Signal Processing Algorithms

PRENTICE-HALL SIGNAL PROCESSING SERIES

Alan V. Oppenheim, Editor

Signal Processing Algorithms

Samuel D. Stearns
Ruth A. David

Sandia National Laboratories
Albuquerque, New Mexico

Prentice-Hall, Inc., Englewood Cliffs, New Jersey 07632

Library of Congress Cataloging-in-Publication Data

Stearns, Samuel D. (date)
 Signal processing algorithms.

 Includes bibliographies.
 1. Signal processing—Digital techniques.
2. Algorithms. I. David, Ruth A. (date)
II. Title.
TK5102.5.S699 1988   621.38′043   86-30493
 ISBN   0-13-809435-7

Editorial/production supervision: *Lisa Schulz Garboski*
Manufacturing buyer: *S. Gordon Osbourne*

If your diskette is defective or damaged in transit, return it directly to Prentice-Hall at
the address below for a no-charge replacement within 90 days of the date of purchase.
Mail the defective diskette together with your name and address.

> Prentice-Hall, Inc.
> Attention: Ryan Colby
> College Operations
> Englewood Cliffs, NJ 07632

LIMITS OF LIABILITY AND DISCLAIMER OF WARRANTY:

The author and publisher of this book have used their best efforts in preparing this
book and software. These efforts include the development, research, and testing of the
theories and programs to determine their effectiveness. The author and publisher make
no warranty of any kind, expressed or implied, with regard to these programs or the
documentation contained in this book. The author and publisher shall not be liable in
any event for incidental or consequential damages in connection with, or arising out of,
the furnishing, performance, or use of these programs.

ISBN   0-13-809435-7   025

Prentice-Hall International (UK) Limited, *London*
Prentice-Hall of Australia Pty. Limited, *Sydney*
Prentice-Hall Canada Inc., *Toronto*
Prentice-Hall Hispanoamericana, S.A., *Mexico*
Prentice-Hall of India Private Limited, *New Delhi*
Prentice-Hall of Japan, Inc., *Tokyo*
Prentice-Hall of Southeast Asia Pte. Ltd., *Singapore*
Editora Prentice-Hall do Brasil, Ltda., *Rio de Janeiro*

This book is dedicated to the cause of peace among the peoples of earth. We hope and trust that its contents may be used toward this end.

Contents

Preface

This book is a text on basic algorithms of digital signal processing. It is written around a set of over 50 subprogram modules, which are listed in Fortran 77 in Appendix A. The book does *not*, however, consist of an extensive system of software with instructions for its use. It differs in this respect from other signal processing software books. The subprograms are simple expressions of the algorithms discussed in the text. Each subprogram is short, typically around 20 Fortran statements long, excluding comment statements, and is written to be easily understood, easily modified to fit the reader's own requirements, and highly portable.

The book is meant to be a text for a one-semester course on algorithms in signal processing as well as a reference text for the practicing engineer. It emphasizes the application as opposed to the theory of digital signal processing. Each chapter covers a particular area, such as discrete transforms, spectral analysis, and so on, and includes only enough theory to allow one to understand how the algorithms work. The emphasis is on applications of the algorithms and interpretation of results in practical situations. Therefore, each chapter has examples of applications as well as exercises that emphasize applications. A typical one-semester course covers about one chapter per week, with exercises being assigned once a week from each chapter.

The algorithms described here have been developed and used over the past several years. Together they comprise a selection of the most used signal processing operations in a general industrial setting that includes data acquisition, data reduction, and real-time applications.

As shown in the Contents, with the exception of Chapters 1 and 2, each chapter covers a set of related signal processing operations and includes descriptive material, calling sequences, examples, and exercises. Chapters 1 and 2 are introductory and do not describe algorithms or include exercises; they provide background and notation for the rest of the text.

Chapters 3 and 4 are on discrete Fourier transforms, including the FFT, and on spectral and coherence estimation with applications to random time-series analysis. Chapter 5 is on computing the frequency and time-domain responses of linear systems.

Chapters 6 through 8 are on digital IIR and FIR filtering and filter design. Filtering algorithms for direct, cascade, parallel, and lattice structures are presented and discussed in Chapter 6. Then algorithms for designing different types of IIR and FIR filters, which are in turn used by the filtering algorithms, are presented in Chapters 7 and 8.

Fast convolution and correlation algorithms are discussed in Chapter 9, and algorithms for decimation and interpolation are presented in Chapter 10. Chapter 11 is on least-squares design and modeling using the Levinson and Durbin algorithms, and Chapter 12 is on adaptive signal processing using the least-mean-squares (LMS) algorithm, which is essentially a time-varying least-squares design process. Next, Chapter 13 includes algorithms for computing statistical parameters as well as other parameters useful in waveform analysis. Finally, Chapter 14 covers some miscellaneous algorithms, including data-window routines, phase unwrapping, a Hilbert transformer, the chirp z-transform, and a Walsh function generator.

The algorithms in these chapters were first developed into a general signal processing library. Then, as the number of users of the library grew, additional descriptive literature was written, and finally a one-semester course was developed and taught at Sandia National Laboratories. The descriptive literature for the program library and the course notes have formed the basis for this text.

During the development of the algorithms, the course, and this text, we have been fortunate to have the help and encouragement of many friends. We are especially grateful to all of our students, who have made countless suggestions to eliminate errors and clarify the text. We also thank James A. Codzow, John O'Donnell, David L. Soldan, and Thomas D. Sullivan for their helpful comments, and Ken Sabish for his support work. We thank Sandia National Laboratories for their encouragement and support, especially for the patient typing services of Patricia G. Willan.

Comments on the Enclosed Disk

The floppy disk included with this text contains the Fortran 77 source code for all of the subprograms listed in Appendix A, that is, all of the signal processing algorithms discussed in this text. The subprograms are in two ASCII files, SPAA.FOR and SPAA.BAK. The latter is a back-up duplicate of the former. The floppy disk has a standard-density, 360kB format and was written by MS-DOS Version 2.11. The subprograms themselves are portable and are meant to compile and run on any system supporting a full version of Fortran.

Samuel D. Stearns
Ruth A. David

Signal Processing Algorithms

chapter 1

Digital Signal Processing

1.1 Introduction

In recent years, tremendous advances have been made in computer hardware as well as in digital technology in general. For many applications, information is now most conveniently recorded, transmitted, and stored in digital form. As a result, digital signal processing is becoming an increasingly important modern tool.

Digital signal processing deals with the representation of signals as ordered sequences of numbers and the processing of those sequences. Typical reasons for signal processing include: estimation of characteristic signal parameters, elimination or reduction of unwanted interference, and transformation of a signal into a form that is in some sense more informative. In some cases, digital signal processing is used as illustrated in Fig. 1.1 to replace or supplement the analysis of signals in the continuous time domain. Three simple tasks involving the use of signal processing techniques are illustrated pictorially in Figs. 1.2 through 1.4.

A common problem in parameter estimation is that of measuring the rise time of a signal. The rise time is generally defined as the time required for a signal to increase from 10% to 90% of its peak value. The solid curve on the plot in Fig. 1.2 represents a continuous time signal, while the asterisks denote the samples that would make up the ordered sequence of numbers used for digital signal processing. Ideally, the computed rise time should equal the difference between times t_1 and t_2, as shown in the figure. In digital signal processing, however, only the sample values are available. Thus, the accuracy of the rise time computation is affected by both the sample interval (elapsed time between samples) and the precision of the digitization process. Rise time estimation is discussed further in Chapter 13.

Reduction of unwanted interference is achieved via filtering in the example illustrated in Fig. 1.3. Once again, both the continuous signal and a corresponding sample sequence are shown, since this task can be accomplished in either the

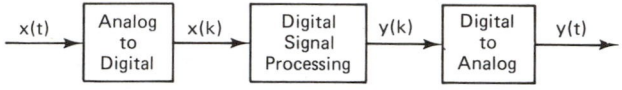

Figure 1.1. Analysis of a continuous time signal $x(t)$ using digital signal processing to operate on the sample sequence $x(k)$. The result is the $y(k)$ sequence whose continuous time counterpart is $y(t)$.

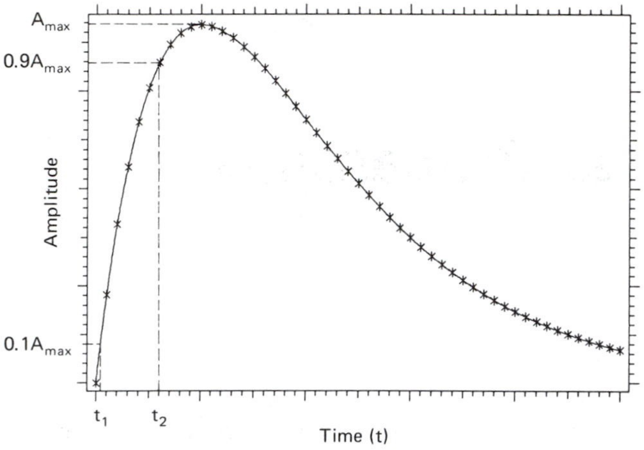

Figure 1.2. An example of parameter estimation involving measurement of the signal rise time. As shown, the rise time is $t_r = t_2 - t_1$.

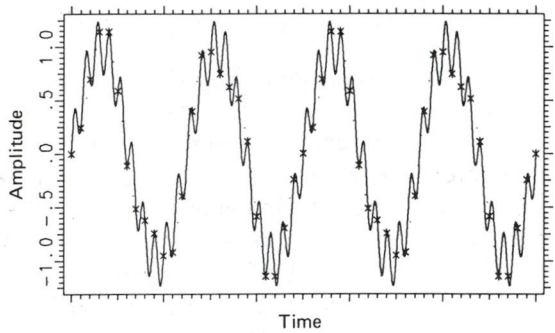

(a) Low frequency signal corrupted by 60 Hz sinusoid.

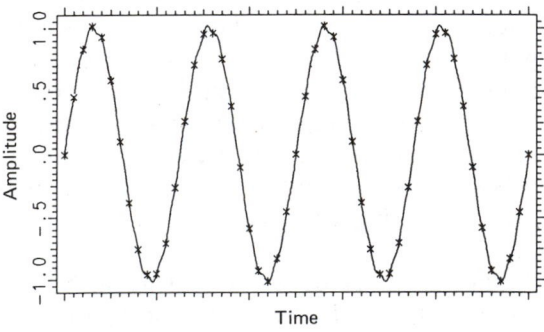

(b) Enhanced signal after filtering.

Figure 1.3. Reduction of interference via filtering.

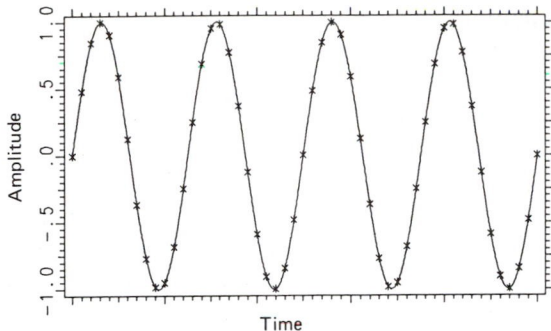

(a) Pure sinusoid in the time domain.

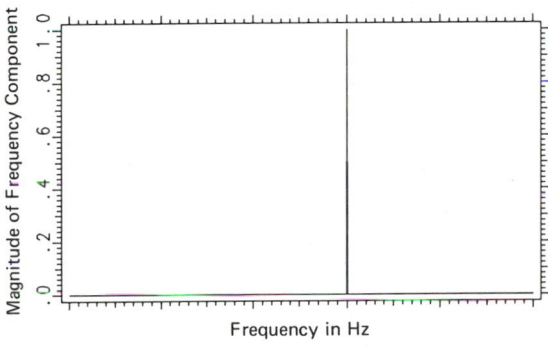

(b) Pure sinusoid in the frequency domain.

Figure 1.4. Signal transformation from time to frequency.

continuous or the discrete time domain. A low-frequency signal corrupted by a 60-Hz sinusoid is shown in Fig. 1.3(a). A simple notch filter was used to reduce the interference by attenuating signal components within a narrow band of frequencies around 60 Hz while passing all other components to the output unchanged in amplitude. The resulting enhanced signal is illustrated in Fig. 1.3(b). Digital filtering is covered in more detail in Chapters 6, 7, and 8.

The most common signal or data transformation used in digital signal processing is the *discrete Fourier transform* (DFT). This algorithm uses the time-domain sample sequence to compute a frequency-domain sequence which describes the spectral content of the signal. Corresponding time- and frequency-domain values for a pure sinusoid are illustrated in Fig. 1.4. Algorithms for computing the DFT are described in Chapter 3.

In later chapters, this book addresses each of the topics just described as well as other cases for which signal processing is useful.

Early techniques for digital signal processing were developed for off-line usage in large, general-purpose computers. When on-line or, equivalently, real-time data processing was necessary, it was typically performed in the analog (continuous) time domain because of speed limitations in digital circuitry. Due to the evolution of digital technology, however, many signal processing tasks are now accomplished in real time using microprocessors or other dedicated digital hardware.

A primary benefit of digital signal processing over analog signal processing is flexibility. In general, a digital processing system is more easily reconfigured as parameters of the problem change. In fact, digital simulations are now frequently used to analyze designs in an attempt to identify potentially costly design errors before the hardware is built. One example of this is the use of computer-aided engineering systems (CAE) to simulate and evaluate the operation of digital logic designs rather than building prototype circuits for this purpose.

Applications for digital signal processing currently exist in diverse fields such as acoustics, sonar, radar, geophysics, communications, and medicine. Examples that follow describe two practical cases in which digital signal processing was used to enhance experimental data.

The results from a task involving a ground-penetration radar system are shown in Figs. 1.5 and 1.6 [A]. The goal of this task was the detection of the hollow steel cylinder, which was buried as illustrated in Fig. 1.5. As the radar system was towed across the surface above the target, scans were recorded at 3-in. intervals. In this application, the data set consisted of a collection of sample sequences. Each sequence was associated with a particular point on the earth's surface, whereas the individual values within a sequence represented time-ordered samples from the radar return. Due to the width of the radar beam, the target was detected even when the recording system was not directly above the steel cylinder. The parabolic pattern appearing in the raw data in Fig. 1.6(a) results from the varying path lengths of these returns. Array-processing techniques were used to focus the radar beam, thus improving the signal-to-noise ratio and simplifying target detection, as demonstrated by the processed data in Fig. 1.6(b). For each point on the surface where a scan was recorded, adjacent radar returns were shifted to account for the differing path delays and then summed in phase. This array processing enhanced the signal component (target return) while attenuating the noise components.

A category of digital signal processing known as *adaptive signal processing* is a relatively new area in which applications are increasing rapidly. When

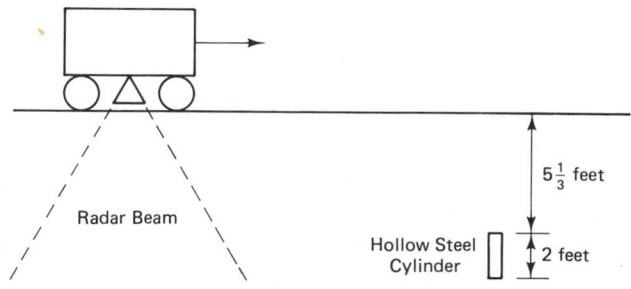

Radar Beam

$5\frac{1}{3}$ feet

Hollow Steel
Cylinder   2 feet

Figure 1.5. Configuration for ground-penetration radar experiment.

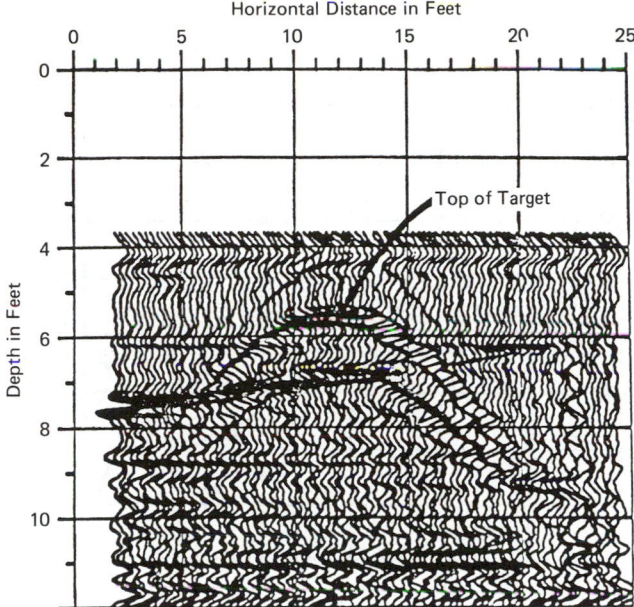

(a) Raw data from ground penetration radar system.

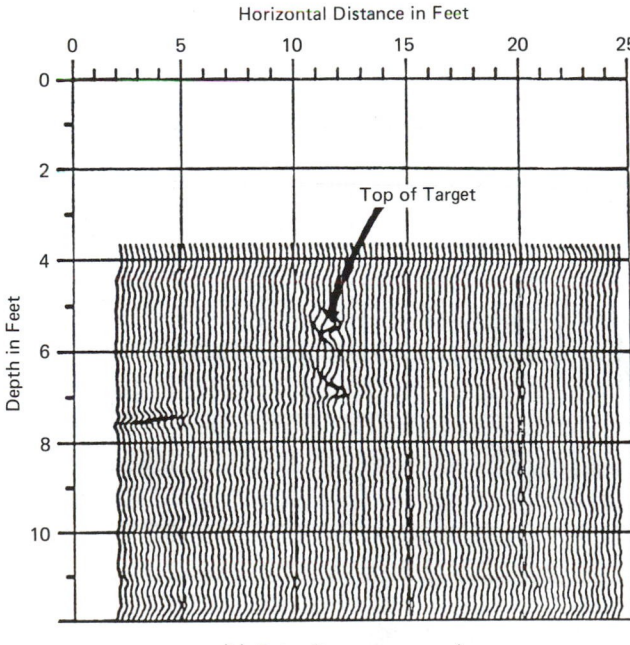

(b) Data after array processing.

Figure 1.6. Data from ground-penetration radar system.

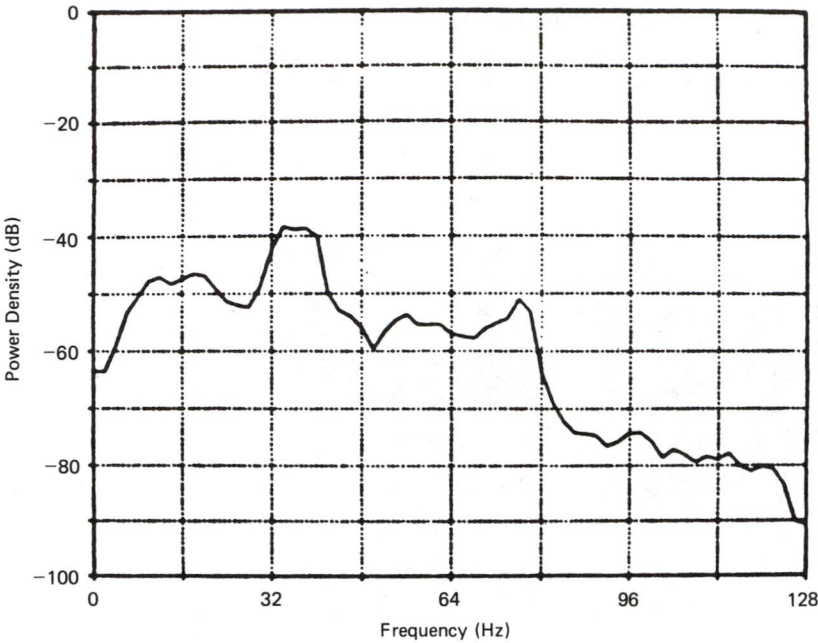

(a) Power density spectrum of noisy unprocessed data.

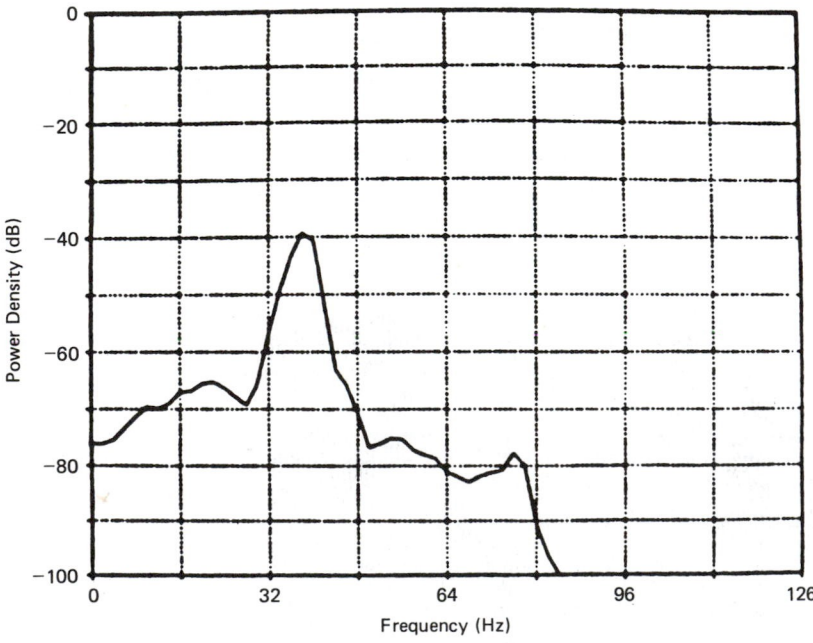

(b) Power density spectrum of enhanced data.

Figure 1.7.   Acoustical data containing corrupted narrow-band signal.

adaptive filters are used to process nonstationary signals, that is, signals whose statistical properties vary in time, the required amount of a priori information is often less than that required for processing via fixed digital filters. The results of an experiment involving an adaptive line enhancer are shown in Fig. 1.7 [B]. The peaks in the power density spectrum of the unprocessed acoustical data shown in Fig. 1.7(a) indicate the presence of narrow-band signals amid wideband noise. The adaptive line enhancer is essentially a time-varying narrow-band filter whose goal is to detect and pass an incoming narrow-band signal while attenuating the surrounding noise. The power density spectrum of the enhanced output is shown in Fig. 1.7(b). Other examples of adaptive signal processing are described in Chapter 12.

Additional new applications for digital signal processing are continually emerging as advances in digital technology allow the processing of new types of signals. While new applications sometimes necessitate an increase in signal processing capabilities, many basic signal processing techniques are already applicable to a wide variety of problems. These general techniques are the subject of this book.

Our presentation is designed primarily for the engineer, scientist, or technician whose principal job is to apply digital signal analysis rather than to study the subject itself. By using the "black box" approach, we present information in a format that circumvents the typical requirement of a sophisticated background in mathematics and computing science. Our goal is to provide an introduction to the subject together with enough working knowledge to enable utilization of the signal processing modules provided herein.

1.2 Algorithm Format

The signal processing algorithms in this book are unique because they are presented as a series of simple, independent modules, which are highly portable. To attain this goal, the algorithms are written in subprogram form, thus eliminating all device-dependent input and output. Descriptive comments are included and calling sequences are standardized in an attempt to simplify the use of the package as a whole. An additional feature of this package is that all subprogram names are six characters in length and begin with the letters SP. Thus, the modules are easily distinguished from other signal processing or math library routines. The modules are written in FORTRAN 77; complete listings are provided in Appendix A.

A variety of digital signal processing examples are presented, thus permitting the reader to gain some insight into the utilization of this collection of algorithms for real-world applications. For each module, simple verification programs are provided which specify a set of input data, the calling sequence, and the corresponding output data. These programs provide users with a means for testing the software on their own computers before progressing to the solution of new signal processing problems.

1.3 Scope

The background material in Chapter 2 provides a brief description of the types of signals commonly encountered in signal processing applications. The consequences of analog-to-digital (A/D) conversion with regard to the sampling theorem are discussed for both the time and the frequency domains. Block diagrams and the concept of a system transfer function are introduced.

Since the DFT is probably the most widely used digital signal processing technique, Chapter 3 is entirely devoted to this topic. Forward and inverse transforms for a general data sequence via the DFT are first described. The fast Fourier transform (FFT), which is a more computationally efficient algorithm for computing the DFT, is then introduced.

The concept of frequency-domain information is extended to spectral analysis in Chapter 4 with a discussion of power density spectra. Algorithms for computing cross spectra and performing coherence analysis are also included.

Techniques for analyzing a digital system transfer function are described in Chapter 5. Frequency-domain analysis yields the system gain and phase response; time-domain analysis provides the output response to a specified input signal.

Digital filtering is the topic of the next three chapters. Chapter 6 introduces the subject and discusses the fundamentals of infinite impulse response (IIR) and finite impulse response (FIR) digital filtering. Direct-form implementations and lattice structures are described, and routines are provided that enable conversion from one form to the other. Other routines enable digital filtering via cascade and parallel filter structures. IIR filter-design routines are provided in Chapter 7. Included are routines that enable Butterworth, Chebyshev, and Bessel filter design. The discussion of FIR filters in Chapter 8 includes routines for filter design via the windowing of an ideal impulse response sequence.

Fast correlation routines are discussed in Chapter 9. Routines for both auto- and cross correlation are described.

Methods for signal decimation and interpolation are described in Chapter 10. In effect, these routines allow the user to change the sampling rate in a digital data sequence.

The discussion of least-squares-system design techniques in Chapter 11 includes descriptions of the Levinson and Durbin algorithms for designing optimum systems. Examples of system identification and modeling as well as data deconvolution and equalization are provided.

A simple routine for adaptive signal processing, called the least-mean-squares (LMS) algorithm, is discussed in Chapter 12. Examples are also included to illustrate several applications of adaptive signal processing.

Simple techniques for time-domain parameter estimation and waveform analysis are described in Chapter 13. Included are routines to enable the computation of basic statistical parameters as well as signal rise and fall times. Several basic curve-fitting routines are also provided.

Chapter 14 contains several miscellaneous routines. Included here are routines for data windows, phase unwrapping, the Hilbert and chirp z-transforms, and Walsh functions.

The routines in this book are not intended to function as a comprehensive digital signal processing package. Instead, our goal is to provide a library of simple routines that are portable and easy to use and yet are applicable to a large variety of signal processing tasks. Thus, our intent is to provide basic algorithms in a modular form so that users can construct their own higher-level signal processing packages.

1.4 References

[A] David, R. A., C. W. Cook, and S. D. Stearns, "Array Processing Techniques Applied to Ground Penetration Radar Data," *Proc. of 16th Asilomar Conf. on Circuits, Systems and Computers* (November 1982).

[B] David, R. A., S. D. Stearns, G. R. Elliott, and D. M. Etter, "IIR Algorithms for Adaptive Line Enhancement," *Proc. of ICASSP 83* (April 1983).

chapter 2

Signals
and Sampled Data

2.1 Introduction

This chapter provides background material and establishes notational conventions that are maintained throughout the book. After a description of the types of signals commonly encountered in digital signal processing, implications of the sampling theorem are discussed. The z-transform is introduced as an alternative means of representing data sequences. The concepts of system transfer functions and block diagrams are described.

2.2 Types of Signals

Signals are typically categorized as being either *deterministic* or *nondeterministic*. For a deterministic signal, each value is uniquely specified by a mathematical expression, a table of data, or a rule of some sort. The class of deterministic signals contains periodic signals, finite-duration signals, transient signals, and almost periodic signals. *Periodic signals,* which are completely described by being specified within a single finite period, can be decomposed and expressed as a sum of sinusoidal components. *Finite-duration signals* are defined during some finite time interval and are undefined for all time outside that interval. *Transient signals* are those that are nonzero only within some finite time interval or vary over a short interval and then decay to a constant value. Signals composed of sums of sinusoids that are not harmonically related are *almost periodic*. This type of signal is not usually encountered in practical engineering applications. One could argue that recorded data can never be truly deterministic due to unforeseen factors that affect the experiment. In many practical applications, however, a mathematical model of the recorded signal may be used to simplify the analysis with very little loss of accuracy. Nondeterministic or random signals are not completely predictable and are therefore described in terms of averages and other statistical properties.

In Fig. 2.1, examples of periodic, transient, and random signals are shown. The signal amplitude, $x(t)$, is plotted versus time, t, with $t = 0$ marking the start of the signal. Although the independent variable, t, may actually denote space or some other physical variable for a specific application, the convention of using time will be maintained throughout this book. In addition, we will restrict the scope of our discussion to one-dimensional signals, that is, signals that are described in terms of one independent and one dependent variable.

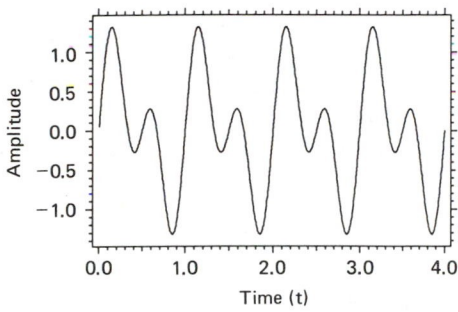

(a) Periodic signal: $x_1(t) = \sin(\omega t) + \sin(2\omega t)$

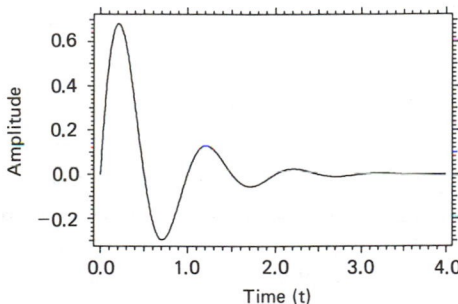

(b) Transient signal: $x_2(t) = e^{-\alpha t} \sin(\omega t)$

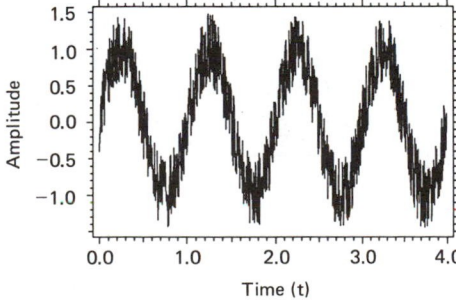

(c) Random signal: $x_3(t) = \sin(\omega t) + n(t)$

Figure 2.1.   Examples of basic types of signals.

The periodic signal illustrated in Fig. 2.1(a) is the sum of two sinusoids, one at twice the frequency of the other. The damped sinusoid in (b) is typically categorized as a transient signal even though as defined, $x_2(t)$ requires an infinite amount of time to decay to zero. The sinusoid corrupted by noise in (c) is an example of a random signal, since the noise, $n(t)$ can be described only in terms of its statistical properties.

The three signals in Fig. 2.1 are called *analog,* or *continuous, signals* since both x and t are continuous variables. To enable digital signal processing, these signals must be converted to sequences of numbers. The conversion is accomplished via analog-to-digital (A/D) conversion, or sampling.

A/D conversion is illustrated in *black box* form in Fig. 2.2. Every T seconds, the continuous signal, $x(t)$, is sampled. This amplitude value is denoted by $x(kT)$; thus, $x(kT) = x(t)$ when $t = kT$. In general, the sampling process introduces a quantization error, which is a function of the number of digital bits available to represent the sample [A–D]. Thus, digital signal processing algorithms actually utilize a quantized sample, $\hat{x}(kT)$, rather than the true sample, $x(kT)$. Although the resolution of the quantizer affects the accuracy of the data, this distinction between $x(kT)$ and $\hat{x}(kT)$ is rarely made in discussions of signal processing algorithms. In this text we use $x(kT)$ to denote the sample value available for processing.

The parameter T is the time step or sample interval, and the integer value k denotes the sample number. A related quantity is the *sampling frequency* or, equivalently, sampling rate, which is equal to $1/T$ hertz. The notation $x(k)$, or equivalently x_k with T included implicitly, is often used interchangeably for $x(kT)$.

A *sample set,* or *data sequence,* is a collection of data samples over some range of k. For the general case, the sample set is defined over the entire range from $k = -\infty$ to $k = \infty$. In digital signal processing, however, we generally deal with finite data sequences. Thus, the sample set is typically defined over the interval from $k = 0$ to $k = N - 1$, so that the total number of data samples is equal to N. Sample sets, or data sequences, may be denoted by either $[x(kT)]$, $[x(k)]$, or $[x_k]$. Since digital signal processing deals primarily with data sequences, details of the sampling process illustrated in Fig. 2.2 are not considered here. The discussion of the sampling theorem in the next section, however, includes

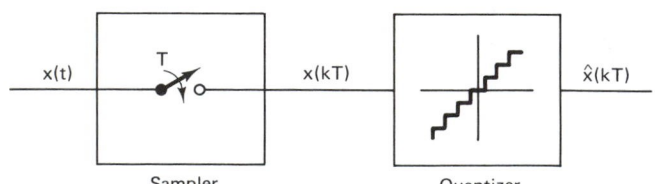

Sampler   Quantizer

Figure 2.2. Black box illustration of the sampling process.

guidelines for the selection of the sample interval T and comments on ways in which the choice of T affects signal processing.

Figure 2.3 shows a sampled, or digital, representation of each of the continuous time signals in Fig. 2.1. Samples are denoted by the bold dots, and straight-line interpolation is used to connect adjacent samples. The horizontal distance between successive data points is the sample interval T, and the horizontal scale is now

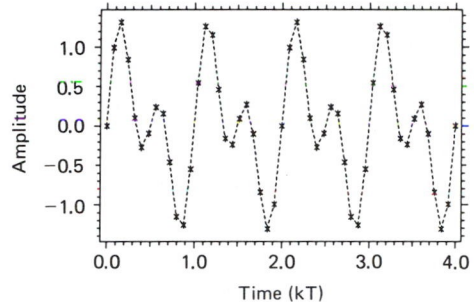

(a) Periodic signal: $x_1(kT) = \sin(\omega kT) + \sin(2\omega kT)$

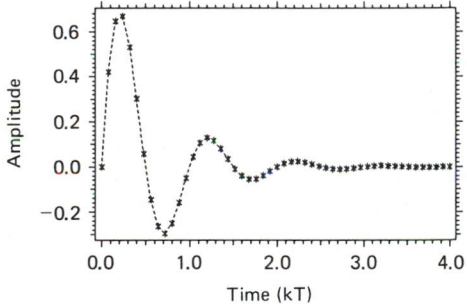

(b) Transient signal: $x_2(kT) = e^{-\alpha kT} \sin(\omega kT)$

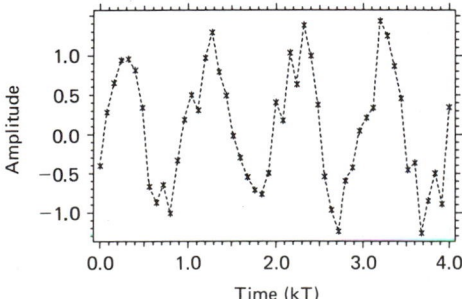

(c) Random signal: $x_3(kT) = \sin(\omega kT) + n(kT)$

Figure 2.3. Sampled representation of signals in Fig. 2.1.

drawn in terms of $t = kT$. Alternatively, data may be plotted versus sample number, so that the integer sequence $k = 0, 1, 2 \ldots$ is used for the horizontal scale. In practical applications, the kT scale is inherently more informative since the data are plotted versus real time.

While data sequences may be exactly represented by plotting discrete points, some method of interpolation is generally used to connect successive data points. Unless otherwise specified, straight-line interpolation is used when continuous plots are generated from discrete sample sets in this book. Other more accurate techniques for data interpolation are discussed in Chapter 10.

2.3 The Sampling Theorem

Rather than present a detailed mathematical derivation of the sampling theorem, which is covered in the references [A–D], examples are provided here to give the reader an intuitive feel for its implications. The sampling theorem, simply stated, requires that an analog signal be sampled at a rate greater than twice its highest frequency. If this condition is satisfied, the continuous signal can theoretically be reconstructed exactly from its sample sequence. In practice this reconstruction will always be approximate due to the infinite duration of the sinc functions $(\sin(\omega t)/\omega t)$ that are required for exact interpolation. In addition, the sampling theorem as stated in this simple form assumes that no quantization error is introduced by the sampling process.

The frequency in hertz (cycles per second) of the sinusoid illustrated in Fig. 2.4 is 1. Suppose first that the signal is sampled as in Fig. 2.4(a). The sampling rate is exactly twice the signal frequency, which does not meet the condition imposed by the sampling theorem. In this case, the analog signal cannot be reconstructed, since the sample values are $x_k = 0$ for all k. Next, suppose that the same sinusoid is sampled at a rate that is less than twice per cycle, as illustrated in Fig. 2.4(b). In addition to the 1-Hz signal, a sinusoid at 0.5 Hz exactly fits the sample sequence. This case demonstrates the ambiguity in signal reconstruction that results when the sampling rate is too low. In Fig. 2.4(c), the 1-Hz sinusoid is sampled three times per cycle. Since the requirement of the sampling theorem is satisfied, the continuous signal can theoretically be reconstructed exactly and unambiguously in this case.

The examples in Fig. 2.4 illustrate the implications of the sampling theorem for time-domain sampling and signal reconstruction. A more interesting and more realistic example of a bandlimited signal is shown in Fig. 2.5. Here, the deterministic signal is a sum of sinusoids, which vary in amplitude and range in frequency from 1 Hz to 10 Hz. According to the sampling theorem, the signal can be exactly and uniquely reconstructed if it is sampled at more than twice the highest frequency, that is, at more than 20 times per second.

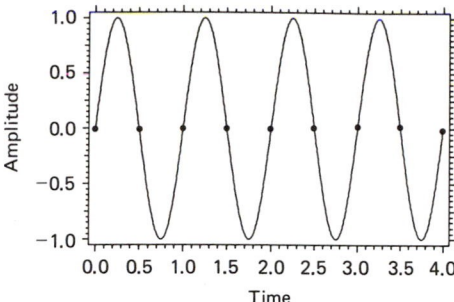

(a) Sample rate equal to 2 samples per cycle.

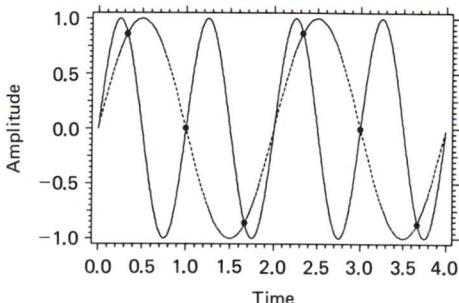

(b) Sample rate less than 2 samples per cycle.

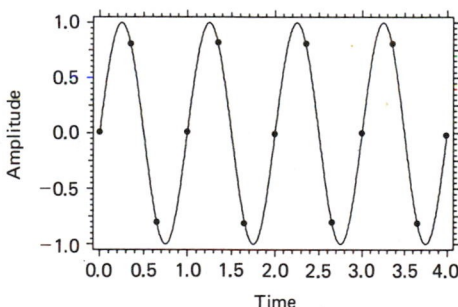

(c) Sample rate greater than 2 samples per cycle.

Figure 2.4. Effects of sample rate on reconstruction of sinusoid.

Bandpass signals, that is, signals whose frequency content is restricted to the range between f_1 and f_2 hertz, where $0 < f_1 < f_2 < \infty$, may be treated as a special case when the sampling theorem is applied. By generating samples of the in-phase and quadrature components (that is, components differing in phase by 90°) of a bandpass signal, the minimum sampling rate per component can be reduced from $f_s > 2f_2$, which is the rate required by the sampling theorem, to $f_s = B$, where $B = f_2 - f_1$ is the signal bandwidth [D]. This technique, which

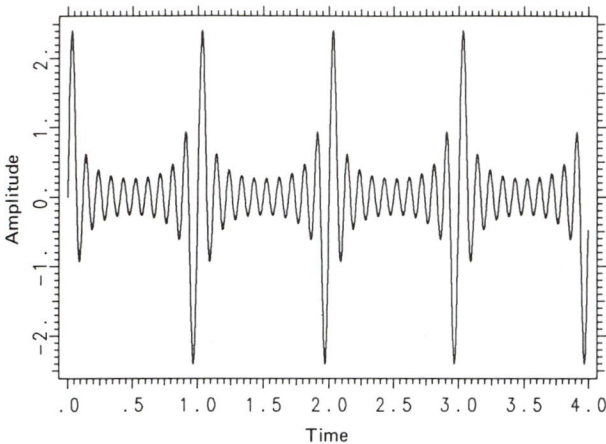

Figure 2.5. Bandlimited signal: $x(t) = \sum\limits_{i=1}^{10} A_i \sin(2\pi\,{}^d it_2)$.

is discussed further in Chapter 14, is used primarily when dealing with narrow-band signals with relatively high center frequencies.

Frequency-domain implications of the sampling theorem can be demonstrated by considering a bandlimited continuous signal with a limited frequency spectrum. Such a spectrum is illustrated in Fig. 2.6(a). Periodic sampling in the time domain results in replication of the spectrum about multiples of the sampling frequency in the frequency domain. In order to allow recovery of the continuous signal, the replicated spectra must not overlap. For this example, the sampling theorem requires a sampling rate of at least 1.2 Hz because the spectrum is contained below 0.6 Hz. Figure 2.6(b) shows the aliasing, or overlap, which occurs when the bandlimited signal in (a) is sampled at 1.0 Hz. The overlapping spectral components add to those of the original spectrum, so the continuous signal cannot be reconstructed from the sample sequence in this case. This effect is illustrated in the time domain in Fig. 2.4(b), where the 1.0-Hz signal that is sampled at 1.5 Hz may be reconstructed as a 0.5-Hz signal. The frequency spectrum in Fig. 2.6(c) results when the signal whose spectrum is shown in (a) is sampled at 1.5 Hz. Since this sampling rate satisfies the sampling theorem, no aliasing of spectral components results. In this case, the original signal can be recovered as in the example of Fig. 2.4(c).

In practical applications, a low-pass analog antialiasing filter is often used ahead of the digitizer to eliminate possible aliasing effects by ensuring that the signal to be sampled is limited in spectral content. Since an ideal rectangular filter cutoff characteristic is unrealizable, actual sampling rates are often selected to be four to five times the filter cutoff frequency. In addition to protecting against aliasing, this oversampling eliminates the need for sophisticated data-interpolation techniques in many cases. When signals are oversampled, straight-line interpolation between samples often produces sufficiently informative data plots. Filtering and signal reconstruction via data interpolation are discussed in more detail in Chapters 6–8 and 10.

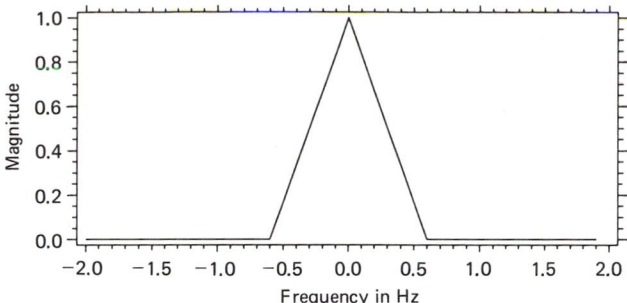

(a) Frequency spectrum of band-limited continuous signal.

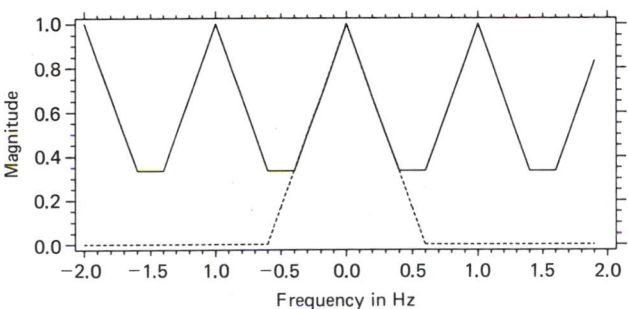

(b) Aliased spectrum of the digitized signal resulting from slow sampling rate = 1.0 Hz.

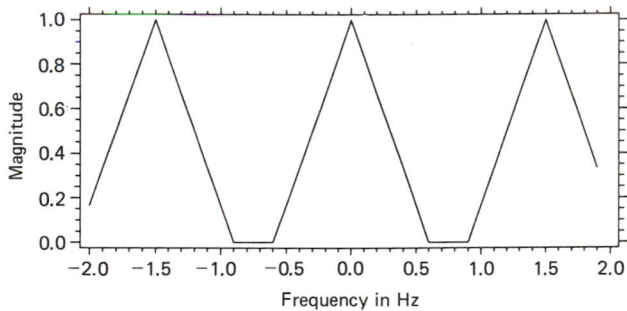

(c) Correct spectrum of the digitized signal resulting from sufficient sample rate = 1.5 Hz.

Figure 2.6. Frequency domain effects of sampling rate.

2.4 The z-Transform

Before proceeding with the discussion of block diagrams and digital system transfer functions, it is necessary to introduce the concept of z-transform representation of sampled signals and systems. Like other mathematical transformations (for example, the Fourier transform), the z-transform exists in both forward and inverse forms [A–D]. Here we review only the forward transform, that is,

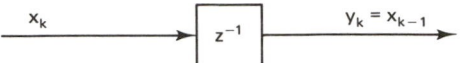

Figure 2.7. The unit delay element.

the transformation of a discrete time sample sequence into the z-domain. A rigorous mathematical description of the transform is not essential to this discussion and is not presented here.

The two-sided z-transform $X(z)$ of a complete data sequence $[x_k]$ is defined as

$$X(z) = \sum_{k=-\infty}^{\infty} x_k z^{-k} \tag{2.1}$$

for the previously assumed condition that the signal x_k is defined for all k. As noted previously, in many practical applications where N samples are involved, the limits on the summation are zero and $N - 1$. A closed form of (2.1) exists for a large number of deterministic signals. Tables of z-transform pairs appear in most digital signal processing texts [A–D].

In general, the variable z in (2.1) is a complex variable, which is expressed in polar form as $z = re^{j\theta}$. This representation is used in Chapter 5, where the z-transform is used to evaluate the frequency response of a linear digital system.

When associated with a particular sample x_k, as in (2.1), the z^{-k} factor represents a time delay of kT seconds, or k sample intervals from time $t = 0$. Thus, each data sample in the sequence $[x_k]$ is associated with a unique power of z, which defines its position in the sequence.

Using this interpretation, we define a unit delay by the symbol z^{-1}. This element, which is illustrated in Fig. 2.7, has the effect of delaying the signal by one sample interval, that is, T seconds in time. Thus, at time $t = kT$, the output y_k of the unit delay element is equal to the delayed input sample x_{k-1}.

The z-domain representation of sampled signals and systems is used extensively throughout this text because it provides notation that relates easily to computer processing of ordered sequences. Additional details relating to the transform are provided as needed.

2.5 Digital Transfer Functions

The block diagram of a general digital system is illustrated in Fig. 2.8. The transfer function $H(z)$ describes how the system operates on the input data sequence $[x_k]$ to produce the output sequence $[y_k]$. The transfer function is given by

$$H(z) = \frac{Y(z)}{X(z)} \tag{2.2}$$

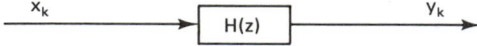

Figure 2.8. Block diagram of a general digital system.

where $X(z)$ and $Y(z)$ are the z-transforms of the data sequences $[x_k]$ and $[y_k]$, respectively.

For the simple example in Fig. 2.7, the transfer function is $H(z) = z^{-1}$, which is obtained by expressing the z-transform of $[y_k]$ in terms of $X(z)$ as follows:

$$Y(z) = \sum_{k=-\infty}^{\infty} x_{k-1} z^{-k} = z^{-1} \sum_{k=-\infty}^{\infty} x_k z^{-k} = z^{-1} X(z) \qquad (2.3)$$

A digital transfer function appearing repeatedly in later chapters is

$$H(z) = \frac{b_0 + b_1 z^{-1} + \cdots + b_L z^{-L}}{1 + a_1 z^{-1} + \cdots + a_L z^{-L}} \qquad (2.4)$$

The coefficient sets $[b_n]$ and $[a_n]$ are constants that determine the system response. In this form, $H(z)$ defines the direct-form transfer function of an Lth-order, linear, time-invariant digital system. Note that some of the b_n coefficients may be equal to zero in the general case, since the order of the numerator must be less than or equal to the order of the denominator.

The time-domain output of the digital system illustrated in Fig. 2.8, whose transfer function $H(z)$ is given in (2.4), is

$$y_k = \sum_{n=0}^{L} b_n x_{k-n} - \sum_{n=1}^{L} a_n y_{k-n} \qquad (2.5)$$

Equation (2.5) is the difference equation, which describes the input-output relationship of $H(z)$. It can be derived from (2.2) and (2.4) by cross multiplication and interpretation of z^{-k} as a k-sample delay.

The transfer function of a digital system need not be realized in direct form as in (2.4). Two alternate forms frequently used are the cascade and parallel structures shown in Fig. 2.9. Note that a transfer function described in either

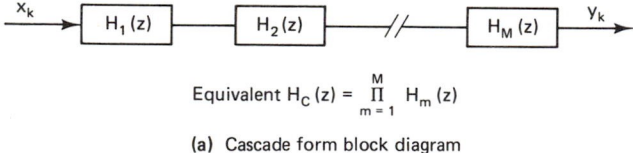

Equivalent $H_C(z) = \prod_{m=1}^{M} H_m(z)$

(a) Cascade form block diagram

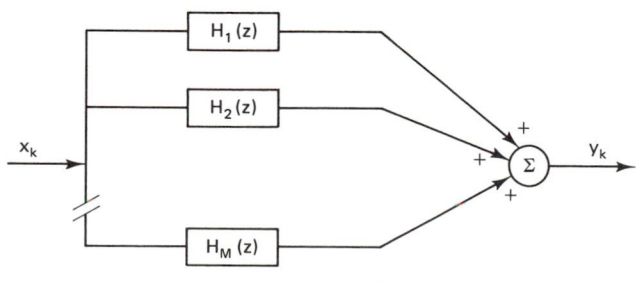

Equivalent $H_P(z) = \sum_{m=1}^{M} H_m(z)$

(b) Parallel form block diagram

Figure 2.9. Structures that are alternate forms of Fig. 2.8.

of these forms could be reduced to the direct form of (2.4); however, there are often numerical advantages to utilizing an alternate structure [A–D].

The term *digital filter* is often used interchangeably with *digital system*. Later chapters provide additional information regarding digital filtering and computation of the response of a digital system in both the time and the frequency domain.

2.6 References

[A] Stearns, S. D., *Digital Signal Analysis* (Rochelle Park, New Jersey: Hayden, 1975).

[B] Ahmed, N. and T. Natarajan, *Discrete-Time Signals and Systems* (Englewood Cliffs, N.J.: Prentice-Hall, Inc., 1983).

[C] Oppenheim, A. V. and R. W. Schafer, *Digital Signal Processing* (Englewood Cliffs, N.J.: Prentice-Hall, Inc., 1975).

[D] Tretter, S. A., *Introduction to Discrete-Time Signal Processing* (New York: John Wiley, 1976).

chapter 3

Discrete Fourier Transform Routines

3.1 The Discrete Fourier Transform

The *discrete Fourier transform* (DFT) is a basic operation used in many different signal processing applications. It is used to transform an ordered sequence of data samples, usually from the time domain into the frequency domain, so that spectral information about the sequence can become known explicitly. We describe the DFT briefly here before discussing the routines for computing it.

As shown later, the DFT is a complex function of frequency, that is, an ordered sequence of complex numbers, each number consisting of a real part and an imaginary part. Usually the data sequence being transformed is real, but it need not be, and in either case the DFT is, in general, complex.

Suppose we have a real data sequence consisting of N real samples of a signal $x(t)$, given by

$$\text{Samples of } x(t) = [x_k]$$
$$= [x_0, x_1, x_2, ..., x_{N-1}] \tag{3.1}$$

In this notation, k is usually a time index and ranges from 0 to $N - 1$. When $[x_k]$ is a real data sequence, the computed DFT of $[x_k]$ consists of $N/2 + 1$ complex samples (we assume here that N is even), given by

$$[X_m] = \text{DFT } [x_k]$$
$$= [X_0, X_1, ..., X_{N/2}] \tag{3.2}$$

The relationship of $[X_m]$ to $[x_k]$ is discussed later; here we simply note that $[X_m]$ is the DFT of $[x_k]$ and that the index m designates the frequency of each component X_m. Also, the DFT is complex, so we might represent each X_m in polar form as

$$X_m = |X_m| \, e^{j\theta_m} \tag{3.3}$$

In this notation, $|X_m|$ is the amplitude of X_m, and a plot of $|X_m|$ versus the frequency index m is called the *amplitude spectrum* of $[x_k]$. Similarly, a plot of θ_m versus m is called the *phase spectrum* of $[x_k]$.

Some examples of the amplitude spectrum, which is usually of more interest than the phase spectrum, are illustrated in Fig. 3.1. In each case the data sequence $[x_k]$ consists of $N = 64$ samples, x_0 through x_{63}, and so the computed DFT has $N/2 + 1 = 33$ components, X_0 through X_{32}. In the first (upper) example, $[x_k]$ is a pure sinusoid and has only one frequency component*, so its amplitude spectrum is nonzero only at a single point, where $m = 4$. (The connection between m and the frequency of $[x_k]$ is discussed in Section 3.3.)

In the center example of Fig. 3.1, a periodicity can be seen in the x_k, and on the right the amplitude spectrum tells us that $[x_k]$ has only two spectral components, one at $m = 4$ and another at $m = 8$. In fact, $x(t)$ is the sum of two sinusoids, one at twice the frequency of the other. In the lower example, $[x_k]$ is a sequence of samples of a decaying sinusoid, and the amplitude spectrum is concentrated near $m = 6$ but is seen to be nonzero for all values of m.

3.2 The Fast Fourier Transform

The *fast Fourier transform* (FFT) is not a new kind of transform different from the DFT. Instead, it is simply an algorithm for computing the DFT, and its output is precisely the same set of complex values expressed in (3.2). The FFT algorithm eliminates most of the repeated complex products in the DFT, however, so its execution time is much shorter. Specifically, the ratio of computing times is approximately [A]

$$\frac{\text{FFT computing time}}{\text{DFT computing time}} = \frac{1}{2N} \log_2 N \qquad (3.4)$$

Using the FFT, one can also do the computation "in place," so that $[X_m]$ in (3.2) replaces $[x_k]$ in (3.1), with only a limited amount of auxiliary storage needed for work space.

On the other hand, the FFT algorithm is more complicated than the DFT and becomes particularly lengthy when N, the number of data samples, is unrestricted. Thus, in many applications it is simpler and preferable to use a simple DFT algorithm instead of an FFT.

3.3 Forward Transform and Frequency Index

The forward transform is the DFT (or FFT) just described, which transforms $[x_k]$ in (3.1) into $[X_m]$ in (3.2). The reverse (or inverse) DFT is used to transform $[X_m]$ back into $[x_k]$ and is discussed in Section 3.5.

---

* To have only one nonzero DFT component, the sinusoid must be sampled over an integral number of cycles. See exercises 30–32 at the end of this chapter.

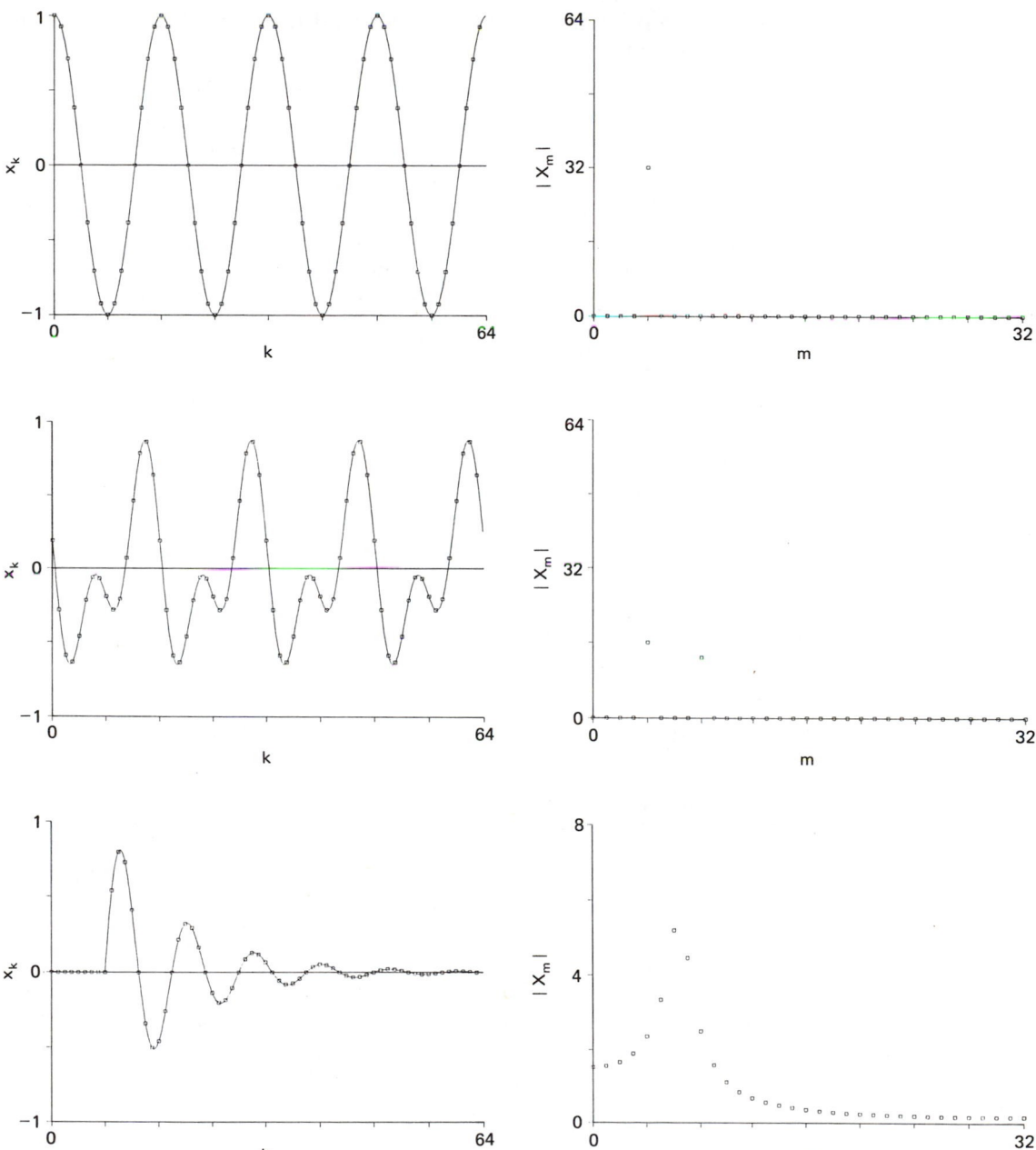

Figure 3.1. Three examples of the DFT amplitude spectrum. In each case the data sequence has $N = 64$ samples.

The relationship implemented by the forward transform between $[x_k]$ and $[X_m]$ can be expressed as

$$X_m = \sum_{k=0}^{N-1} x_k e^{-j(2\pi mk/N)}; \qquad m = 0, 1, ..., \frac{N}{2} \qquad (3.5)$$

Again we are assuming here for convenience that N, the number of data samples, is even. In this formula for X_m, the exponential function, $\exp(-j2\pi mk/N)$, is a *complex sinusoid* and is periodic. If we think of $\exp(-j2\pi mk/N)$ as a function of k, the time index, then its period is seen to be N/m; that is, when k goes through a range of N/m, $\exp(-j2\pi mk/N)$ goes through one cycle. We can see this even more clearly by separating the real and imaginary parts of (3.5):

$$X_m = \sum_{k=0}^{N-1} x_k \cos\left(\frac{2\pi m}{N} k\right) - j \sum_{k=0}^{N-1} x_k \sin\left(\frac{2\pi m}{N} k\right); \qquad m = 0, 1, ..., \frac{N}{2} \qquad (3.6)$$

Thus, each part (real or imaginary) of each DFT component X_m is a correlation (summed product) of the data sequence $[x_k]$ with a cosine or sine sequence having a period of N/m data samples.

The corresponding frequencies of the DFT components are shown in Fig. 3.2 for $m = 1$ and 2. (Note that X_0 is the zero-frequency component, that is, the sum of all the data samples.) If T is the time step between data samples, then the first component X_1 is the summed product of $[x_k]$ times a complex sinusoid with one cycle over N samples or NT seconds, and in general X_m is the summed product of $[x_k]$ times a complex sinusoid with m cycles over the same period. The frequency corresponding to each component is usually designated in one of four ways, as summarized in Table 3.1.

TABLE 3.1   FREQUENCIES OF DFT COMPONENTS IN DIFFERENT UNITS

	Hz-s	rad	Hz	rad/s
Symbol	ν	ω	f	Ω
Frequency of X_1	$1/N$	$2\pi/N$	$1/NT$	$2\pi/NT$
Frequency of X_m	m/N	$2\pi m/N$	m/NT	$2\pi m/NT$
Frequency of $X_{N/2}$	0.5	π	$1/2T$	π/T
Sampling frequency	1.0	2π	$1/T$	$2\pi/T$

Note that the frequency of the final component, $X_{N/2}$, of the DFT of any real time series is exactly one-half the sampling frequency in accordance with the sampling theorem (Chapter 2). The period of X_m is just the reciprocal of the frequency in all cases; thus, from row 2 of Table 3.1,

$$\text{Period of } X_m = \frac{N}{m} \text{ samples} = \frac{NT}{m} \text{ seconds} \qquad (3.7)$$

3.4 Redundancy in the DFT

The periodicity of the DFT just discussed is a very important property. From it we can show that when $[x_k]$ is a sequence of real data samples, the range of

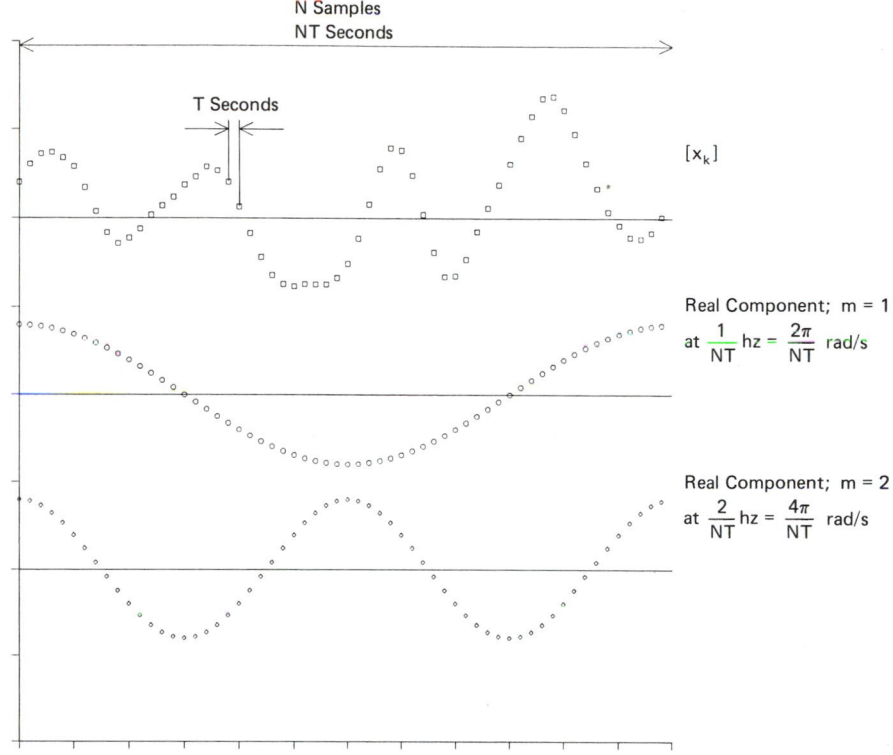

Figure 3.2. Real components of the complex exponential in the DFT, equation (3.6), illustrating frequencies with $m = 1$ and 2, that is, $1/NT$ and $2/NT$ hertz. The real part of X_1 is the summed product of x_k times the component with $m = 1$, the real part of X_2 is the summed product of x_k times the component with $m = 2$, and so on.

m from 0 through $N/2$ in (3.5) covers all of the independent values of X_m. When m is outside of this range, X_m is redundant, that is, equivalent to some single component within the range, as shown in the following argument.

First, from (3.5), we can see that adding N to the index of X_m does not change the value of X_m, that is,

$$X_{m+N} = \sum_{k=0}^{N-1} x_k e^{-j[2\pi(m+N)k/N]}$$

$$= \sum_{k=0}^{N-1} x_k e^{-j(2\pi mk/N)} e^{-j2\pi k} = X_m \qquad (3.8)$$

Furthermore, this rule holds whether $[x_k]$ is real or complex; X_0 through X_{N-1} is a complete set of DFT components.

Secondly, when $[x_k]$ is real, we can also show from (3.5) that X_m and X_{N-m} are complex conjugates. From (3.5) we have

$$X_{N-m} = \sum_{k=0}^{N-1} x_k e^{-j[2\pi(N-m)k/N]}$$

$$= \sum_{k=0}^{N-1} x_k e^{+j(2\pi mk/N)} e^{-j2\pi k} = X_m^* \qquad (3.9)$$

The star (*) denotes the complex conjugate, resulting from the change of sign in the exponential function. Note that if $[x_k]$ is complex, (3.9) does not apply.

Thus, when $[x_k]$ is a sequence of real data, (3.8) and (3.9) show that all values of X_m outside of the set X_0, X_1, ..., $X_{N/2}$ are redundant. When $[x_k]$ is a sequence of complex data, (3.8) shows that all values outside of the set X_0, X_1, ..., X_{N-1} are redundant.

3.5 The Inverse Transform

The reverse (or inverse) DFT is used to obtain a data sequence $[x_k]$ from its complex spectrum $[X_m]$. The formula for the inverse DFT, similar to (3.5), is

$$x_k = \frac{1}{N}\sum_{m=0}^{N-1} X_m e^{j(2\pi mk/N)}; \qquad k = 0, 1, ..., N-1 \qquad (3.10)$$

By substituting (3.5) into (3.10), one can prove that (3.10) is correct [Q].

Thus the inverse DFT is the same as the forward DFT except for the sign of the exponential and the scaling factor $1/N$.

In (3.10), we can see that N values of X_m, not just $N/2 + 1$ values, are required for the inverse transformation. Thus, given X_0 through $X_{N/2}$, one could generate $X_{N/2+1}$ through X_{N-1} using (3.9) before applying (3.10). Alternately, (3.10) can be modified as follows, assuming $[x_k]$ is real and using (3.9):

$$\begin{aligned}
x_k &= \frac{1}{N}\sum_{m=0}^{N-1} X_m e^{j(2\pi mk/N)} \\
&= \frac{1}{N}\left[X_0 + \sum_{m=1}^{N/2-1} X_m e^{j(2\pi mk/N)} + X_{N/2}\, e^{j\pi k} + \sum_{m=N/2+1}^{N-1} X_{N-m}^* e^{j(2\pi mk/N)} \right] \\
&= \frac{1}{N}\left[X_0 + (-1)^k X_{N/2} + \sum_{m=1}^{N/2-1} X_m e^{j(2\pi mk/N)} + \sum_{n=1}^{N/2-1} X_n^* e^{-j(2\pi nk/N)} \right] \qquad (3.11) \\
&= \frac{1}{N}\left\{ X_0 + (-1)^k X_{N/2} + 2\,\text{Real}\left[\sum_{m=1}^{N/2-1} X_m e^{j(2\pi mk/N)} \right] \right\} \\
&= \frac{2}{N}\sum_{m=0}^{N/2}\left(R_m' \cos\frac{2\pi mk}{N} - I_m \sin\frac{2\pi mk}{N} \right)
\end{aligned}$$

where $[R_m]$ and $[I_m]$ are, respectively, the real and imaginary parts of $[X_m]$, and $[R'_m]$ is the same as $[R_m]$ but with R_0 and $R_{N/2}$ reduced to half-strength. We note from the original definition in (3.5) that both X_0 and $X_{N/2}$ are real, so only the two real components in (3.11) need to be modified.

One final but important point about the inverse transform is that most computer routines, in order to preserve symmetry between (3.5) and (3.10), omit the factor $1/N$ from (3.10) and (3.11). Thus, the sequence $[x_k]$ is scaled to N times its correct amplitude by most inverse DFT and FFT algorithms, including those described below.

3.6 DFT and FFT Algorithms

Several DFT and FFT programs, listed in Fortran-77 in Appendix A, are described here. Each has different constraints on its use in accordance with the foregoing discussion, that is, restrictions on the type of input data, on allowable values of N, and so on. The brevity and portability of the routines, along with the absence of references to input-output devices mentioned in Chapter 1, Section 1.2, are properties shared by all of the routines described in this book.

The Fourier transform programs are listed in Table 3.2, along with their restrictions and special properties. As indicated by the properties in the table, we recall that the DFT routines are very slow compared with the FFT routines as described by (3.4) but are also comparatively short in length. Thus the DFTs should be used for convenience with small amounts of data, where storage and computing time are negligible factors.

In the subroutine names, the letter I stands for the inverse transform, the letters DFT or FFT designate the type of algorithm, and the final letter, R or C, designates real or complex data. The final name, SPCOMP, stands for *component,* because the routine computes a single DFT component.

The *Time* column in Table 3.2 lists measured execution times for each of the six subroutines executing a 128-point transform. We note that the FFT times are not quite as low as predicted by (3.4). The discrepancy is due to extra, or overhead, computations in the FFT programs not considered in the derivation of (3.4).

3.7 Storage Requirements

The DFT routines require twice as much storage as the FFT routines because the FFT routines all replace the original data values with the transform values, whereas the DFT routines do not. Also, the complex routines SPDFTC and SPFFTC require complex storage locations, whereas the other routines require only real storage, so these two routines require about twice as much storage as their real counterparts. The function SPCOMP requires less storage because it only computes a single DFT component.

TABLE 3.2 CHARACTERISTICS OF FOURIER TRANSFORM PROGRAMS

Program[a]	Algorithm	Direction	Type of Time Series	Number of Samples (N)	Storage	Time[b]
SPDFTR	DFT (real data)	Forward	Real	Any	$2N + 2$ real	256.0
SPIDTR	Inverse DFT (real output)	Reverse	Real	Even	$2N + 2$ real	268.0
SPDFTC	DFT (complex data)	Both	Complex	Any	$2N$ complex	534.0
SPFFTR	FFT (real data)	Forward	Real	Power of 2	$N + 2$ real	9.7
SPIFTR	Inverse FFT (real output)	Reverse	Real	Power of 2	$N + 2$ real	9.9
SPFFTC	FFT (complex data)	Both	Complex	Power of 2	N complex	18.0
SPCOMP[a]	DFT (real data)	Forward	Real	any	N real	

[a] The first six programs are subroutines. SPCOMP is a function.

[b] Approximate number of CPU milliseconds per transform with $N = 128$ data points on a CDC Cyber 170/855 computer running the NOS timesharing system.

3.8 Data Sequence Length Restrictions

As shown in Table 3.2, the data sequence $[x_k]$ can be any length (N) for the DFT routines but must have length equal to a power of 2 for the FFT routines. This restriction on N is not as severe as it might at first appear to be, because zeros can always be appended to the sequence to cause N to become a power of 2 without changing the spectrum of $[x_k]$. For example, if $[x_k]$ contained 1000 samples originally, we could append 24 zero-samples to the end of the sequence, giving $[x_0, x_1, \ldots, x_{999}, 0, 0, \ldots, 0]$, with 1024 samples in all, and not change the spectrum in the following sense.

Suppose the original real data sequence is $x_0, x_1, \ldots, x_{M-1}$, and that M is not a power of 2. We now append $N - M$ zeros to bring the length (N) up to a power of 2, giving

$$[x_k] = [x_0, x_1, \ldots, x_{M-1}, \underbrace{0, 0, \ldots, 0}_{N - M \text{ zeros}}] \tag{3.12}$$

When we compute the DFT of x_k using (3.5), the last $N - M$ terms in the sum are all zero, so the DFTs before and after appending the zeros are similar:

$$\text{Before: } X_m = \sum_{k=0}^{M-1} x_k e^{-j(2\pi mk/M)}; \qquad m = 0, 1, \ldots, \frac{M}{2} \tag{3.13}$$

$$\text{After: } X_m = \sum_{k=0}^{M-1} x_k e^{-j(2\pi mk/N)}; \qquad m = 0, 1, \ldots, \frac{N}{2} \tag{3.14}$$

In fact, since only the exponents differ in (3.13) and (3.14), we can say that the DFT is giving components of the same spectrum in both cases, but that the components are computed at different frequencies. Notice from (3.7) that the maximum frequency in both cases is one-half the sampling frequency, that is,

$$\text{Maximum frequency} = \frac{M/2}{MT} = \frac{N/2}{NT} = \frac{1}{2T} \text{ hertz} \tag{3.15}$$

Thus, with the zeros appended to x_k, we can see in (3.14) that the DFT components are equally spaced over the same frequency range as in (3.13) before the zeros were appended, that is, from 0 to $1/2T$ hertz, but that the components are now more densely spaced.

An example, similar to the third example in Fig. 3.1, is shown in Fig. 3.3. We can see that the zeros appended to $[x_k]$ do not change the form, amplitude, or frequency range of the spectrum, but only the points in the frequency domain at which the spectral values are computed.

3.9 Calling Sequences

The seven routines in Table 3.2 have similar calling sequences, although the variable definitions differ depending on the properties listed in the table. The calling sequences are:

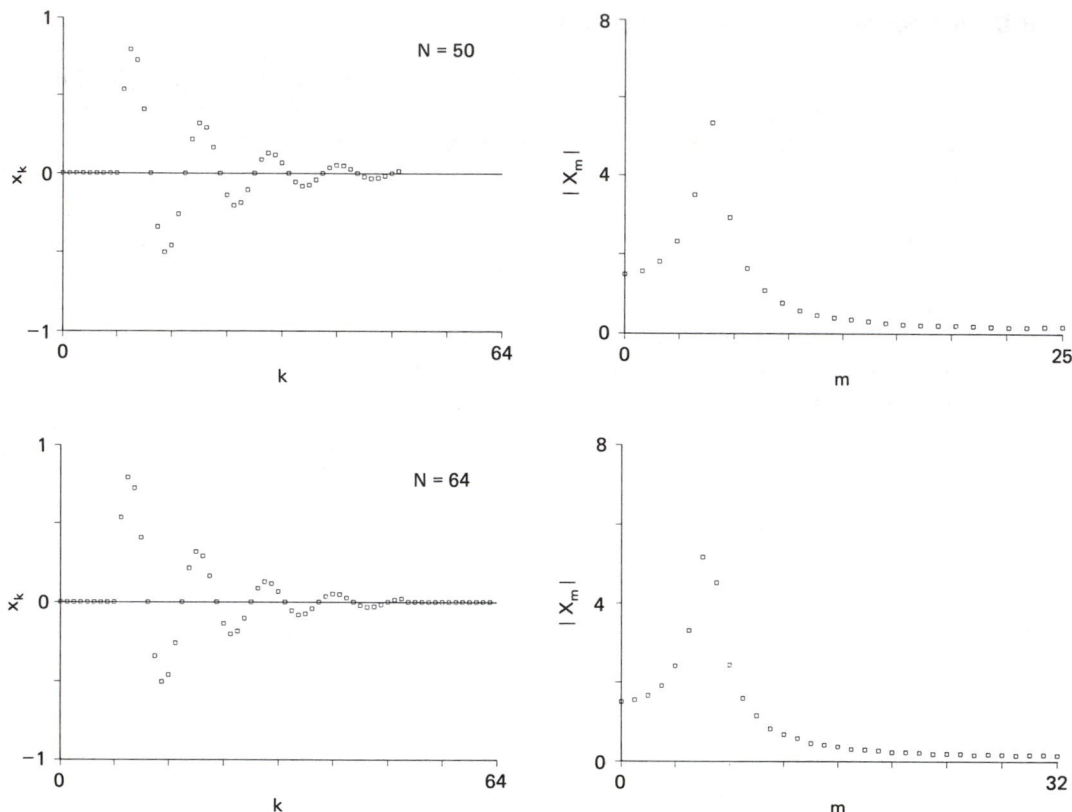

Figure 3.3. An example showing the effect of appending zeros to $[x_k]$. Upper plots show $[x_k]$ with $N = 50$ and corresponding DFT. Lower plots show $[x_k]$ with 14 zeros appended and corresponding DFT. In both cases the DFT frequency range is 0 to $1/2T$ hertz, but the individual DFT component spacing is $1/50T$ in the upper plot and $1/64T$ in the lower plot.

```
CALL SPDFTR (X, Y, N)
CALL SPIDTR (Y, X, N)
CALL SPDFTC (X, Y, N, ISIGN)
CALL SPFFTR (X, N)
CALL SPIFTR (X, N)
CALL SPFFTC (X, N, ISIGN)
COMPLEX FUNCTION SPCOMP (X, N, F)
```

For all six routines, the variables are defined as follows:

X   = input data vector, real or complex depending on routine.

Y   = output data vector in the DFT routines. In the FFT routines, the output replaces the input.

N   = number of samples in the time series data sequence. (Power of 2 for FFTs.)

ISIGN = −1 for forward transform, +1 for inverse.

F   = frequency in Hz-s of single DFT component, where sampling frequency equals 1.0.

Thus, X and Y are the data vectors for the three DFT subroutines, and in the three FFT subroutines, the output replaces the input in X. The types of data in X and Y depend on the routine, as summarized in Table 3.3. For the function SPCOMP, the output is SPCOMP, a single complex DFT component.

Carefully note the use of the square brackets in Table 3.3; these brackets have the following connotation. From the table we can easily see how to specify the X and Y arrays in the calling program for most of the six transform subroutines. For example, if we were using SPDFTR on a sequence of N voltage samples, we might have DIMENSION VOLTS(0:99) with N = 100 and COMPLEX SPECT(0 : 50) with N/2 = 50. The exceptions are SPFFTR and SPIFTR, where we have complex data replacing real data, and vice-versa. In these two cases, we specify the arrays in accordance with the input data, and the output in square brackets is in effect an "equivalent" array replacing the input data. All of this becomes clear when we consider the examples in the following section.

To conclude this section, Table 3.4 lists the array index ranges that are needed to accommodate the arrays shown in Table 3.3. As in Table 3.3, N is the number of elements in the time-domain data array $[x_k]$. The elements may be real or complex (or both), depending on the routine. In Table 3.4, note especially that $X(0:N+1)$ is required for SPFFTR and SPIFTR in order to accommodate the complex FFT components X_0 through $X_{N/2}$, which require $2(N/2 + 1) = N + 2$ real storage locations, that is, $X(0)$ through $X(N+1)$.

3.10 Examples

We now consider a series of examples to show how each subroutine is used. The examples are also useful for checking for correct operation of the routines.

TABLE 3.3   DATA ARRAYS IN FOURIER TRANSFORM PROGRAMS

Routine	Input	Output
SPDFTR	Real X(0), X(1), . . . , X(N-1)	Complex Y(0), Y(1), . . . , Y(N/2)[a]
SPIDTR	Complex Y(0), Y(1), . . . , Y(N/2)	Real X(0), X(1), . . . , X(N-1)
SPDFTC	Complex X(0), X(1), . . . , X(N-1)	Complex Y(0), Y(1), . . . , Y(N-1)
SPFFTR	Real X(0), X(1), . . . , X(N-1)	[Complex X(0), X(1), . . . , X(N/2)]
SPIFTR	Complex X(0), X(1), . . . , X(N/2)	[Real X(0), X(1), . . . , X(N-1)]
SPFFTC	Complex X(0), X(1), . . . , X(N-1)	Complex X(0), X(1), . . . , X(N-1)
SPCOMP	Real X(0), X(1), . . . , X(N-1)	Complex SPCOMP

[a] In SPDFTR, if N is odd, the final Y index is (N-1)/2 instead of N/2.

TABLE 3.4   REQUIRED INDEX RANGES FOR X AND Y ARRAYS

Routine	X Type[a]	X Index Range	Y Type[a]	Y Index Range
SPDFTR	Real	$0:N-1$	Complex	$0:N/2$
SPIDTR	Real	$0:N-1$	Complex	$0:N/2$
SPDFTC	Complex	$0:N-1$	Complex	$0:N-1$
SPFFTR	Real	$0:N+1$	—	—
SPIFTR	Real[b]	$0:N+1$	—	—
SPFFTC	Complex	$0:N-1$	—	—
SPCOMP	Real	$0:N-1$	—	—

[a] Any of the complex arrays could be declared real in the main program, but then of course twice as many elements would be required.

[b] In SPIFTR, X is made real to receive the real output time sequences shown in Table 3.3.

First, consider the forward transform (3.5) of the following sequence of 8 data points:

$$x_k = [-6, -2, 0, 2, 4, 6, 3, -1] \qquad (3.16)$$

Since $N = 8$, we can use either SPDFTR or SPFFTR to obtain the spectrum. We could also use SPDFTC or SPFFTC by making $[x_k]$ a complex sequence with real components as in (3.16) and imaginary components equal to zero.

The following program demonstrates the use of SPDFTR on $[x_k]$ in (3.16). The results of program execution are given immediately after the RUN SPA0301 command to the computer. In this result, notice that, in accordance with (3.5), X_0 is real and is the sum of the time series. Also, $X_{N/2}$ is real and equal to the alternating time series sum. The remaining spectral values, X_1, X_2, and X_3, are complex.

```
            PROGRAM SPA0301
C-DEMONSTRATION OF THE USE OF SUBROUTINE SPDFTR.
            DIMENSION X(0:7)
            COMPLEX Y(0:4)
            DATA X/-6.,-2.,0.,2.,4.,6.,3.,-1./
            CALL SPDFTR(X,Y,8)
            PRINT 1, (M,Y(M),M=0,4)
          1 FORMAT(' M     REAL        IMAGINARY'/(I2,2F10.4))
            STOP
            END
$ RUN SPA0301
   M     REAL        IMAGINARY
   0     6.0000      0.0000
   1   -17.7782      6.5355
   2    -5.0000     -3.0000
   3    -2.2218      0.5355
   4    -4.0000      0.0000
FORTRAN STOP
```

The next program, similar to SPA0301, demonstrates the use of SPFFTR on the same data. The main difference here is that the computation is done in place, with the spectrum replacing (and extending beyond) the time series. The results are, of course, the same for the two programs.

```
        PROGRAM SPA0302
C-DEMONSTRATION OF THE USE OF SUBROUTINE SPFFTR.
        DIMENSION X(0:9)
        DATA (X(M),M=0,7)/-6.,-2.,0.,2.,4.,6.,3.,-1./
        CALL SPFFTR(X,8)
        PRINT 1,(M,X(2*M),X(2*M+1),M=0,4)
      1 FORMAT(' M     REAL      IMAGINARY'/(I2,2F10.4))
        STOP
        END
$ RUN SPA0302
  M     REAL      IMAGINARY
  0     6.0000    0.0000
  1   -17.7782    6.5355
  2    -5.0000   -3.0000
  3    -2.2218    0.5355
  4    -4.0000    0.0000
FORTRAN STOP
```

Next, we demonstrate the use of SPIDTR in SPA0303. The time series used above is stored originally in X1, then transformed into Y using SPDFTR, then reverse-transformed to X2 using SPIDTR, and then printed. (If we had wished, we could have transformed from Y back to X1.) We note that the printed time series is $N = 8$ times the original series in X1, in accordance with the earlier discussion.

```
        PROGRAM SPA0303
C-DEMONSTRATION OF THE USE OF SPIDTR.
        DIMENSION X1(0:7),X2(0:7)
        COMPLEX Y(0:4)
        DATA X1/-6.,-2.,0.,2.,4.,6.,3.,-1./
        CALL SPDFTR(X1,Y,8)
        CALL SPIDTR(Y,X2,8)
        PRINT 1,(K,X2(K),K+4,X2(K+4),K=0,3)
      1 FORMAT(' K   X(K)',11X,'K   X(K)'/(I2,F6.0,10X,I2,F6.0))
        STOP
        END
$ RUN SPA0303
  K   X(K)          K   X(K)
  0   -48.          4    32.
  1   -16.          5    48.
  2     0.          6    24.
  3    16.          7    -8.
FORTRAN STOP
```

A similar demonstration of SPIFTR now follows in SPA0304. Again, the differences here are that the transformations are in place and that N is required to be a power of 2. The results are, of course, identical for SPA0303 and SPA0304.

```
        PROGRAM SPA0304
C-DEMONSTRATION OF THE USE OF SUBROUTINE SPIFTR.
        DIMENSION X(0:9)
        DATA (X(M),M=0,7)/-6.,-2.,0.,2.,4.,6.,3.,-1./
        CALL SPFFTR(X,8)
        CALL SPIFTR(X,8)
        PRINT 1,(K,X(K),K+4,X(K+4),K=0,3)
      1 FORMAT(' K   X(K)',11X,'K   X(K)'/(I2,F6.0,10X,I2,F6
        STOP
        END
$ RUN SPA0304
  K   X(K)          K   X(K)
  0   -48.          4    32.
  1   -16.          5    48.
  2     0.          6    24.
  3    16.          7    -8.
FORTRAN STOP
```

The two routines intended primarily for use with complex time series, SPDFTC and SPFFTC, are demonstrated in the next two examples, SPA0305 and SPA0306. In both cases we begin with a minimal-length ($N = 4$) complex time series, then transform the series forward using ISIGN = -1, then transform it in reverse using ISIGN = $+1$. In the output, the spectral components are listed on the right, and the output time series, which is N times the original data, is on the left. Again, we note that SPFFTC performs the computation in place and requires that N be a power of 2. Notice also that the rounding errors in the spectral output of SPDFTC in SPA0305 are slightly worse than those of SPFFTC in the output of SPA0306. This result is more pronounced with longer data sequences because the FFT algorithm requires fewer products, as shown in (3.4), and therefore tends to have smaller rounding errors.

```
        PROGRAM SPA0305
C-DEMONSTRATION OF THE USE OF SUBROUTINE SPDFTC.
        COMPLEX X1(0:3),X2(0:3)
        DATA X1/(2.,1.),(0.,2.),(1.,1.),(-1.,0.)/
        CALL SPDFTC(X1,X2,4,-1)
        CALL SPDFTC(X2,X1,4,+1)
        PRINT 1,(I,X1(I),X2(I),I=0,3)
      1 FORMAT(' I    X1(I)',19X,'X2(I)'/(I2,2F5.0,7X,2F12.7))
        STOP
        END
$ RUN SPA0305
   I    X1(I)                 X2(I)
   0  8.   4.        2.0000000   4.0000000
   1  0.   8.        3.0000000  -1.0000000
   2  4.   4.        4.0000000  -0.0000002
   3 -4.   0.       -0.9999999   1.0000000
FORTRAN STOP

        PROGRAM SPA0306
C-DEMONSTRATION OF THE USE OF SUBROUTINE SPFFTC.
        COMPLEX X1(0:3),X2(0:3)
        DATA X1/(2.,1.),(0.,2.),(1.,1.),(-1.,0.)/
        CALL SPFFTC(X1,4,-1)
        DO 1 M=0,3
      1 X2(M)=X1(M)
        CALL SPFFTC(X1,4,+1)
        PRINT 2,(I,X1(I),X2(I),I=0,3)
      2 FORMAT(' I    X1(I)',19X,'X2(I)'/(I2,2F5.0,7X,2F12.7))
        STOP
        END
$ RUN SPA0306
   I    X1(I)                 X2(I)
   0  8.   4.        2.0000000   4.0000000
   1  0.   8.        3.0000000  -1.0000001
   2  4.   4.        4.0000000   0.0000000
   3 -4.   0.       -1.0000000   1.0000001
FORTRAN STOP
```

The foregoing examples used short data sequences with no physical significance in order to provide simple, repeatable tests of the routines. The next example, in Fig. 3.4, illustrates the use of SPFFTR or SPDFTR on $N = 1024$ samples of a hypothetical voltage waveform, given by

$$x_k = \begin{cases} 0, & k < 100 \\ \exp\left[-\dfrac{k-100}{176}\right]\sin\left[\dfrac{2\pi(k-100)}{160}\right], & k = 100, 101, ..., 1023 \end{cases} \qquad (3.17)$$

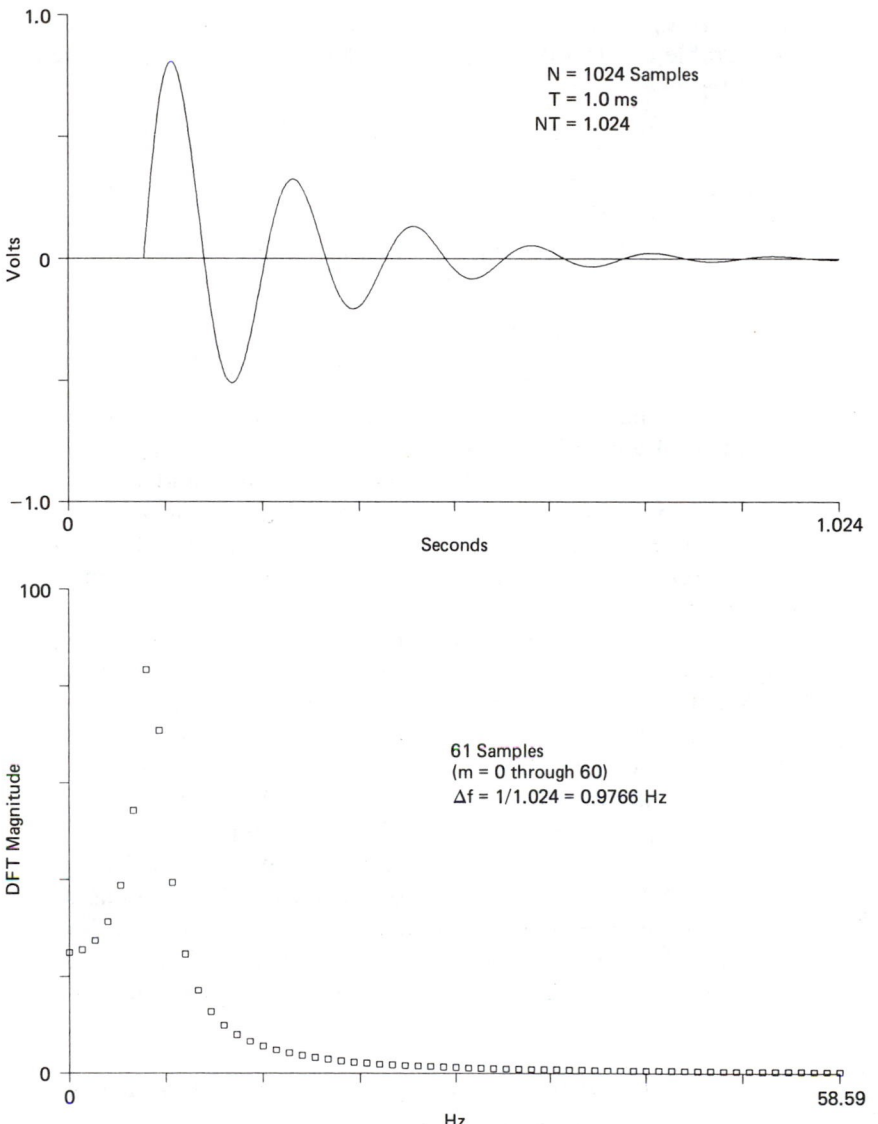

Figure 3.4.   Illustration of the use of SPFFTR or SPDFTR on a sampled voltage waveform. The lower plot shows the first 61 (out of 513) values of the DFT.

If the time step between samples is $T = 1.0$ ms, then we may say that the kth sample occurs at $t = 10^{-3}k$ seconds, so the continuous waveform may be expressed as

$$x(t) = \begin{cases} 0, & t < 0.1 \text{ s} \\ \exp\left[-\dfrac{t - 0.1}{0.176}\right] \sin\left[\dfrac{2\pi(t - 0.1)}{0.160}\right], & 0.1 \text{ s} \leq t < 1.024 \text{ s} \end{cases} \tag{3.18}$$

Since N is a power of 2, either SPDFTR or SPFFTR is appropriate, but SPFFTR is preferable due to the data sequence length. The calling sequence for SPFFTR would be

CALL SPFFTR(X, 1024)

assuming that the sequence $[x_k]$ is originally in the array dimensioned, say, X(0:LX), with LX $\geq$ 1025.

Having called SPFFTR, the following code would compute the amplitude spectrum in X(0) through X(512):

DO 1 M = 0, 512
1 X(M) = SQRT (X(2*M)**2 + X(2*M + 1)**2)

The first 61 values of the amplitude spectrum are plotted in Fig. 3.4. These may be compared with Fig. 3.3, in which a similar waveform was used.

The values of the DFT components in this and other examples do not, unfortunately, have a direct or simple physical significance. They do, however, give a relative indication of the distribution of the amplitude of $x(t)$ over frequency. Thus, in Fig. 3.4, we see that the amplitude of $x(t)$ is concentrated in the neighborhood of 6 Hz.

In fact, the DFT values depend on the sampling rate as well as the number of samples, as one can see by comparing Figs. 3.3 and 3.4. If the time step in Fig. 3.4 were decreased from 1 ms to, say, 0.5 ms and N was increased from 1024 to 2048 samples, the DFT values would be about twice what they are in the figure. The significance of DFT values is discussed further in Chapter 4.

Our final examples demonstrate SPCOMP, which is a complex function routine designed to compute a single DFT component of a real data sequence. To test SPCOMP we could run the following program, SPA0307, to get the same results as in SPA0301. We note particularly in the argument list of SPCOMP in SPA0307 the frequency value $m/8$, which is called F in the calling sequence. This value gives the frequency in hertz-seconds of each successive DFT component in accordance with Table 3.1, that is, m/N hertz-seconds, with $N = 8$ in this case.

```
            PROGRAM SPA0307
C-DEMONSTRATION OF THE USE OF COMPLEX FUNCTION SPCOMP
            DIMENSION X(0:7)
            COMPLEX Y(0:4),SPCOMP
            DATA X/-6.,-2.,0.,2.,4.,6.,3.,-1./
            DO 1 M=0,4
          1 Y(M)=SPCOMP(X,8,M/8.)
            PRINT 2,(M,Y(M),M=0,4)
          2 FORMAT(' M     REAL       IMAGINARY'/(I2,2F10.4))
            STOP
            END
    $ RUN SPA0307
     M     REAL       IMAGINARY
     0     6.0000     0.0000
     1   -17.7782     6.5355
     2    -5.0000    -3.0000
     3    -2.2218     0.5355
     4    -4.0000     0.0000
    FORTRAN STOP
```

This frequency specification in SPCOMP gives us the ability to "zoom in" on a part of the spectrum of a time series. The DFT values are not really independent of each other unless m is an integer; nevertheless, it is sometimes convenient to let m slide between the integers and thereby focus on a part of the frequency domain for plotting purposes.

As an example, suppose we wish to look in detail at the spectrum in Fig. 3.4 in the frequency range from 0 to 30 Hz. Given that the 1024 time series samples are stored in X(0) through X(1023), the following code computes, in Y(0) through Y(200), 201 equally spaced samples of the DFT in the frequency range from 0 to 30 Hz. Note that as the program is executed, the F argument in SPCOMP goes from 0 to $30T$ hertz-seconds, where T is the 1-ms time step. Thus, the frequency range (0, 30) Hz corresponds with (0, 0.03) Hz-s in accordance with Table 3.1.

```
DIMENSION X(0:1023), Y(0:200)
COMPLEX SPCOMP



        .
        .
        .


DO 1 I = 0, 200
 F = 30.*0.001*I/200
1 Y(I) = ABS(SPCOMP(X,1024,F))
```

Note also that the function ABS is generic in Fortran 77 and so computes the correct magnitude of the complex quantity returned by SPCOMP. The 201 values of this DFT magnitude are plotted in Fig. 3.5.

Thus, the SPCOMP function gives us the ability to compute arbitrary spectral components or to compute a portion of a spectrum in arbitrarily fine detail. A related routine, called SPCHRP, is described in Chapter 14. SPCHRP implements the chirp z-transform. It also is useful for zooming in on a part of the spectrum and is faster than SPCOMP but is more restricted in use and more complicated to use than SPCOMP, so we leave its description to Chapter 14.

3.11 Theory of Operation

In this section we discuss briefly the theory of operation of each of the Fourier transform routines. First, the three DFT routines follow exactly the transform definitions already given. SPDFTR and SPCOMP are exact implementations of (3.5), SPIDTR is an implementation of (3.11) with the factor $1/N$ omitted, and SPDFTC is described by either (3.5) or (3.10) without the factor $1/N$, depending on whether ISIGN is -1 or $+1$, denoting the forward or the inverse transform. In the case of SPDFTC, note that N transform values are computed in either the forward or the inverse case, because the time series is complex.

SPFFTC is the basic member of the set of three FFT routines. The other

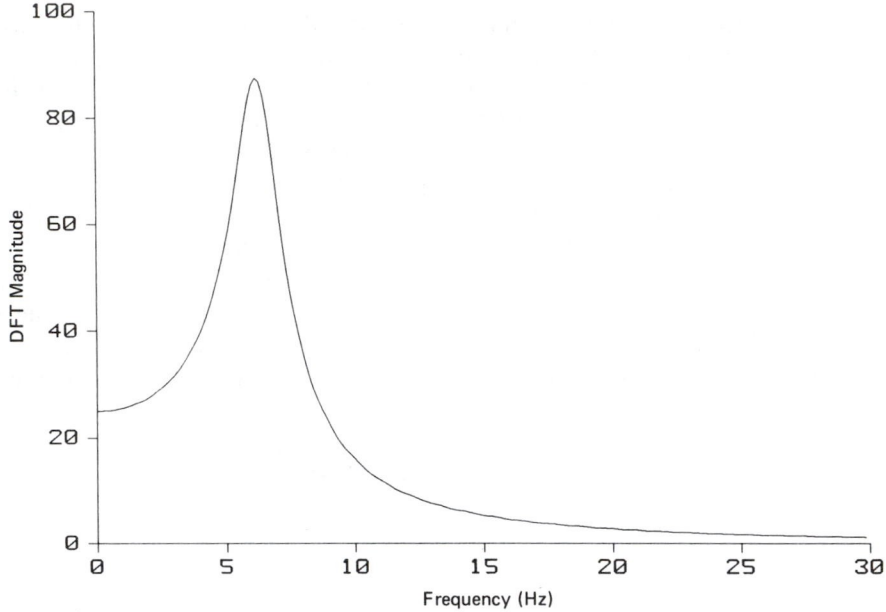

Figure 3.5. An expansion of the spectrum in Fig. 3.4 using the complex function SPCOMP to obtain 200 spectral values from 0 to 30 Hz. The ordinary DFT algorithms would produce only 30 points in this range.

two routines both call SPFFTC. The algorithm implemented in SPFFTC is the symmetric Cooley-Tukey algorithm [C] based on time decomposition with input bit reversal. This algorithm is discussed at length in the references [A–P], and its operation will not be covered here in detail. The result of SPFFTC is exactly the same as that of SPDFTC, except that the computation is in place and the sequence length, N, must be a power of 2.

SPFFTR accomplishes the in-place transformation of a real sequence of length N in half the storage required by SPFFTC using a method given by Brigham [Q]. In Brigham's method, the original real sequence of N points is viewed as a complex sequence of $N/2$ points, that is,

$$[x_0, x_1, ..., x_{N-1}] = [v_0, v_1, ..., v_{N/2-1}] \tag{3.19}$$

In the complex sequence $[v_k]$, the real part of v_0 is x_0, the imaginary part of v_0 is x_1, the real part of v_1 is x_2, and so on. The complex routine SPFFTC is then used to take the $N/2$-point transform of $[v_k]$, replacing $[x_k]$ with

$$[V_m] = [V_0, V_1, ..., V_{N/2-1}] \tag{3.20}$$

The desired transform $[X_m]$ can then be obtained from $[V_m]$ in accordance with the following formulas, which are expressions of Brigham's formulas:

$$\begin{bmatrix} X_m \\ X^*_{N/2-m} \end{bmatrix} = \frac{1}{2} \begin{bmatrix} (1 - U_m) & (1 + U_m) \\ (1 + U_m) & (1 - U_m) \end{bmatrix} \begin{bmatrix} V_m \\ V^*_{N/2-m} \end{bmatrix} \tag{3.21}$$

where:
$$U_m = j \exp(-j2\pi m/N) = \sin(2\pi m/N) + j\cos(2\pi m/N)$$

In this result both formulas are needed together in SPFFTR because $[X_m]$ replaces $[V_m]$ during the computation. Thus, X_0 and $X_{N/2}$ replace V_0 and $V_{N/2}$, then X_1 and $X_{N/2-1}$ replace V_1 and $V_{N/2-1}$, and so on. Also, since $[V_m]$ is an $N/2$-point transform, $V_{N/2}$, which is not given in (3.20), must equal V_0 in accordance with (3.8). Thus, SPFFTR accomplishes the transformation of $[x_k]$ using (3.21).

SPIFTR reverses the effect of SPFFTR, except for the scale factor N. In other words, executing the two statements

CALL SPFFTR(DATA, N)
CALL SPIFTR(DATA, N)

would leave the DATA array exactly as it was originally, except that each element would be N times its original value. To accomplish the in-place inverse transform, SPIFTR essentially reverses the operation of SPFFTR. Equation (3.21) can be written

$$\begin{bmatrix} V_m \\ V_{N/2-m}^* \end{bmatrix} = \frac{1}{2}\begin{bmatrix} (1 - U_m^{-1}) & (1 + U_m^{-1}) \\ (1 + U_m^{-1}) & (1 - U_m^{-1}) \end{bmatrix}\begin{bmatrix} X_m \\ X_{N/2-m}^* \end{bmatrix} \tag{3.22}$$

so that now $[V_m]$ can be computed in terms of the spectrum, $[X_m]$. Then, when $[V_m]$ is computed, it can be inverted to N times $[v_k]$ using SPFFTC with ISIGN $= +1$, and $[Nv_k]$ is equivalent to $[Nx_k]$ as in (3.19). The inversion using SPFFTC is actually an $N/2$-point transform, so $[2V_m]$ is used instead of $[V_m]$ in order to achieve a final scaling factor of N.

The scaling factor N could easily be included in any of the inverse transforms, but its omission for the sake of symmetry between (3.5) and (3.10) has become standard practice. For example, the user could easily modify SPIDTR or SPIFTR by inserting $1/N$ into the computation of the output.

3.12 Summary of Calling Sequences

To conclude this chapter we summarize the calling sequences listed in Section 3.9. Here we use a slightly different format to list the variables used in all of the routines. The calling sequences are given in the following statements:

CALL SPDFTR(X, Y, N)
CALL SPIDTR(Y, X, N)
CALL SPDFTC(X, Y, N, ISIGN)

CALL SPFFTR(X, N)
CALL SPIFTR(X, N)
CALL SPFFTC(X, N, ISIGN)

COMPLEX FUNCTION SPCOMP(X, N, F)

Variable	Definition	Input-Output	Remarks
X	Data vector	I,O	See Table 3.3.
Y	Data vector	I,O	Not used with FFTs. See Table 3.3.
N	Transform size	I	Power of 2 for FFTs.
ISIGN	−1 for forward +1 for inverse	I	Used only in complex transforms.
F	Frequency in hertz-seconds, normally in (0, 0.5).	I	Specifies single frequency for SPCOMP.

In this tabulation of the variables, the symbols I and O indicate whether the variable is an input from the user to the routine or an output computed by the routine during execution.

3.13 Exercises

3.1. Demonstrate subroutine SPDFTR by using it to compute the DFT of the following sets of data. Check some of your results using equation (3.5).

	(a)		(b)		(c)
k	$x(k)$	k	$x(k)$	k	$x(k)$
0	1.0	0	0.0	0	0.0
1	2.0	1	0.0	1	−1.0
2	3.0	2	1.0	2	0.0
3	2.0	3	0.0	3	1.0
				4	0.0
				5	−1.0

3.2. Write a program to demonstrate SPFFTR on the following sets of data.

	(a)		(b)		(c)
k	$x(k)$	k	$x(k)$	k	$x(k)$
0	1.0	0	−1.0	0	−2.0
1	2.0	1	−2.0	1	−4.0
2	3.0	2	−1.0	2	−2.0
3	2.0	3	0.0	3	0.0
		4	1.0	4	2.0
		5	2.0	5	4.0
		6	1.0	6	2.0
		7	0.0	7	0.0

3.3. Using the data in exercise 3.1, write a program that uses SPDFTR to transform the data, then uses SPIDTR to transform back, and then scales the output to obtain the original data.

3.4. Using the data in exercise 3.2, write a program that uses SPFFTR to transform the data, then uses SPIFTR to transform back, and then scales the output to obtain the original data.

3.5. Using the data in exercise 3.1, write and run a program to compute the amplitude spectra.

3.6. Using the data in exercise 3.2, write and run a program that computes the amplitude spectra.

3.7. Use SPFFTR with zero extension to compute the amplitude spectra in exercise 3.1(b) and (c).

3.8. Using SPDFTC, compute the amplitude spectra of the following sets of data.

	(a)		(b)
k	$x(k)$	k	$x(k)$
0	$1.0 + j0.0$	0	$0.0 + j1.0$
1	$0.5 + j0.5$	1	$0.0 - j1.0$
2	$0.0 + j1.0$	2	$1.0 + j0.0$
3	$-1.0 + j0.0$	3	$-1.0 + j0.0$
4	$0.0 - j1.0$	4	$0.0 + j0.0$
		5	$0.5 + j0.5$
		6	$0.5 - j0.5$

3.9. Using SPFFTC with zero extension, repeat exercise 3.8.

3.10. Write and run a program that computes the correct amplitude spectra for exercise 3.1 using SPDFTC.

3.11. Write and run a program that transforms the data in exercise 3.1 using SPDFTR, then uses SPDFTC to recover the original sequences.

3.12. Write and run a program that transforms the data in exercise 3.2 using SPFFTC and then recovers the original data using SPFFTC.

3.13. Write a program to compute 2048 sequential samples of the function $x(t) = t \exp(-t/3)$. Use time step $T = 0.015$ s, with the first sample at $t = 0$. Then run the program and plot the function by drawing straight lines between sample points.

3.14. Using the data sequence in exercise 3.13, compute and plot the amplitude spectrum using SPFFTR. Plot the spectrum only over the range from 0 to 1.0 Hz.

3.15. Using the data sequence in exercise 3.13, compute and plot the phase spectrum, $[\theta_k]$, over the interval from 0 to 1.0 Hz.

3.16. Examine the effects on the spectrum of truncation of the data sequence in exercise 3.13. Using SPFFTR, plot the complete amplitude spectrum based on 512 samples and then 256 samples. Plot both spectra over the same frequency scale from zero to 1.0 Hz.

3.17. Repeat exercise 3.16 for the phase spectrum.

3.18. Using SPFFTR on the data sequence in exercise 3.13, compute the amplitude spectrum. Then, plot just the spectral values $|X_m|$ for $m = 0$ through 24. Label the frequency axis of your plot in terms of m, also in hertz, in Hz-s, in rad, and in rad/s.

3.19. Using SPCOMP, examine in detail the spectrum of the waveform in exercise 3.13.

Make a continuous plot in the range from 0 to 0.8 Hz, using 100 values of the amplitude spectrum spaced evenly over this frequency range. Compare with the result of exercise 3.18.

3.20. Do exercise 3.18 for the phase spectrum.

3.21. Do exercise 3.19 for the phase spectrum.

Note: Exercises 3.22 through 3.29 involve the effects of *shifting* a data sequence in the time domain.

3.22. Using SPFFTR, show that the amplitude spectrum does not change when the data in exercise 3.2(a) is shifted circularly (that is, end-around) one place to the right. In other words, show that (1, 2, 3, 2) and (2, 1, 2, 3) have the sample amplitude spectra.

3.23. Using SPFFTR, show that the amplitude spectrum is invariant when the data in exercise 3.2(b) is shifted circularly two places to the right.

3.24. Show that the amplitude spectrum is unchanged when the sequence in exercise 3.2(c) is shifted circularly three places to the right.

3.25. Prove the following shifting theorem: If a data sequence $[x_k]$ is rotated n places, with $n > 0$ for a right shift, producing a shifted sequence $[x_k']$, the new amplitude and phase spectra are

$$|X_m'| = |X_m| \tag{3.23}$$

$$\theta_m' = \theta_m + \frac{2\pi nm}{N} \tag{3.24}$$

where N is the sequence length. (*Hint:* Use equation (3.5).)

3.26. Verify equation (3.24) by shifting the data in exercise 3.2(a) one place to the right, that is, toward increasing k.

3.27. Verify (3.24) by shifting the data in exercise 3.2(c) three places to the left.

3.28. Plot the amplitude spectra before and after shifting the data in exercise 3.13 4.5 s to the right, *without rotation*. Explain why the two spectra are so similar, using the theorem in exercise 3.25.

3.29. Verify (3.24) by plotting the phase spectra before and after shifting the data in exercise 3.13 4.5 s to the right.

Note: Exercises 3.30–3.32 examine a phenomenon known as *leakage*. If a pure sinusoidal waveform is sampled evenly over an integral number of periods, the DFT in (3.5) will contain a single nonzero component at the exact frequency of the sinusoid. If the sample set does not cover an integral number of cycles, the DFT will consist of nonzero components with larger amplitudes in the vicinity of the frequency of the sinusoid. This result can be shown by letting $x_k = A \sin(2\pi k/p + \alpha)$, when p is any real period, in (3.6).

3.30. Compute and plot the amplitude spectrum of

$$x_k = \cos\left(\frac{2\pi k}{16}\right), \qquad k = 0, 1, \ldots, 63$$

Compare with the upper DFT in Fig. 3.1.

3.31. Compute and plot the amplitude spectrum of

$$x_k = 0.7\sin\left[\frac{2\pi(k+7)}{16}\right] + \sin\left[\frac{4\pi(k+4)}{16}\right], \qquad k = 0, 1, \ldots, 63$$

Compare with the center DFT in Fig. 3.1.

3.32. Compute and plot the amplitude spectrum of

$$x_k = \cos\left(\frac{2\pi k}{17}\right), \qquad k = 0, 1, \ldots, 63$$

Compare the result with the spectrum in exercise 3.30.

3.14 References

[A] Bergland, G. D., "A Guided Tour of the Fast Fourier Transform," *IEEE Spectrum,* v. 6 (July, 1969): 41.

[B] Cochran, W. T., et al., "What Is the Fast Fourier Transform?" *IEEE Trans.,* AU-15, no. 2 (June, 1967): 45.

[C] Cooley, J. W., and J. W. Tukey, "An Algorithm for the Machine Calculation of Complex Fourier Series," *Math. Comput.,* 19 (April, 1965): 297.

[D] Dejongh, H. R., and E. DeBoer, "The Fast Fourier Transform and Its Use," *1971 DECUS Proceedings,* Maynard, Mass.: Digital Equipment Corp.

[E] Glisson, T. H., C. I. Black, and A. P. Sage, "The Digital Computation of Discrete Spectra Using the Fast Fourier Transform," *IEEE Trans.,* AU-18 no. 3 (September, 1970): 271.

[F] Gentleman, W. M., and G. Sande, "Fast Fourier Transforms—for Fun and Profit," *1966 Fall Joint Computer Conf. AFIPS Proc.,* 29 (Washington, D. C.: Spartan, 1966): 563.

[G] Gold, B., and C. M. Rader, *Digital Processing of Signals* (New York: McGraw-Hill, 1969), Chapter 6.

[H] Good, I. J., "The Interaction Algorithm and Practical Fourier Series," *J. Roy. Statist. Soc. Ser. B,* 20 (1958): 361; 22 (1960): 372.

[I] Hovanessian, S. A., and L. A. Pipes, *Digital Computer Methods in Engineering* (New York: McGraw-Hill, 1969), Chapter 4.

[J] *IEEE Transactions on Audio and Electroacoustics* (Special Issues on the Fast Fourier Transform), AU-15, no. 2 (June, 1967): AU-17, no. 2 (June, 1969).

[K] Kahaner, D. K., "Matrix Description of the Fast Fourier Transform," *IEEE Trans.,* AU-18, no. 4 (December 1970): 442.

[L] Oppenheim, A. V., and R. W. Schafer, *Digital Signal Processing* (Englewood Cliffs, N.J.: Prentice-Hall, 1975), Chapter 6.

[M] Stearns, S. D., *Digital Signal Analysis* (Rochelle Park, N.J.: Hayden, 1975), Chapter 6.

[N] Rabiner, L. R., and B. Gold, *Digital Signal Processing* (Englewood Cliffs, N.J.: Prentice-Hall, 1975), Chapter 6.

[O] Ahmed, N., and K. R. Rao, *Orthogonal Transforms for Digital Signal Processing* (New York: Springer-Verlag, 1975).

[P] Elliott, D. F., and K. R. Rao, *Fast Transforms* (New York: Academic Press, 1983).

[Q] Brigham, E. O., *The Fast Fourier Transform* (Englewood Cliffs, N.J.: Prentice-Hall, 1974), Chapter 10.

chapter 4

Routines for Random Numbers, Spectral Analysis, and Coherence Estimation

4.1 Introduction

The basic routines for spectral analysis were introduced in Chapter 3. Now we wish to incorporate these routines into higher-order routines that compute spectral periodograms and to incorporate these, in turn, into programs that compute power spectra, cross spectra, and coherence. The subject of spectral analysis, particularly the spectral analysis of random time series, is very broad and has many facets. There are excellent texts devoted to spectral analysis and related subjects, for example, [A–D]. Nevertheless, with the routines introduced in this chapter, the reader will be able to estimate spectra in practice and to interpret the results of spectral estimation.

To estimate power spectra we will concentrate in this chapter on the average periodogram method described by Welch [F]. This method is the simplest to understand and is quite general in application. In Chapter 11, we introduce a different approach to spectral estimation.

We first describe the periodogram and a routine, SPPOWR, for computing it and then show how this routine can be used in programs to compute or estimate power and energy spectra. After that we describe the computation of cross spectra and coherence.

4.2 The Periodogram

Given a real sample set $[x_k]$ as in equation (3.1) of the last chapter, the periodogram is defined as the squared magnitude of the discrete Fourier transform (DFT) of $[x_k]$, usually scaled by N, the number of samples in $[x_k]$. Thus, from equation (3.5), assuming N is even, we define the periodogram as a collection of $N/2 + 1$ real values:

$$P_m \triangleq \text{Periodogram of } [x_k]$$

$$\triangleq \frac{1}{N}|X_m|^2, \qquad m = 0, 1, \ldots, \frac{N}{2} \tag{4.1}$$

where
$$X_m = \sum_{k=0}^{N-1} x_k e^{-j(2\pi km/N)}$$

From equation (3.9), we note that when $[x_k]$ is a real time series, $P_m = P_{N-m}$ for the rest of the values of m from $N/2 + 1$ to $N - 1$. This is the most common definition of the periodogram, although sometimes a different scaling factor is used [A, F, I, J, K].

Assume for discussion that the real data sequence $[x_k]$ is embedded in a long time series. Then an element P_m of the periodogram can be viewed as a measure of the *power density* of the time series in the vicinity of $[x_k]$. The entire periodogram is therefore a measure of the *power density spectrum* of $[x_k]$. We can demonstrate this by using a sampled sine wave with amplitude A for $[x_k]$. Let

$$x_k = A \sin\left(\frac{2\pi k}{K}\right), \qquad k = 0, 1, \ldots, MK - 1 \tag{4.2}$$

This sequence is seen to have K samples per cycle and we assume that the sampling theorem is satisfied, that is, $K > 2$. Also the number of samples (N) in this case is MK, so that the sample set covers an integral number of cycles. As stated in Chapter 3, the DFT of $[x_k]$ has therefore exactly one nonzero component, that is, the DFT is

$$X_m = \begin{cases} 0, & m \neq M \\ \dfrac{AMK}{2j}, & m = M \end{cases} \tag{4.3}$$

This result is easy to obtain from the DFT definition, equation (3.5). See exercise 4.1. Combining (4.3) and (4.1) with $N = MK$, we find that the periodogram for this example is

$$P_m = \begin{cases} 0, & m \neq M \\ \dfrac{A^2 N}{4}, & m = M \end{cases} \tag{4.4}$$

Again, note that $N = MK$ is the total number of data samples over exactly M cycles of the sine wave in this example. This periodogram, or power density spectrum, is plotted in Fig. 4.1.

The units, that is, the frequency and power density scales in Fig. 4.1, are especially important. They are the key to understanding how the periodogram is related to power density. The periodogram is usually plotted in the form of a bar graph, as in Fig. 4.1. The amplitude of each bar is the power density given by (4.1), or by (4.4) in this example. Since there are N bars in the range from 0 to 1.0 Hz-s on the normalized frequency scale, the width of each bar is

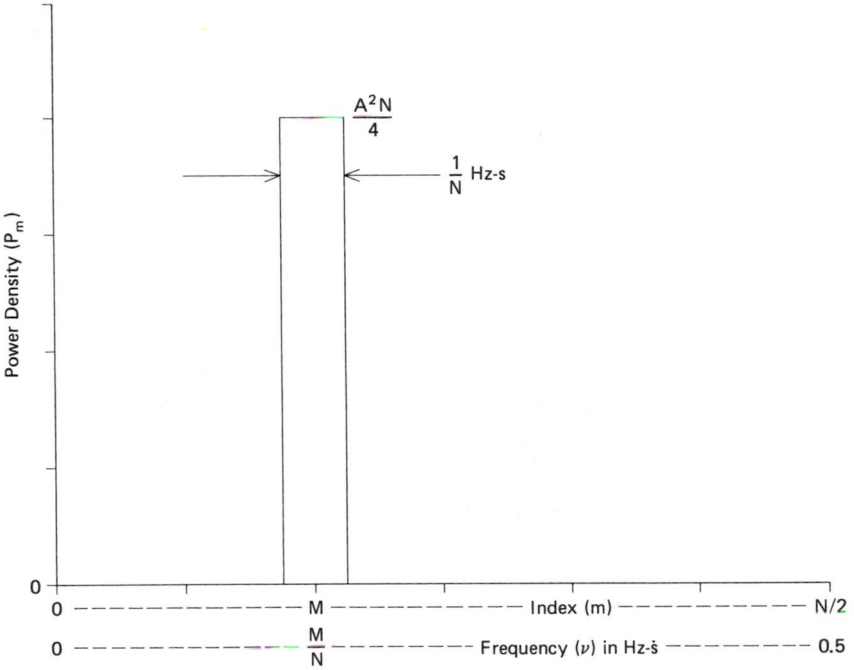

Figure 4.1. Periodogram (power density spectrum) of $N = MK$ samples of the sequence $x_k = A \sin (2\pi k/K)$, $k = 0, 1, \ldots, MK - 1$. Plot is made as if $N = 24$ and $M = 4$. Note that there is a similar bar, not usually plotted, at $M = N - M = 20$.

$$\text{Bar width} = \frac{1}{N} \text{ hertz-seconds} \qquad (4.5)$$

Let us assume now that there are N bars in the power density spectrum at $m = 0$ through $N - 1$. Since $P_{N-m} = P_m$, only the first $N/2 + 1$ bars are normally plotted as in Fig. 4.1. The power concentrated in each bar with index m, that is, the power in the range from $(m - 0.5)/N$ to $(m + 0.5)/N$ hertz-seconds, is then equal to the area of the bar. Thus, generalizing from Fig. 4.1 and using equation (4.1), the power at each frequency index m is

$$\begin{array}{l} \text{Power in frequency} \\ \text{range from } (m - 0.5)/N \\ \text{to } (m + 0.5)/N \text{ hertz-seconds} \end{array} = \frac{P_m}{N} = \frac{1}{N^2}|X_m|^2, \qquad 0 \leq m \leq \frac{N}{2} \qquad (4.6)$$

The *total power* in the data sequence $[x_k]$ is the integral of the power density spectrum, that is, twice the area of the bar graph of the entire periodogram, from $m = 0$ through $m = N - 1$.

$$\begin{aligned} \text{Total power} &= \frac{1}{N^2}\sum_{m=0}^{N-1}|X_m|^2 \\[2mm] &= \frac{1}{N}\sum_{m=0}^{N-1} P_m \end{aligned} \qquad (4.7)$$

In Fig. 4.1, for example, the total power is twice the area of the single nonzero bar:

$$\begin{matrix} \text{Total power} \\ \text{in Fig. 4.1} \end{matrix} = 2\left(\frac{1}{N}\right)\left(\frac{A^2N}{4}\right) = \frac{A^2}{2} \tag{4.8}$$

This total power is seen to be the average squared value (or average power) of the data sequence $[x_k]$ in (4.2). Thus, in general, the following rule relates the periodogram to the power in a data sequence that is assumed, as above, to be extracted from a long time series:

> The total power, or average power, or average squared value, of a data sequence is equal to the area of the periodogram at positive frequencies from $m = 0$ through $m = N - 1$.
> The total power is therefore given by
>
> $$E[x_k^2] = \frac{1}{N^2}\sum_{m=0}^{N-1}|X_m|^2 \tag{4.9}$$
>
> $$= \frac{1}{N}\sum_{m=0}^{N-1}P_m$$

The units of total power are the squared units of $[x_k]$. If x_k is velocity in meters per second, power is measured in (meters per second)^2, and so on. The units of power density are therefore power per hertz-second in Fig. 4.1. Power density is also sometimes plotted as power per hertz instead of hertz-second. In this case the bar width in Fig. 4.1 changes to $1/NT$ hertz, and the ordinate values must therefore be multiplied by T to give the result in (4.9). See exercise 4.2.

4.3 Averaged Periodograms of Overlapped Data Segments

We defined the periodogram in (4.1) as if the entire data sequence $[x_k]$ were transformed to obtain $[X_m]$ and the squared magnitude of $[X_m]$ were then to be used to obtain the periodogram. Alternately, with random or noisy data, an improved estimate of the power spectrum can be computed by taking an *average periodogram*. The average periodogram is obtained by breaking the data sequence into segments of equal length, applying a data window to each segment, then computing the periodogram of each windowed segment, and finally averaging the periodograms together.

The segments of data may be overlapped or they may be disjoint. If the segments are made to overlap, more segments and, therefore, more periodograms can be taken from the same sequence; however, the periodograms are not independent. Disjoint and overlapping segments are illustrated in Fig. 4.2 for a sequence of 25 samples and a segment size of $N = 8$ samples. There are only three disjoint, nonoverlapping segments, but there are five half-overlapping segments. Other degrees of overlap could have been used. Note that the last sample of the sequence is not used in a segment in this example. A residue of unused samples is typical.

Figure 4.2. A data sequence with 25 samples shown broken into segments with 8 samples each. We can obtain 3 nonoverlapping segments, 5 half-overlapping segments, and so forth.

The power spectral estimate for a sequence of random or noisy data is obtained typically by averaging the periodograms of half-overlapping segments [F]. When the spectra are used in coherence estimates (described in Section 4.9), an overlap of 62.5% has been recommended by Carter et al. [O]. We give some examples of spectral estimation with overlapping segments after first introducing a routine for computing the average periodogram.

4.4 Average Periodogram Routine

We now describe a routine, SPPOWR, for computing the average periodogram, or estimated power spectrum, of a real data sequence. The routine uses SPFFTR to compute the DFT of each segment of the sequence, so each segment length must be a power of 2. This restriction is usually easy to accommodate. Another restriction is that the entire data sequence must be in the data array X(0:LX) at execution. If the data sequence is too long for the maximum array size, the routine must be modified and called repeatedly.

SPPOWR itself is simple and quite easy to understand and use. It is listed in Appendix A. It uses SPFFTR to compute the segment periodograms in accordance with (4.1) and accumulates the average periodogram by summing the segment periodograms. The calling sequence is

CALL SPPOWR(X, Y, WORK, LX, LY, IWINDO, OVRLAP, NSGMTS, IERROR)

X(0:LX)	= input data sequence, containing LX + 1 samples
Y(0:LY)	= output spectrum, where LY is a power of 2; corresponding segment size is $N = 2*LY$
WORK(0:2*LY + 1)	= work array
LX	= last index in data sequence X(0), X(1), . . . , X(LX); must be at least $N - 1$
LY	= last frequency index. Corresponds to half sample rate; must be a power of 2.
IWINDO	= indicator of type of data window: 1 (rectangular), 2 (tapered rectangular), 3 (triangular), 4 (Hanning), 5 (Hamming), or 6 (Blackman)
OVRLAP	= fraction that each data segment overlaps its predecessor. Must be at least 0 and less than 1
NSGMTS	= number of overlapping segments averaged together, computed internally
IERROR	= output error indicator, 0 if no error is detected, 1 if IWINDO is out of range, 2 if LX is too small for one segment, or 3 if LY is not a power of 2

All the arguments except IWINDO should be easy to understand. IWINDO indicates the type of data window to use on each data segment. A data window is a sequence of numbers that is multiplied by the data segment before taking the FFT, and its purpose is to smooth or otherwise shape the periodogram of the data segment. We will leave the discussion of data windows to Chapters 8 and 14, except to note here that windows 1 and 2 are generally used with relatively short or nonstationary data files, windows 4 through 6 with long, stationary files, and window 3 in special applications.

Note that we now have two data sequence lengths: LX, the overall sequence length, and $N = 2*LY$, the segment length. Typically, N and LY are small compared with LX. We reserve the symbol N to denote the segment length, that is, the DFT size as in Chapter 3, in (4.1), and so forth.

In the calling sequence for SPPOWR, notice also that the upper frequency index, LY, is a power of 2. Thus, as described in Chapter 3, the corresponding data segment size (N) is 2 times LY, also a power of 2, and the power density spectrum computed using SPFFTR has LY + 1 components, that is, P_0 through $P_{N/2}$ in equation (4.1), with $N = 2*LY$.

A very simple example of the use of SPPOWR is given in SPA0401 below. The DO loop at the beginning of the program generates $N = 32$ samples of the sinusoidal sequence in equation (4.2) with $A = 1.0$ and $K = 8$. The call to

```
        PROGRAM SPA0401
C-GENERATES A PERIODOGRAM SIMILAR TO FIG. 4.1.
        DIMENSION X(0:31),Y(0:16),WORK(0:33)
        PI=4.*ATAN(1.)
        DO 1 K=0,31
        X(K)=SIN(2.*PI*K/8.)
     1  CONTINUE
        CALL SPPOWR(X,Y,WORK,31,16,1,0.,NSGMTS,IERROR)
        PRINT 2,NSGMTS,IERROR,Y
     2  FORMAT(' NSGMTS,IERROR=',2I4/' Y=',8F7.3/2X,9F7.3)
        CALL HP(Y,17,0,.1,.1,.9,.9)
        STOP
        END
$ RUN SPA0401
  NSGMTS,IERROR=    1    0
  Y=  0.000  0.000  0.000  0.000  8.000  0.000  0.000  0.000
      0.000  0.000  0.000  0.000  0.000  0.000  0.000  0.000  0.000
FORTRAN STOP
```

SPPOWR then computes $[P_m]$ in accordance with (4.1) by using the rectangular data window and treating the entire sequence of 32 samples as a single data segment. Note that the values of NSGMTS and IERROR, computed at execution and printed below the program listing, are 1 and 0, as they should be. The periodogram, which is plotted by HP (see Appendix B) in Fig. 4.3, is therefore

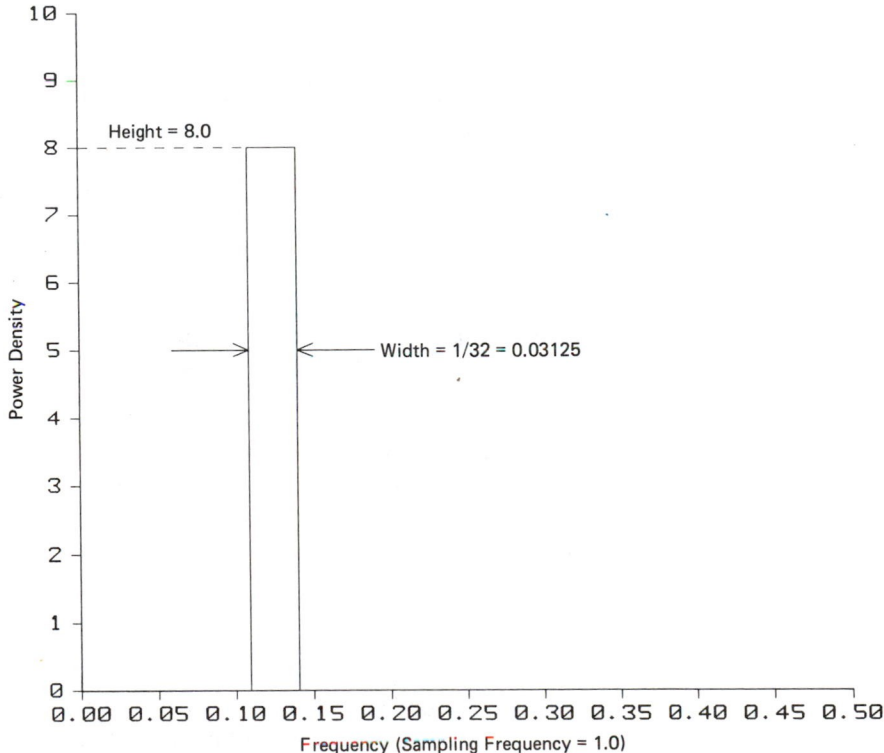

Figure 4.3. Power density spectrum of a single sinusoid at $\frac{1}{8}$ of the sampling frequency, plotted by SPA0401. Similar to Fig. 4.1 but with $N = 32$, $M = 4$, and $K = 8$. Nonzero bar is centered at $m = M = 4$, or $\nu = 0.125$.

similar to Fig. 4.1, only in this case $N = 32$ and $M = \frac{32}{8} = 4$. Notice also that the call to SPPOWR in SPA0401 causes a single segment to be taken by having the data sequence length LX + 1 equal to 32, and the segment size 2*LY, also equal to 32. In this case, with only one segment, the OVRLAP argument is superfluous.

Again, as in equation (4.8), the total power in the sine wave can be found as twice the integral of the periodogram, that is, twice the area of the nonzero bar in Fig. 4.3, or $2(8)(0.03125) = 0.5$. This is seen to be the mean-squared value of the unit sine wave in this example.

The power density spectrum is often estimated for data containing noise or other random components. This type of computation provides more typical uses of SPPOWR, with multiple segments being required to provide a reliable estimate in the form of an average periodogram. To demonstrate the use of SPPOWR in this type of spectral estimation, we need first to discuss the generation of random numbers and random noise sequences.

4.5 Random Number Routine

To generate a random noise sequence, we need a random number routine. Most computer libraries have routines for generating random numbers, but often these routines are not portable in the sense that they work only on the specific computers for which they were designed. Here we present a function SPRAND, which is quite portable in that it can be written in a high-level language like Fortran (as it is in Appendix A) and will work on systems with fixed-point words of 32 bits or longer. The function is used in statements such as

$$\text{Statement: } \text{X} = \text{SPRAND (ISEED)} \qquad (4.10)$$
$$\text{Result: } \quad 0.0 < \text{X} < 1.0$$

In the statement, X is the function output and is always a random number between 0 and 1. ISEED is the "seed" of the random number routine and must be set initially by the user to an integer (any positive nonzero integer) and then left alone for repeated executions of (4.10). For example, the following instructions would generate a random sequence in X(0: 9999):

$$\begin{aligned} &\text{ISEED} = 12357 \\ &\text{DO 1 K} = 0, 9999 \\ 1 \quad &\text{X(K)} = \text{SPRAND (ISEED)} \end{aligned} \qquad (4.11)$$

The sequence is not really random in the sense that if these same instructions were repeated, the same sequence would be obtained, but it is random in the sense that an outside observer could not predict a particular sample in the sequence, given all the other samples. Thus we have in SPRAND a routine for generating repeatable random sequences, so that the reader can repeat exactly the examples in this book.

To illustrate the characteristics of SPRAND, we first plot in Fig. 4.4 the histogram of the results from (4.11). The bars in the histogram give the relative frequencies of occurrence of X(K) in the 10,000-sample sequence, in the ten ranges from 0.0 to 0.1, 0.1 to 0.2, . . . , and 0.9 to 1.0. We note that the histogram is nearly flat, and thus we infer that essentially all values of X between 0 and 1 are equally likely. Therefore, X is called a *uniformly distributed* (or uniform) random number. Note that the mean value of X in Fig. 4.4 is 0.5, and that a zero-mean sequence is therefore obtained with

$$X(K) = \text{SPRAND (ISEED)} - 0.5 \tag{4.12}$$

Also, the variance, or mean-squared deviation from the mean of X, is $\frac{1}{12}$ in either (4.11) or (4.12) [E]. Thus, to generate a *unit-power* random sequence with zero mean and unit mean-squared value, we would multiply X(K) by $\sqrt{12}$. In general,

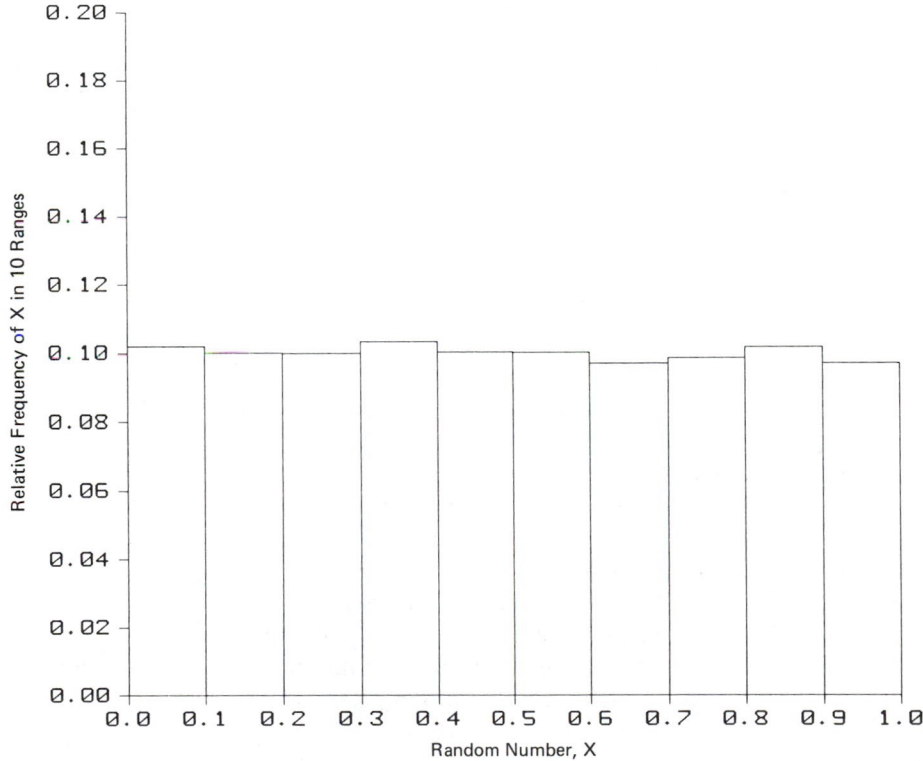

Figure 4.4. Histogram of the first 10,000 outputs of X = SPRAND (ISEED), initiated with ISEED = 12357. Each bar gives the relative frequency of occurrence of X in the range of the bar—for example, 0.0 to 0.1, 0.1 to 0.2, and so on.

then, we have

> For a uniform random sequence with zero mean and power P:
>
> Statement: $X(K) = SQRT(12.*P)*(SPRAND (ISEED) - 0.5)$
> Result: $E[X(K)] = 0$, $E[X^2(K)] = P$

(4.13)

An approximation to a *normally distributed* or *Gaussian* random number can be obtained by summing uniformly distributed random numbers. If a random number Y is formed by summing n values of X, the variance of Y is then n times the variance of X. It is therefore simple and convenient to obtain the Gaussian approximation by summing 12 values of X(K) in (4.10), thus eliminating the factor of 12 in (4.13). In general

> For an approximately Gaussian random sequence with zero mean and power P:
>
> Statements: SUM = 0.0
>   DO 1 I = 1, 12
>   1   SUM = SUM + SQRT (P) * (SPRAND (ISEED) - 0.5)
>   Y(K) = SUM
>
> Result: $E[Y(K)] = 0$, $E[Y^2(K)] = P$

(4.14)

Note that the loop in (4.14) is over I, not K. As in (4.13), the statements are assumed to be embedded in another loop over K, with ISEED initialized as in (4.11). The result of using (4.14) with P = 1 in a loop similar to (4.11) is plotted in the form of a histogram in Fig. 4.5. Comparing with Fig. 4.4, we see that the values Y(K) are now approximately normally distributed. The actual mean and variance computed for Fig. 4.5 were $E[Y(K)] = -0.010$ and $E[Y^2(K)] = 0.986$, which are close to the values predicted by (4.14), namely, 0.0 and 1.0.

Finally, another important property of any random number routine like SPRAND is the cycle length, which is the number of iterations before the sequence begins to repeat itself. The cycle length of SPRAND is

$$\text{Cycle length} = 1,048,576 \qquad (4.15)$$

This cycle length is relatively short but still more than long enough for the purposes of this book and for many signal processing experiments. Having the cycle length this short allows us to obtain the portability and simplicity shown in the listings in Appendix A.

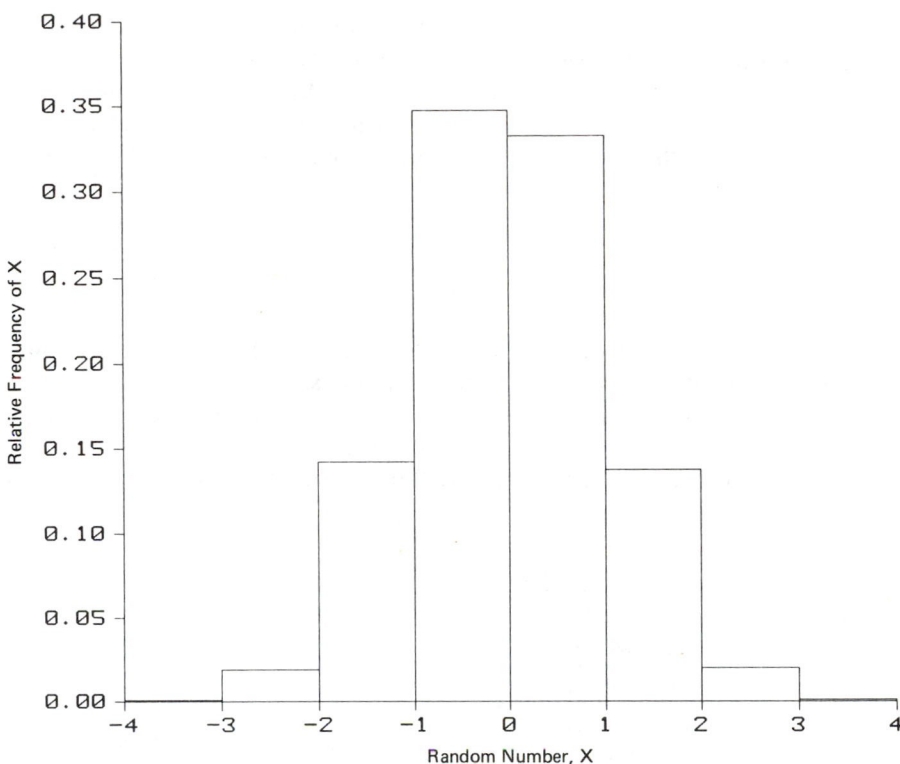

Figure 4.5. Histogram of the first 10,000 outputs, Y(0) through Y(9999) in (4.14), with P = 1.0 and initialized with ISEED = 12357. The distribution is approximately Gaussian with mean = 0 and variance (power) = 1.

4.6 Power Spectra of Random Sequences

We can now discuss the use of SPPOWR to estimate power spectra of random sequences. For power spectral estimates, unless there is an unlimited amount of data, we will generally use half-overlapping data segments, as discussed in section 4.3 above. The segment length, which is specified as 2 LY in the calling sequence of SPPOWR, is generally chosen to provide the desired frequency resolution, which, with $N = 2$ LY in equation (4.5), is seen to be

$$\text{Frequency resolution} = \Delta\nu = \frac{1}{2\,\text{LY}}\ \text{hertz-seconds} \qquad (4.16)$$

However, the segment length should also be as small as possible so that the largest possible number of segments can be averaged to produce the most reliable power estimate. So there is a trade-off between frequency resolution and reliability of the power density estimates. This trade-off is rather complex and is discussed at length in the references ([C] and so on). We do not discuss the theory here, but we can illustrate the trade-off in some examples and exercises.

The first example, in SPA0402, illustrates the estimation of the power spectrum of a uniform random sequence of 1000 samples. The sequence is created in the first DO loop in SPA0402. Relating the computation of X(K) to equation (4.13), note that the theoretical mean value of X(K) is E[X(K)] = 0.0 and that the theoretical power (P) in the sequence is $E[X^2(K)] = 3.0$. The actual power in the sequence is also found as EX2 in the first DO loop and is printed below the program as ACTUAL POWER.

The call to SPPOWR in SPA0402 has the last X index as LX = 999, and LY = 16 indicates a frequency resolution $\Delta\nu$ = 0.03125 Hz-s in accordance with equation (4.16). The corresponding segment length is 2∗LY = 32 which, with a sequence length of 1000, allows a total of 31 nonoverlapping segments to be averaged. This total is calculated in the subroutine and is printed as NSGMTS below the program. We use nonoverlapping segments, that is, OVRLAP = 0.0 in this example, to make the calculation of NSGMTS simple. For the short data sequence of 1000 samples, the rectangular data window is used, that is, IWINDO = 1.

```
      PROGRAM SPA0402
C-POWER DENSITY SPECTRUM OF WHITE UNIFORM NOISE.
      DIMENSION X(0:999),Y(0:16),WORK(0:33)
      ISEED=12357
      EX2=0.
      DO 1 K=0,999
        X(K)=SQRT(12.*3.)*(SPRAND(ISEED)-.5)
        EX2=EX2+X(K)**2/1000.
    1 CONTINUE
      CALL SPPOWR(X,Y,WORK,999,16,1,0.0,NSGMTS,IERROR)
      TPOWER=(Y(0)+Y(16))/32.
      DO 2 M=1,15
        TPOWER=TPOWER+2.*Y(M)/32.
    2 CONTINUE
      PRINT 3,IERROR,NSGMTS,Y
    3 FORMAT(' IERROR,NSGMTS=',2I4/' Y=',8F7.3/2X,9F7.3)
      PRINT 4,EX2,TPOWER
    4 FORMAT(' ACTUAL POWER, ESTIMATED POWER:',2F7.3)
      CALL HP(Y,17,0,.1,.2,.9,.9)
      STOP
      END
$ RUN SPA0402
 IERROR,NSGMTS=    0  31
 Y=  2.813  3.257  3.159  3.931  3.077  2.585  2.736  3.125
     3.342  3.435  2.529  2.039  2.624  3.576  3.330  2.554  2.884
 ACTUAL POWER, ESTIMATED POWER:  3.001  3.009
FORTRAN STOP
```

After the call to SPPOWR in SPA0402, the total power is estimated in the second DO loop as TPOWER and printed below the program as ESTIMATED POWER. Note that the computation in the second DO loop is in accordance with equation (4.9) and that the estimated power from the average periodogram agrees fairly well with the actual power $E[x_k^2]$. The periodogram values (Y_m) are printed below the program listing and are plotted in Fig. 4.6.

The sequence generated in SPA0402 is called a *uniform white noise* sequence. The samples in $[x_k]$ are independent uniform random numbers, and a white, or flat, spectrum is one of the properties of such a sequence. Because the total power in this example is 3.0, the theoretical value of each periodogram magnitude

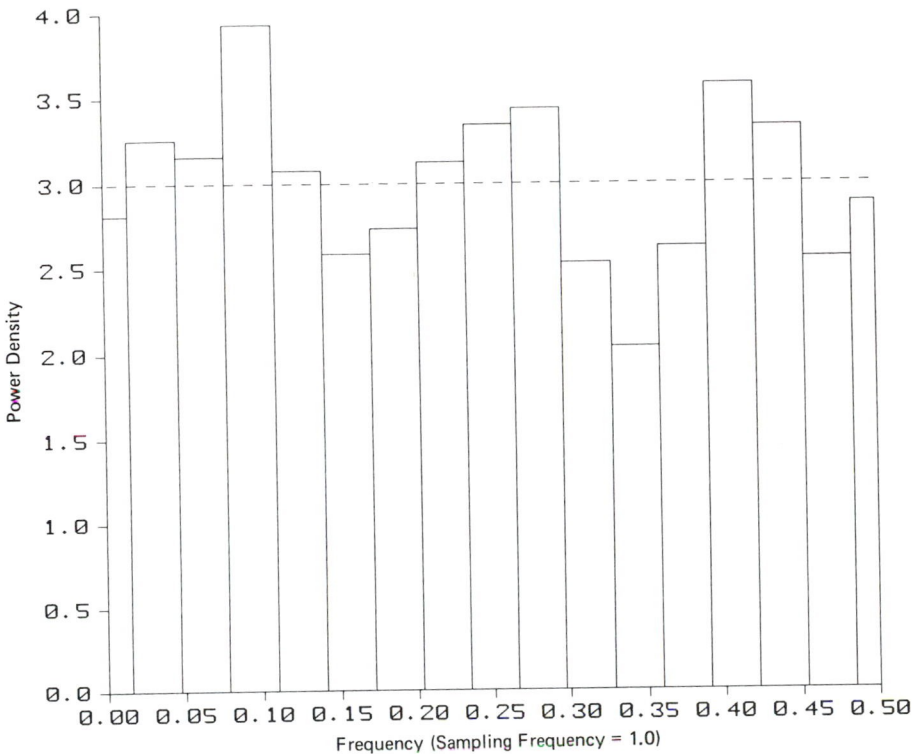

Figure 4.6. Estimated power density spectrum from 1000 samples of a uniform white noise sequence, computed with SPA0402 (average of 31 periodograms from nonoverlapping segments of length 32; total power = 3.0).

in this case is also 3.0, so that the integral of the power density spectrum, from 0.0 to 1.0 Hz-s, is also 3.0.

We can see in Fig. 4.6 that the periodogram values differ noticeably from the "correct" value of 3.0. The difference is due to the relatively small number (31) of nonoverlapping segments and is typical for this number of segments of a white noise sequence. For the same sequence length (1000 samples) and data window, Fig. 4.7 shows the result with half-overlapping segments, that is, with the calling statement in SPA0402 changed to

CALL SPPOWR (X, Y, WORK, 999, 16, 1, 0.5, NSGMTS, IERROR).

Here we see some improvement, but not much, in the accuracy of the estimate. The number of segments in this case is NSGMTS = 61.

Of course, we can always improve the spectral estimate by taking a shorter segment length and sacrificing frequency resolution, as discussed in connection with equation (4.16). Figure 4.8 shows the result of changing the calling statement to

CALL SPPOWR (X, Y, WORK, 999, 8, 1, 0.5, NSGMTS, IERROR)

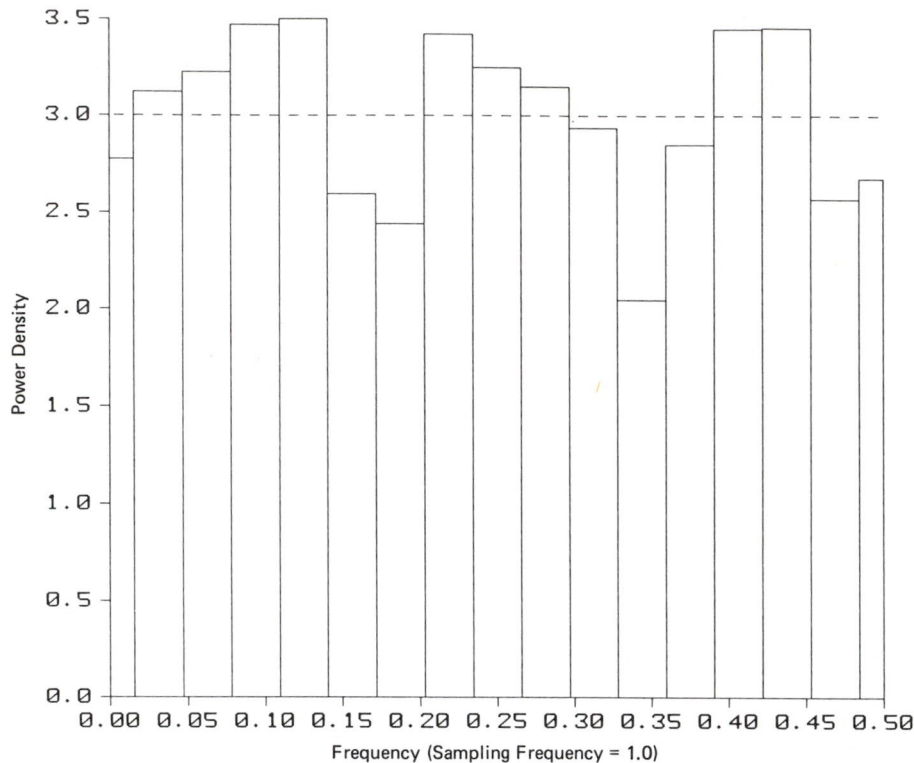

Figure 4.7. Estimated power density spectrum from 1000 samples of a uniform white noise sequence, similar to Fig. 4.6, but in this case with 61 half-overlapping segments of length 32.

In this case, by halving the frequency resolution to $\Delta \nu = \frac{1}{16}$ Hz-s in accordance with (4.16), the number of half-overlapping segments is increased to NSGMTS = 124, and we notice a further improvement in the accuracy of the spectral estimate. Taken together, Figs. 4.6–4.8 give an indication of the accuracy to be expected when estimating the power spectrum of a stationary random time series. Another example, showing the effect of different data windows, is given in Fig. 14.2.

A final example of spectral estimation is given in SPA0403, with the results

```
      PROGRAM SPA0403
C-POWER DENSITY SPECTRUM OF SINE WAVE IN WHITE UNIFORM NOISE.
      DIMENSION X(0:999),Y(0:16),WORK(0:33)
      PI=4.*ATAN(1.)
      ISEED=12357
      DO 1 K=0,999
        SIGNAL=SIN(3.*PI*K/16.)+0.5*SIN(7.*PI*K/16.)
        X(K)=SIGNAL+SQRT(12.*0.01)*(SPRAND(ISEED)-0.5)
    1 CONTINUE
      CALL SPPOWR(X,Y,WORK,999,16,1,0.5,NSGMTS,IERROR)
      DO 2 M=0,16
        Y(M)=10.*ALOG10(Y(M))
    2 CONTINUE
      CALL HP(Y,17,0,.1,.2,.9,.9)
      STOP
      END
```

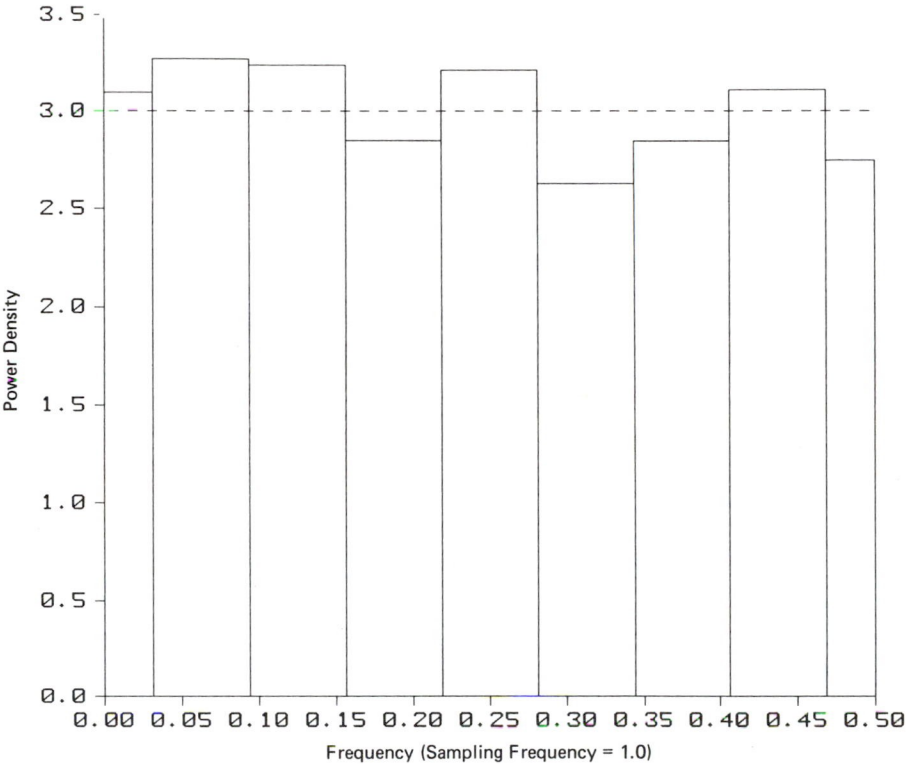

Figure 4.8.   Estimated power density spectrum from 1000 samples of a uniform white noise sequence, similar to Fig. 4.7, but in this case with 124 half-overlapping segments of length 16.

plotted in Fig. 4.9. In loop 1 we have 1000 data samples generated as before, but here the data has the following components:

Sine wave at $\frac{3}{32}$ Hz-s with power 0.5

Sine wave at $\frac{7}{32}$ Hz-s with power 0.125

Uniform white noise with power 0.01

The power spectral estimate computed by SPPOWR has the same parameters as used for Fig. 4.7, but of course the results differ due to the sine components. In cases like this it is often preferable to measure the power density in *decibels*. The decibel (dB) measure is defined as

$$\text{Power density in decibels} = 10\log_{10}P_m, \qquad m = 0, 1, ..., \frac{N}{2} \qquad (4.17)$$

where P_m is the power density (averaged-periodogram value). In SPA0403, the power density (Y) values are converted to decibels in the second DO loop and

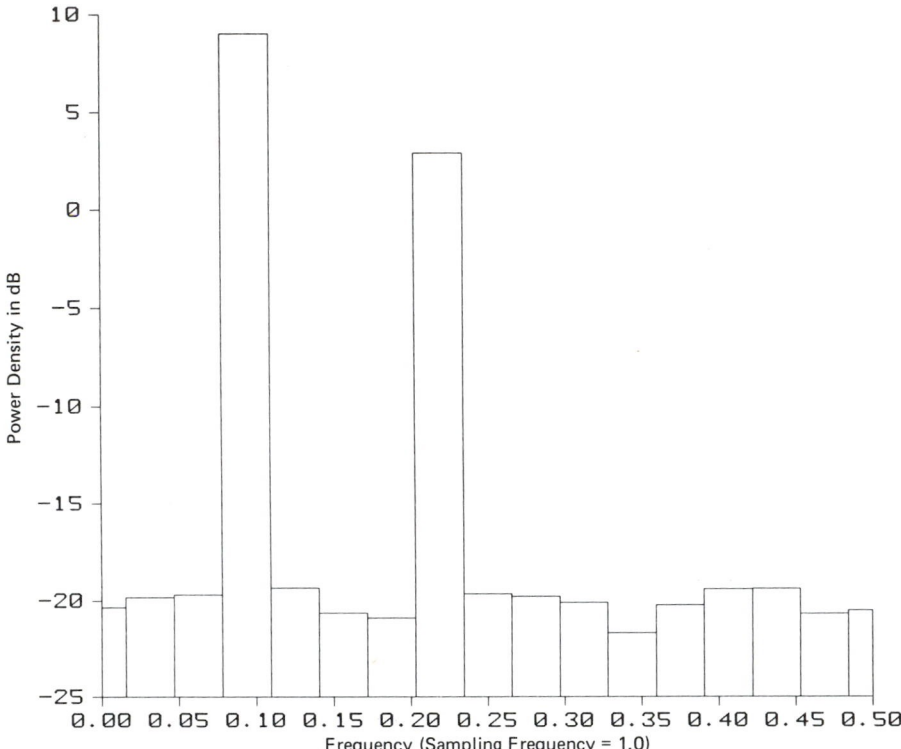

Figure 4.9. Power spectrum of two sine waves plus noise, estimated in SPA0403. Sine components are at approximately 9.0 and 3.0 dB and white noise is at approximately -20 dB.

plotted in Fig. 4.9 by calling HP. Since the total white noise power is 0.01, we know that the theoretical noise power density must also be 0.01, or

$$\text{Noise power density in decibels} = 10 \log_{10}(0.01) = -20 \text{ dB} \qquad (4.18)$$

The total power of the sine wave at $\frac{3}{32}$ Hz-s is 0.5, and all the power in this case is concentrated in component P_3 of the periodogram. Therefore, the additional signal power density in P_3 is the total power, 0.5, divided by the bar width, $\frac{1}{32}$, and divided by 2 because half the power is in P_{29}, as described above, which gives 8. Thus the signal plus noise power density in P_3 is

$$\text{Power density in } P_3 \text{ in decibels} = 10 \log_{10}(8.01) = 9.0 \text{ dB} \qquad (4.19)$$

Similarly, the second signal component is in P_7, and the signal-plus-noise-power density there is

$$\text{Power density in } P_7 \text{ in decibels} = 10 \log_{10}(2.01) = 3.0 \text{ dB} \qquad (4.20)$$

These values are seen to match the estimates computed by SPA0403 and plotted in Fig. 4.9.

4.7 Cross-Spectrum Routine

We have seen that the power density spectrum of $[x_k]$ is the average squared magnitude of the DFT, that is, the average of periodograms in the form of (4.1). The cross-power density spectrum, or cross spectrum for short, is similar but involves two time series, $[x_{1k}]$ and $[x_{2k}]$, instead of just $[x_k]$. The estimated cross-power spectrum is the average of DFT products of the form

$$Q_m \triangleq \text{cross periodogram of } [x_{1k}, x_{2k}]$$

$$\triangleq \frac{1}{N} X_{1m}^* X_{2m}, \qquad m = 0, 1, ..., \frac{N}{2} \qquad (4.21)$$

As in Chapter 3, the star signifies the complex conjugate. Unlike P_m in (4.1), Q_m in (4.21) is complex, with $Q_m = Q_{N-m}^*$ for $m = N/2 + 1$ through $N - 1$. It is a measure of the cross power between $[x_{1k}]$ and $[x_{2k}]$, or in other words the power that $[x_{1k}]$ and $[x_{2k}]$ have in common at specific frequencies. Also, the definition of Q_m implies an ordering of the two sequences; that is,

$$Q_m^* = \frac{1}{N} X_{1m} X_{2m}^* = \text{cross periodogram of } [x_{2k}, x_{1k}] \qquad (4.22)$$

All the remarks about the periodogram in Section 4.2 above apply also to the cross periodogram, except that the cross periodogram is complex instead of real. In particular, when the segments of $[x_{1k}]$ and $[x_{2k}]$ both have N data samples, there are $N/2$ unique cross-periodogram components, as in (4.21), and the frequency resolution is defined by (4.5) or (4.16), that is, if N is the segment length, then

$$\text{Frequency resolution} = \Delta\nu = \frac{1}{N} \text{ hertz-second} \qquad (4.23)$$

Equations similar to (4.6) and (4.9) are also applicable to the cross-power spectrum, as follows:

$$\begin{array}{l} \text{Cross power in} \\ \text{frequency range} \\ \text{from } (m - 0.5)/N \\ \text{to } (m + 0.5)/N \text{ Hz-s} \end{array} = \frac{1}{N} \text{Re}[Q_m] = \frac{1}{N^2} \text{Re}[X_{1m}^* X_{2m}]; \qquad 0 \le m \le \frac{N}{2} \qquad (4.24)$$

$$\text{Total cross power} \triangleq E[x_{1k} x_{2k}]$$

$$= \frac{1}{N^2} \sum_{m=0}^{N-1} \text{Re}[X_{1m}^* X_{2m}] \qquad (4.25)$$

$$= \frac{1}{N} \sum_{m=0}^{N-1} \text{Re}[Q_m]$$

Note that here we have used (3.8) and (3.9) and that Q_0 as well as $Q_{N/2}$ are real components (see exercise 4.27). As before, the frequency corresponding with $m = N/2$ is half the sampling rate, as illustrated in the following examples.

With random or noisy data, we usually estimate the cross-power spectrum by taking an average of cross periodograms of overlapped data segments. The process is just as discussed in Section 4.3 for the periodogram, except that there are two data sequences.

We now introduce a routine similar to SPPOWR, called SPCROS, for estimating the cross power spectrum. This routine is very similar to SPPOWR in structure as well as use. The calling sequence for SPCROS is

CALL SPCROS(X1, X2, Y, WORK, LX, LY, IWINDO, OVRLAP, NSGMTS, IERROR)

X1(0:LX)	= first input data sequence, containing LX+1 samples
X2(0:LX)	= second input data sequence, similar to X1
Y(0:LY)	= output cross spectrum, (complex) where LY is a power of 2; corresponding segment size is 2*LY
WORK(0:2*LY-1)	= work array (complex)
LX	= last index in data sequences; must be at least 2*LY-1
LY	= last frequency index (power of 2), corresponding to half the sampling frequency; segment length = 2*LY
IWINDO	= indicator of type of data window: 1 (rectangular), 2 (tapered rectangular), 3 (triangular), 4 (Hanning), 5 (Hamming), or 6 (Blackman)
OVRLAP	= fraction that each data segment overlaps its predecessor; at least 0 and less than 1
NSGMTS	= number of overlapping segments averaged together, computed internally
IERROR	= output error indicator, 0 if no error is detected, 1 if IWINDO is out of range, 2 if LX is too small for one segment, or 3 if LY is not a power of 2

All the arguments are similar to the corresponding arguments in SPPOWR described in section 4.4, except that the Y array is complex to accommodate the complex cross spectrum, and WORK is a complex array also.

Our first example of the use of SPCROS is in SPA0404 below. Here we make $[x_{1k}]$ and $[x_{2k}]$ both the same as $[x_k]$ in SPA0401, and thus the cross periodogram in this case becomes just the periodogram of $[x_k]$, as we can see by comparing (4.1) and (4.21) with $x_{1k} = x_{2k} = x_k$. As in SPA0401, the rectangular data window is used, and there is only one data segment of length 32, which covers exactly 4 cycles of the sine wave in each sequence. In the printout below SPA0404, we see 17 complex-pair values of the cross periodogram Y(0) through Y(16), listed in succession. We note that the only nonzero value is Y(4), which

```
          PROGRAM SPA0404
C-TEST OF SPCROS, WITH RESULT SAME AS FOR SPA0401.
          DIMENSION X1(0:31),X2(0:31)
          COMPLEX Y(0:16),WORK(0:31)
          PI=4.*ATAN(1.)
          DO 1 K=0,31
             X1(K)=SIN(2.*PI*K/8.)
             X2(K)=SIN(2.*PI*K/8.)
        1 CONTINUE
          CALL SPCROS(X1,X2,Y,WORK,31,16,1,0.,NSGMTS,IERROR)
          PRINT 2,NSGMTS,IERROR,Y
        2 FORMAT(' NSGMTS,IERROR=',2I4/' Y=',8F7.3/(3X,8F7.3))
          STOP
          END
$ RUN SPA0404
 NSGMTS,IERROR=    1    0
  Y=  0.000   0.000   0.000   0.000   0.000   0.000   0.000   0.000
      8.000   0.000   0.000   0.000   0.000   0.000   0.000   0.000
      0.000   0.000   0.000   0.000   0.000   0.000   0.000   0.000
      0.000   0.000   0.000   0.000   0.000   0.000   0.000   0.000
      0.000   0.000
 FORTRAN STOP
```

is real and represents the power at $\frac{1}{8}$ of the sampling frequency. This component is equal to the Y(4) output of SPA0401, and thus we see that (4.25) gives the same total power as (4.6) with $N = 32$ in this case, namely, 0.5.

To obtain the second example in SPA0405, we have made three changes in SPA0404. First, we changed the first sample sequence, $[x_{1k}]$, from a sine function to a cosine function, that is, we shifted $[x_{1k}]$ ahead (to the left) one-quarter cycle, or two samples in this case. Secondly, we lowered the segment length, 2*LY, from 32 to 16, and finally we changed from nonoverlapping to half-overlapping segments by setting OVRLAP = 0.5. The last two changes result in NSGMTS = 3 segments being averaged. The second change means that the resolution, $\Delta\nu$ in (4.23), has now doubled from $\frac{1}{32}$ to $\frac{1}{16}$ of the sampling frequency and that the data frequency is thus now $2\Delta\nu$ instead of $4\Delta\nu$. We see this in the printout of Y below SPA0405, where Y(2) is now the nonzero complex periodogram value. We also note that Y(2) is imaginary. We could have anticipated this from (3.5), where $[X_{1m}^*]$ in this case is seen to be real and $[X_{2m}]$ is imaginary, so the product, $[Q_m]$ in (4.21), must be imaginary. See exercise 4.25(c).

```
          PROGRAM SPA0405
C-TEST OF SPCROS:  ONE IMAGINARY COMPONENT.
          DIMENSION X1(0:31),X2(0:31)
          COMPLEX Y(0:8),WORK(0:15)
          PI=4.*ATAN(1.)
          DO 1 K=0,31
             X1(K)=COS(2.*PI*K/8.)
             X2(K)=SIN(2.*PI*K/8.)
        1 CONTINUE
          CALL SPCROS(X1,X2,Y,WORK,31,8,1,0.5,NSGMTS,IERROR)
          PRINT 2,NSGMTS,IERROR,Y
        2 FORMAT(' NSGMTS,IERROR=',2I4/' Y=',8F7.3/(3X,8F7.3))
          STOP
          END
$ RUN SPA0405
 NSGMTS,IERROR=    3    0
  Y=  0.000   0.000   0.000   0.000   0.000   4.000   0.000   0.000
      0.000   0.000   0.000   0.000   0.000   0.000   0.000   0.000
      0.000   0.000
 FORTRAN STOP
```

Thus we have seen in SPA0405 the illustration of a general rule, which is a direct result of Chapter 3, exercise 3.25:

> Rotating $[x_{1k}]$ n places to the right relative to $[x_{2k}]$ does not alter the magnitude of the cross periodogram, $[Q_m]$, and adds $2\pi nm/N$ radians to the phase of $[Q_m]$, where N is the length of the segment used to compute $[Q_m]$.

The shifting of $[x_{1k}]$ relative to $[x_{2k}]$ becomes important when the sequences are nonstationary, nonrandom, or only approximately stationary. Shifting without rotation generally changes both the magnitude and phase of $[Q_m]$. Before computing $[Q_m]$, the sequences are sometimes shifted with respect to each other to obtain a peak in the cross-correlation function of $[x_{1k}]$ and $[x_{2k}]$, as discussed in Chapter 9.

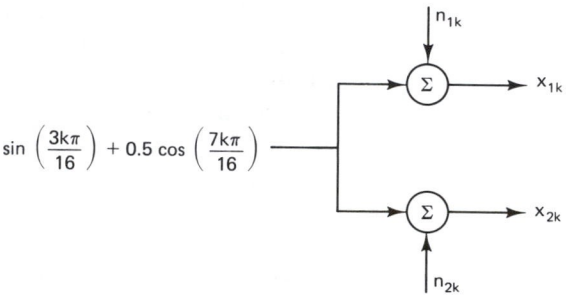

Figure 4.10. Scheme for generating the signals in SPA0406. n_1 and n_2 are independent uniform white noises, each with unit power.

For the third example of a cross-spectrum estimation, we use the scheme shown in Fig. 4.10. Both $[x_{1k}]$ and $[x_{2k}]$ contain the same signal, which is the sum of two sine waves, exactly as in SPA0403. Then uniform white noise n_{1k} is added to produce x_{1k}, and a similar but independent noise, n_{2k}, is added to produce x_{2k}. In program SPA0406, we can see in the "DO 1" loop that the

```
        PROGRAM SPA0406
C-TEST OF SPCROS:  SAME SIGNAL PLUS INDEPENDENT NOISE.
        DIMENSION X1(0:999),X2(0:999),YABS(0:16)
        COMPLEX Y(0:16),WORK(0:31)
        PI=4.*ATAN(1.)
        ISEED=123
        DO 1 K=0,999
          SIGNAL=SIN(3.*PI*K/16.)+0.5*COS(7.*PI*K/16.)
          X1(K)=SIGNAL+SQRT(12.)*(SPRAND(ISEED)-0.5)
          X2(K)=SIGNAL+SQRT(12.)*(SPRAND(ISEED)-0.5)
      1 CONTINUE
        CALL SPCROS(X1,X2,Y,WORK,999,16,1,0.5,NSGMTS,IERROR)
        DO 2 M=0,16
          YABS(M)=REAL(Y(M))
      2 CONTINUE
        CALL HP(YABS,17,0,.1,.2,.9,.9)
        STOP
        END
```

noise power, $E[n_{1k}^2]$ and $E[n_{2k}^2]$, is 1.0 in accordance with (4.13), which is 50 times the noise power in SPA0403. The SNR in this case is 0.5 for the sine signal and 0.125 for the cosine signal. However, in the cross spectrum plotted in Fig. 4.11, which is the real part of $[Q_m]$ used in (4.23) and (4.24), the noise power is close to zero, because the noises are independent (uncorrelated) and therefore do not appear in the cross spectrum. The relations between correlation functions and power spectra are discussed further in Section 4.8.

As a final point on the power spectral routines, we note that SPCROS is just a generalized version of SPPOWR and could be used to compute the power spectrum of a single signal simply by making X1 = X2 in the calling sequence. This was illustrated in SPA0404, where X1 and X2 were the same sequence. However, the output, Y, of SPCROS is complex, and the computation is not as efficient as in SPPOWR, so the latter is always preferred for computing the power density spectrum of a single signal.

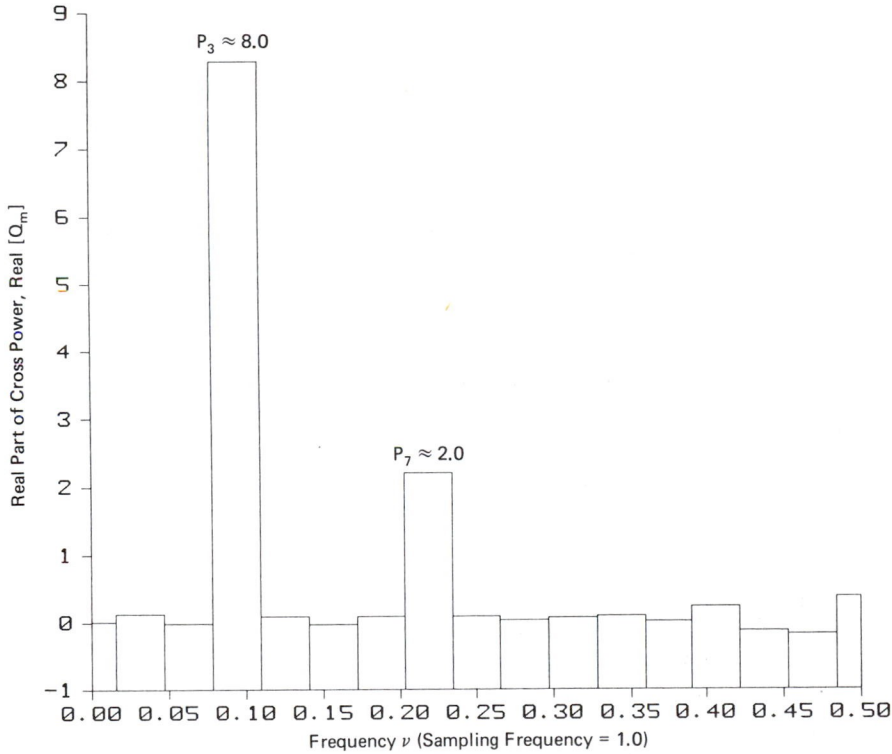

Figure 4.11. Real part of the cross power computed in SPA0406, showing noise level close to zero and signal components at $\nu = \frac{3}{32}$ and $\frac{7}{32}$. Segment length = 16. $\Delta\nu = 0.0625$, power in signal components = 0.5 and 0.125. Total data sequence length = 1000 samples.

4.8 Power Spectral Relationships

In this section we briefly summarize some important relationships between power spectra, transfer functions (which are discussed further in Chapter 5), and correlation functions. These relationships provide further insight into the nature of the power spectrum and will be useful in the discussion of coherence in the next section.

For this discussion we assume two stationary random sequences, $[x_k]$ and $[y_k]$, with k ranging from minus infinity to infinity. The *correlation functions* of $[x_k]$ and $[y_k]$ are the following expected products:

$$\text{Autocorrelation:} \quad \phi_{xx}(n) = E[x_k x_{k+n}] \tag{4.26}$$

$$\text{Cross corelaton:} \quad \phi_{xy}(n) = E[x_k y_{k+n}] \tag{4.27}$$

Similar products hold for $\phi_{yy}(n)$ and $\phi_{yx}(n)$, with the expectation going over all values of k, that is, over all time. With stationary sequences the averages do not change with k, so $x_k y_{k+n}$ and $x_{k-n} y_k$ represent the same relative shift of $[x_k]$ and $[y_k]$, and therefore

$$\begin{aligned} \phi_{yx}(n) &= E[y_k x_{k+n}] \\ &= E[y_{k-n} x_k] = \phi_{xy}(-n) \end{aligned} \tag{4.28}$$

Thus the autocorrelation function is an even function, that is,

$$\phi_{xx}(n) = \phi_{xx}(-n) \tag{4.29}$$

Using the z-transform notation introduced in Chapter 2, we define the *discrete power spectra* of $[x_k]$ and $[y_k]$ to be the transforms of the correlation functions:

$$\text{Power spectrum of } [x_k]: \quad \Phi_{xx}(z) = \sum_{n=-\infty}^{\infty} \phi_{xx}(n) z^{-n} \tag{4.30}$$

$$\text{Cross spectrum of } [x_k] \text{ and } [y_k]: \quad \Phi_{xy}(z) = \sum_{n=-\infty}^{\infty} \phi_{xy}(n) z^{-n} \tag{4.31}$$

By letting $z = e^{j\omega}$, as described in Chapter 2, and letting $\omega = 2\pi m/N$ rad as in Table 3.1, we can make (4.30) and (4.31) equivalent to the previous power spectral expressions in (4.6) and (4.24). To do this we must let m and N increase without bound in (4.6) and (4.24) so that the power spectral estimates become the true power spectra of infinite-length sequences.

These transform relationships between correlation functions and power spectra are important in signal processing. In Chapter 9, finite versions of (4.30) and (4.31) are used to derive fast algorithms for computing correlation functions.

Besides the transform relations in (4.30) and (4.31), the formulas listed next are important relationships involving power spectra.

$$z \triangleq e^{j\omega}; \qquad \omega = 2\pi m/N \tag{4.32}$$

$$\Phi_{yx}(z) = \Phi_{xy}(z^{-1}) = \Phi_{xy}^{*}(z) \tag{4.33}$$

$$\phi_{xy}(n) = \frac{1}{2\pi} \int_{0}^{2\pi} \Phi_{xy}(e^{j\omega}) e^{j\omega n} \, d\omega \tag{4.34}$$

$$\phi_{xx}(n) = \frac{1}{2\pi} \int_{0}^{2\pi} \Phi_{xx}(e^{j\omega}) e^{j\omega n} \, d\omega \tag{4.35}$$

$$E(x_k^2) = \phi_{xx}(0) = \frac{1}{2\pi} \int_{0}^{2\pi} \Phi_{xx}(e^{j\omega}) \, d\omega \tag{4.36}$$

$$Y(z) \triangleq X(z)H(z) \tag{4.37}$$

$$\Phi_{xy}(z) = \Phi_{xx}(z)H(z) \tag{4.38}$$

$$\Phi_{yy}(z) = \Phi_{xx}(z) |H(z)|^2 \tag{4.39}$$

We obtain these formulas as follows. In (4.32) we define z to be a point on the unit circle on the complex z-plane, with the frequency ω and the index m being related as discussed previously. Then (4.33) follows by using (4.28) in (4.31) and is similar to (4.22). The inverse relationships (4.34) and (4.35) can be proved by substituting (4.31) and (4.30) for the spectral functions and reaching an identity. Then (4.36), the formula for total power, follows by letting $n = 0$ in (4.35). Finally, the transfer-function relations in (4.38) and (4.39) follow from the definition in (4.37), where $[y_k]$ is made dependent on $[x_k]$. These relations are not proved in detail here; however, they are similar to the transfer-function relations presented in Chapter 2.

In summary, the auto- and cross-spectral estimates in (4.6) and (4.24) are considered to be approximations to the power spectra in (4.30) and (4.31), as discussed earlier. The additional relationships in (4.32) through (4.39) are aids to understanding power spectra and are used in the next section on coherence.

4.9 The Coherence Function

When two or more signals from the same source are processed, as in array processing, coherence analysis is often useful. The coherence function essentially gives a measure of the similarity between signals and is related to the cross-correlation function discussed in the preceding section and again in Chapter 9. More specifically, coherence is a measure of the linear dependence of one signal on another.

The magnitude-squared coherence (MSC) function [O, P, Q, R] of two sampled signals x_k and y_k is a normalized version of the cross power between x_k and y_k and is defined as follows:

$$\text{MSC function: } \Gamma_{xy}^2(z) \triangleq \frac{|\Phi_{xy}(z)|^2}{\Phi_{xx}(z)\Phi_{yy}(z)} \tag{4.40}$$

To examine the significance of this function, we begin with the situation shown in Fig. 4.12. A source signal, s_k, is measured by two sensors. We assume that the source signal itself is not known explicitly but only by observing the sensor outputs x_k and y_k, which are noisy. This situation exists in making measurements in biological or seismological situations, for example. We assume that x_k and y_k are related to s_k through linear systems H_1 and H_2 but that they also contain additive independent noise components n_{1k} and n_{2k}. The MSC in (4.40) is a measure of the linear dependence between x_k and y_k and thus a measure of the amount of noise, the signal-to-noise ratio, and so on.

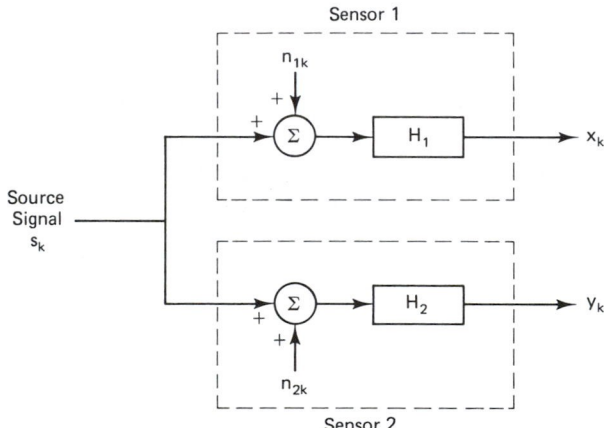

Figure 4.12.   Source signal passing through two sensors.

More specifically, suppose H_1 and H_2 are specified as linear digital transfer functions $H_1(z)$ and $H_2(z)$ and that these functions have no zeros on the unit circle $z = e^{j\omega}$, so that the two sensors are not "dead" to any part of the spectrum. Then we can reduce Fig. 4.12 to its equivalent in Fig. 4.13, in which $H(z) = H_2(z)/H_1(z)$ and n_k is the total noise found by passing the difference $(n_{2k} - n_{1k})$ through $H_1(z)$, as may be seen from the following transform relationships:

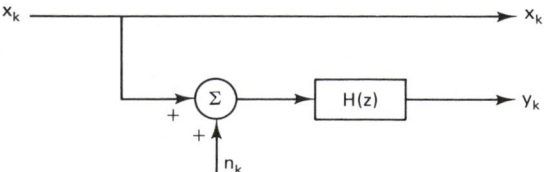

Figure 4.13.   A reduction of Fig. 4.12.

$$X(z) = H_1(z)[N_1(z) + S(z)]$$
$$Y(z) = H_2(z)[N_2(z) + S(z)] \tag{4.41}$$

Therefore,

$$Y(z) = \frac{H_2(z)}{H_1(z)}[X(z) + H_1(z)(N_2(z) - N_1(z))] \tag{4.42}$$

$$H(z) = \frac{H_2(z)}{H_1(z)} \qquad N(z) = H_1(z)[N_2(z) - N_1(z)] \tag{4.43}$$

So in Fig. 14.13 we have

$$Y(z) = H(z)[X(z) + N(z)] \tag{4.44}$$

We assume that the noise n_k is independent of x_k and y_k, so that the cross-power terms $\Phi_{xn}(z)$ and $\Phi_{yn}(z)$ are zero. Then, when (4.38) and (4.39) are applied to Fig. 4.13, we have

$$\Phi_{xy}(z) = H(z)\Phi_{xx}(z) \tag{4.45}$$

$$\Phi_{yy}(z) = |H(z)|^2[\Phi_{xx}(z) + \Phi_{nn}(z)] \tag{4.46}$$

Now $\Phi_{xy}(z)$, $\Phi_{xx}(z)$, and $\Phi_{yy}(z)$ are the spectral terms that we estimate using the routines described previously. From (4.40), (4.45), and (4.46), we can see that the MSC, the transfer function $H(z)$, the noise, and the signal-to-noise ratio (SNR) spectra are obtained from these spectral terms as follows:

$$\Gamma_{xy}^2(z) = \frac{|\Phi_{xy}(z)|^2}{\Phi_{xx}(z)\Phi_{yy}(z)} = \frac{1}{1 + 1/\mathrm{SNR}(z)} \tag{4.47}$$

$$H(z) = \frac{\Phi_{xy}(z)}{\Phi_{xx}(z)} \tag{4.48}$$

$$\Phi_{nn}(z) = \Phi_{xx}(z)[\Gamma_{xy}^{-2}(z) - 1] \tag{4.49}$$

$$\mathrm{SNR}(z) \triangleq \frac{\Phi_{xx}(z)}{\Phi_{nn}(z)} = \frac{\Gamma_{xy}^2(z)}{1 - \Gamma_{xy}^2(z)} \tag{4.50}$$

From these results, we see in Fig. 4.13 that the MSC between x_k and y_k approaches 1 as the noise power density approaches 0 and $H(z)$ remains finite and that the MSC approaches 0 as the noise power density increases or as $H(z)$ decreases toward 0. Thus the MSC indicates, at all frequencies, whether the signals x_k and y_k in Fig. 4.12 are really from the source s_k or are just artificial signals coming from independent sensor noise.

The other functions, $H(z)$, $\Phi_{nn}(z)$, and $\mathrm{SNR}(z)$ in (4.48) through (4.50), are often preferred to the MSC. In any case, the three power spectra $\Phi_{xy}(z)$, $\Phi_{xx}(z)$, and $\Phi_{yy}(z)$ are the basic functions that must be estimated from the data.

To illustrate a computation of the MSC function, we conducted a test using the system shown in Fig. 4.14. For the source signal s_k, we passed a white signal with unit power density (and therefore unit total power) through the digital filter shown in the figure, so that s_k has a nonwhite spectrum with a peak at approximately $\frac{1}{6}$ of the sampling frequency. To obtain y_k, we add independent white noise with power equal to 3.0. Having $H(z) = 1$ is the same as specifying that the two sensor-transfer functions, H_1 and H_2 in Fig. 4.12, are the same— that is, that the two sensors are identical but generate independent broadband noise.

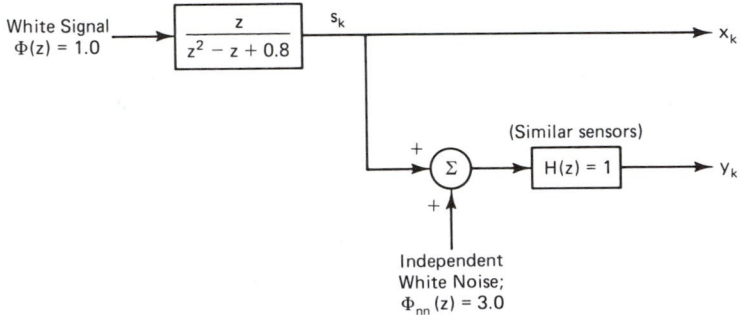

Figure 4.14. System used for coherence computations.

Using (4.32) and (4.39), we determine the power density spectrum of x_k in Fig. 4.14 to be

$$\Phi_{xx}(e^{j\omega}) = \left| \frac{e^{j\omega}}{e^{2j\omega} - e^{j\omega} + 0.8} \right|^2$$

$$= [(1.8 \cos \omega - 1)^2 + (0.2 \sin \omega)^2]^{-1} \tag{4.51}$$

The SNR is therefore

$$\text{SNR}(e^{j\omega}) = \frac{\Phi_{xx}(e^{j\omega})}{\Phi_{nn}(e^{j\omega})}$$

$$= \frac{1}{3}[(1.8 \cos \omega - 1)^2 + (0.2 \sin \omega)^2]^{-1} \tag{4.52}$$

The MSC function is then given by (4.47) as

$$\Gamma_{xy}^2 (e^{j\omega}) = \left[1 + \frac{1}{\text{SNR}(e^{j\omega})} \right]^{-1}$$

$$= \{1 + 3[(1.8 \cos \omega - 1)^2 + (0.2 \sin \omega)^2]\}^{-1} \tag{4.53}$$

In Fig. 4.15 we have a plot of the MSC function in (4.53) superimposed on two estimated MSC functions, first using 1000 data samples and then using 4000 data

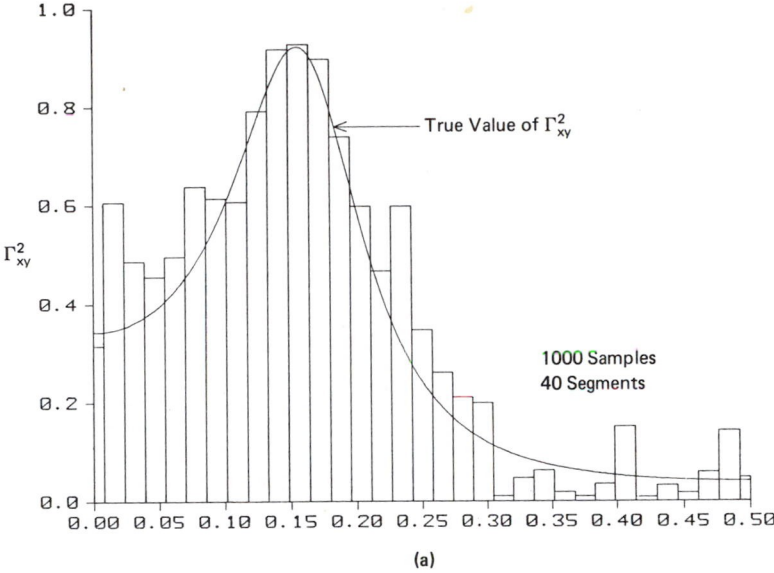

(a)

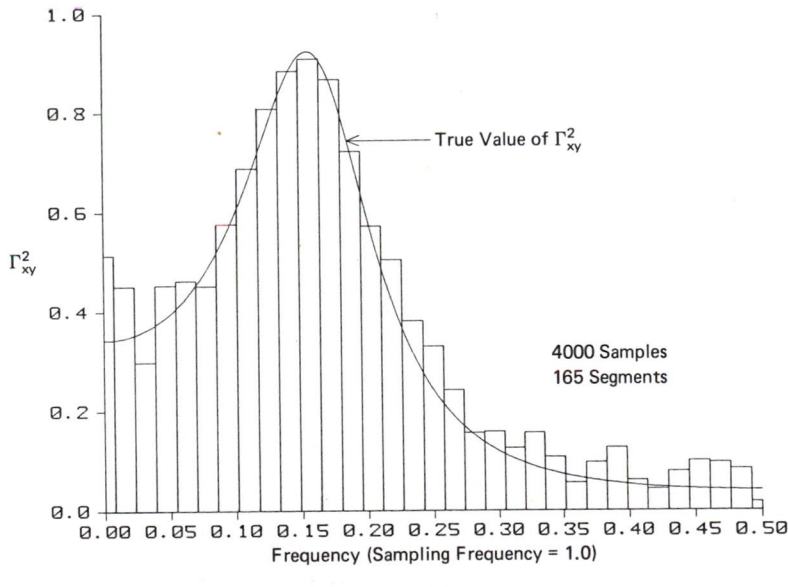

Frequency (Sampling Frequency = 1.0)

(b)

Figure 4.15.   Mean-squared coherence function Γ_{xy}^2 for Fig. 4.14, estimated using (a) 1000 samples and (b) 4000 samples.

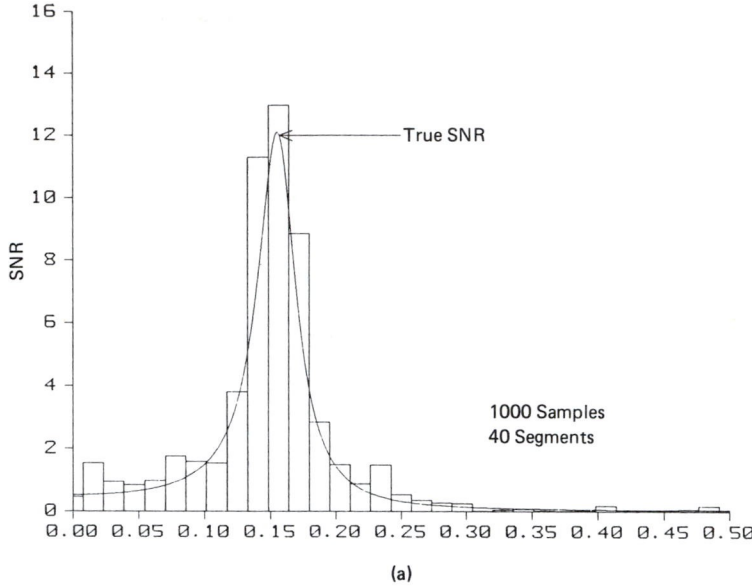

(a)

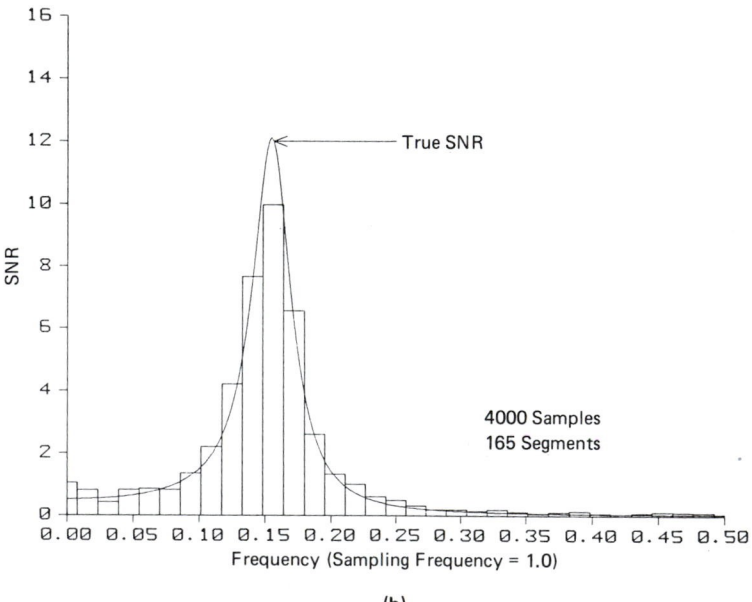

(b)

Figure 4.16. SNR for Fig. 4.14, estimated using (a) 1000 samples and (b) 4000 samples.

samples. The following section of code shows the computation of the MSC estimate in the array GXY for both cases, with NP = 0 for 1000 samples and NP = 1 for 4000 samples. Initially, the arrays X and Y contain the two data sequences, $[x_k]$ and $[y_k]$. We note that the data length, LX + 1, is either 1000 or 4000 samples and that the power spectra are computed in 33 frequency bands, with LY = 32. The overlap is 62.5%, as recommended by Carter, Knapp, and Nuttall [O], resulting in 40 segments for the first estimate and 165 segments for the second estimate in Fig. 4.15.

```
        LX=999+3000*NP
        CALL SPPOWR(X1,PXX,WORK,LX,32,1,0.625,NSEG,IE1)
        CALL SPPOWR(X2,PYY,WORK,LX,32,1,0.625,NSEG,IE2)
        CALL SPCROS(X1,X2,CPXY,CWORK,LX,32,1,0.625,NSEG,IE3)
        IF(IE1+IE2+IE3.NE.0) PAUSE 1
        DO 153 M=0,32
          GXY(M)=ABS(CPXY(M))**2/(PXX(M)*PYY(M))
    153 CONTINUE
```

From the two estimates the reader can get an impression of the amount of averaging needed to obtain a reasonably accurate MSC estimate without going into a detailed analysis of confidence intervals.* Also, in the given section of code, we note that GXY, PXX, PYY, and WORK are real arrays, whereas PXY and CWORK must be complex arrays.

To conclude our example of coherence computation, we plot the true and estimated SNRs in Fig. 4.16. The true SNR is found from (4.50) and the estimations use 1000 and 4000 samples, as earlier. Again, we obtain an impression from Fig. 4.16 of the relative accuracy achieved with different amounts of data. Note that the SNR reaches a peak where the spectrum of s_k is at a peak, that is, at about one-sixth of the sampling frequency.

4.10 Summary of Calling Sequences

The algorithms introduced in this chapter were the two spectral estimation routines, SPPOWR and SPCROS, and the random number function, SPRAND. The calling sequences and variable definitions are given in the following statements and list. As in Chapter 3, the symbols I and O indicate whether the variable is an input to the routine, or an output computed by the routine.

```
CALL SPPOWR (X,Y,WORK,LX,LY,IWINDO,OVRLAP,NSGMTS,IERROR)
CALL SPCROS (X1,X2,Y,WORK,LX,LY,IWINDO,OVRLAP,NSGMTS,IERROR)
FUNCTION SPRAND (ISEED)
```

---

* With broadband stationary data sequences, averaging with smaller numbers of segments can produce misleading results. In the extreme case with just one segment, the MSC function, estimated using (4.40), is identically 1.

Variable	Definition	Input-Output	Remarks
X,X1,X2	Input data sequences	I	X(0:LX), X1(0:LX), X2(0:LX)
Y	Output spectrum	O	Y(0:LY), complex with SPCROS
WORK	Work array	—	SPPOWR: WORK(0:2*LY + 1) SPCROS: WORK(0:2*LY − 1) Complex with SPCROS
LX	Last index of input sequence(s)	I	At least $N - 1$
LY	Last index of Y	I	Power of 2 Segment size = 2*LY Spectrum size = LY
IWINDO	Data window indicator	I	1. Rectangular 2. Tapered rectangular 3. Triangular 4. Hanning 5. Hamming 6. Blackman
OVRLAP	Fraction of segment overlap	I	$0 \leq \text{OVRLAP} < 1$
NSGMTS	Number of segments averaged during execution	O	
ISEED	Random number seed	I	Initialized to ISEED > 0; then left alone
IERROR	Error indicator	O	0. No error 1. IWINDO out of range 2. LX < 2*LY−11 3. LY ≠ power of 2

4.11 Exercises

4.1 Using equation (3.5), prove the result in equation (4.3). One approach is to use the relationship $\sin \alpha = (e^{j\alpha} - e^{-j\alpha})/(2j)$ and then sum the resulting geometric series.

4.2. Show how to convert Fig. 4.1 to a plot of power density versus frequency (f) in hertz. Draw the plot to scale for $N = 24$, $M = 4$, $A = 0.365$, and time step $T = 0.1$ s. Label the scales on the plot. Verify that the total power, $A^2/2$ is the area of the periodogram.

4.3. Given the sample set $[x_k]$ with

$$x_k = 2 \cos\left(\frac{2\pi k}{8}\right), \qquad k = 0, 1, ..., 31$$

plot the periodogram of the entire sample set without the aid of SPPOWR. Show that the total power in $[x_k]$ is the integral of the periodogram.

4.4. Verify your answer to exercise 4.3 by using SPPOWR and printing the results, including the total power in $[x_k]$.

4.5. Write a program to reproduce Fig. 4.4 but with five different values of ISEED. Plot all five histograms on the same graph.

4.6. Write a program to reproduce Fig. 4.5 but with five different values of ISEED. Plot all five histograms on the same graph.

4.7. A periodogram has $LY = 16$ frequency intervals. What is the frequency resolution in hertz-seconds? If the data sampling interval is 1 ms, what is the frequency resolution in hertz?

4.8. A data sequence is sampled at 40 samples per second. What is the minimum segment length (power of 2) that will allow a frequency resolution in the periodogram of $\Delta f = 0.2$ Hz?

4.9. A data sequence is sampled at 10^6 samples per second, and a power density spectrum is computed using segments that are 128 samples in length. What is the frequency resolution, Δf, in hertz?

4.10. Write a DO loop around program SPA0402 so that it is executed five times with $ISEED = 101, 201, \ldots, 501$. Plot all five power density spectra together on a graph similar to Fig 4.6.

4.11. Using a program like SPA0402 with $ISEED = 123$, make a plot similar to Fig. 4.6. Then make the same plot with power density measured in decibels.

4.12. Using $ISEED = 2345$, generate 5000 samples of a uniform white noise sequence as in (4.13) with power $P = 100$. Assume that the time step between samples in this sequence is 5 ms. Using SPPOWR and half-overlapping segments, plot power density (not in decibels) versus hertz with frequency resolution $\Delta f = 6.25$ Hz.

4.13. Repeat exercise 4.12 with the sequence length decreased to 500 samples. Comment on the decreased accuracy of the spectral estimate.

4.14. Repeat exercise 4.12 with eight times the frequency resolution, that is, $\Delta f = 0.78125$ Hz. Comment on the decreased accuracy of the spectral estimate.

4.15. Generate 2000 samples of the sequence given by $x_k = \sqrt{2} \cos(2\pi k/16)$. Using SPPOWR, plot the power density spectrum (not in decibels) using half-overlapping segments 64 samples in length. Then repeat the exercise with $x_k = \sqrt{2} \cos(2\pi k/20)$. Explain the leakage of the spectrum in this second case. (See exercises 3.30–3.32.)

4.16. Examine the leakage phenomenon in the following way. Plot seven power density spectra (not in decibels) using SPPOWR with half-overlapping segments 32 samples in length. For each spectrum, use a time series with 1000 samples of a sine wave at frequency f, with f going from 0.125 Hz-s to 0.15625 Hz-s in equal steps of 0.0052083 Hz-s. Compare the seven spectra and explain the results.

4.17. Repeat exercise 4.12 using a Gaussian white noise sequence in place of the uniform white noise sequence. There should be no significant differences in the result. What, specifically, is necessary for the power spectrum to be flat (white)?

4.18. Repeat exercise 4.12 with a sinusoidal signal component added to the uniform white noise. The signal should be at 20 Hz and the SNR should be 1.0. The power density should be plotted in decibels.

4.19. Generate 5000 samples of the sequence given by

$$x_k = r_k + 0.9\, r_{k-1} + 0.9\, x_{k-1}, \qquad k = 0, 1, \ldots, 4999$$

where r_k is a uniform random number with $E[r_k] = 0$ and $E[r_k^2] = 1.0$. Assume $r_{-1} = x_{-1} = 0$ and use ISEED = 123. Use SPPOWR with half-overlapping segments to plot a power spectrum (not in decibels) with 33 frequency intervals. Also, on the same graph plot the theoretical power spectrum, given by

$$P(\nu) = \frac{1.81 + 1.8\,\cos(2\pi\nu)}{1.81 - 1.8\,\cos(2\pi\nu)}$$

and compare the results.

4.20. Modify SPA0401 so that it uses SPCROS instead of SPPOWR but produces exactly the same printout.

4.21. Modify SPA0402 so that it uses SPCROS instead of SPPOWR but produces exactly the same printout.

4.22. Generate 5000 samples of u_k and v_k as in the accompanying diagram. Let $[n_k]$ be uniform white noise such that the SNR at v_k is 1.0. Using SPCROS in a program with data segments of length 32, plot the real cross power in decibels versus frequency. Comment on the comparison of your result with the theoretical cross-power spectrum.

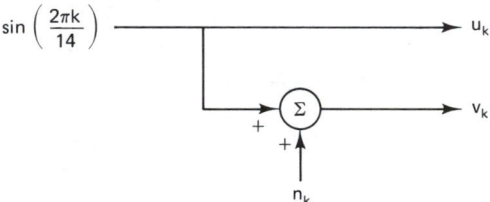

4.23. Plot the power spectrum of $[v_k]$ in decibels in exercise 4.22, again using segment length = 32. Comment on the difference between this spectrum and the cross spectrum.

4.24. Suppose two signals, u_k and v_k, are generated, as shown. H is a scalar gain parameter and n_k is a noise sequence with $E[n_k^2] = 0.1$. The sequences in this case have 2000 samples each. Compute and print the cross spectrum using 16 samples per segment (a) with $H = 1$, (b) with $H = 0.5$, and (c) with $H = 0.25$.

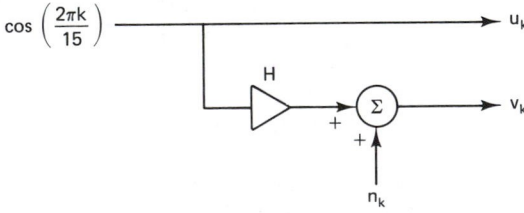

4.25. Do exercise 4.24 with v_k changed to $\cos(2\pi k/15 + \alpha)$, (a) with $\alpha = 0$, (b) with $\alpha = \pi/4$, (c) with $\alpha = \pi/2$, and (d) with $\alpha = \pi$. Explain the result.

4.26. In equation (4.21), prove that $Q_m = Q_{N-m}^*$.

4.27. Using the concept of area of the complex periodogram, derive equations (4.24) and (4.25). Use the redundancy properties of the DFT in equations (3.8) and (3.9).

4.28. Refer to Fig. 4.14. Calculate the estimated MSC function and plot it along with the true MSC, as in the plots in Fig. 4.15. Use data sequences of length 2000. Generate the uniform white signal in Fig. 4.14 beginning with ISEED = 60. Then generate the white noise beginning with ISEED = 78.* Use 32 frequency intervals as in Fig. 4.15, and use 62.5% overlap in the spectral estimations.

4.29. Do exercise 4.28, but plot the estimated SNR instead of the estimated MSC. Make your plot similar to those in Fig. 4.16.

4.30. Compute (a) the estimated MSC, (b) the estimated transfer function, (c) the estimated noise spectrum, and (d) the estimated SNR for the situation shown below. Use sequence length = 1000 samples, segment length = 32 samples (that is, 16 frequency intervals), and 62.5% overlap. Begin with ISEED = 123 for x_k and then use ISEED = 321 for n_k. Plot the results.

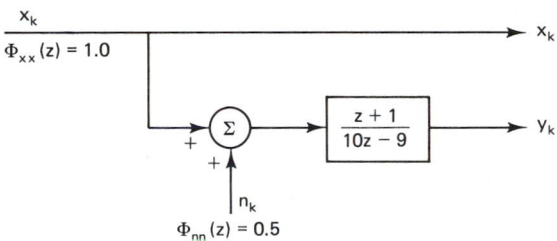

4.12 References

[A] Koopmans, L. H., *The Spectral Analysis of Time Series* (New York: Academic Press, 1974).

[B] Jenkins, G. M., and D. G. Watts, *Spectral Analysis and Its Applications* (San Francisco: Holden-Day, 1968).

[C] Bendat, J. S., and A. G. Piersol, *Random Data: Analysis and Measurement Procedures* (New York: John Wiley, 1971), Chapter 9.

[D] Otnes, R. K., and L. Enochson, *Digital Time Series Analysis* (New York: John Wiley, 1972).

[E] Stearns, S. D., *Digital Signal Analysis* (Rochelle Park, N.J.: Hayden, 1975), Chapters 13 and 14.

[F] Welch, P. D., "The Use of the Fast Fourier Transform for the Estimation of Power Spectra," *IEEE Trans.*, AU-15 (June, 1967): 70.

[G] Richards P. I., "Computing Reliable Power Spectra," *IEEE Spectrum*, V. 4 (January, 1967): 83.

[H] Bingham, C., M. D. Godfrey, and J. W. Tukey, "Modern Techniques of Power Spectrum Estimation," *IEEE Trans.*, AU-15 (June, 1967): 56.

[I] Oppenheim, A. V., and R. W. Schafer, *Digital Signal Processing* (Englewood Cliffs, N.J.: Prentice-Hall, 1975), Chapter 11.

---

* *Note:* These two values of ISEED are about 3000 samples apart in the sequence of SPRAND, so the two random sequences do not overlap.

[J] Rabiner, L. R., and B. Gold, *Theory and Application of Digital Signal Processing* (Englewood Cliffs, N.J.: Prentice-Hall, 1975), Chapter 6.

[K] Gold, B., and Rader, C. M., *Digital Processing of Signals* (New York: McGraw-Hill, 1969), p. 124.

[L] Carter, G. C., and A. H. Nuttall, "On the Weighted Overlapped Segment Averaging Method for Power Spectral Estimation," *Proc. IEEE,* V. 68 (1980): 1352–4.

[M] Nuttall, A. H., and G. C. Carter, "A Generalized Framework for Power Spectral Estimation," *IEEE Trans.,* ASSP-28 (June, 1980): 334.

[N] Yuen, C. K., "A Comparison of Five Methods for Computing the Power Spectrum of a Random Process Using Data Segmentation," *Proc. IEEE,* V. 65 (June, 1977): 984.

[O] Carter, G. C., C. H. Knapp, and A. H. Nuttall, "Estimation of the Magnitude-Squared Coherence Function via Overlapped FFT Processing," *IEEE Trans.,* AU-21 (August, 1973): 337.

[P] Roth, P. R., "Effective Measurements Using Digital Signal Analysis," *IEEE Spectrum,* V. 8 (April, 1971): 62.

[Q] Benignus, V. A., "Estimation of the Coherence Spectrum and Its Confidence Interval Using the Fast Fourier Transform," *IEEE Trans.,* AU-17 (June, 1969): 145.

[R] Hinich, M. J., and C. S. Clay, "The Application of the Discrete Fourier Transform in the Estimation of Power Spectra, Coherence, and Bispectra of Geophysical Data," *Reviews of Geophysics,* V. 6 (August, 1968): 347.

chapter 5

Frequency
and Time-Domain
Response Routines

5.1 Introduction

In this chapter we describe two simple routines for computing the frequency response and the time-domain response of a linear system. The system must be known in terms of its z-transfer function, which was described in Chapter 2. In this chapter the transfer function must be in direct, parallel, or cascade form. These terms are illustrated below. The transfer function can also be in lattice form, but lattice forms are not discussed until Chapter 6.

We begin by describing the function SPGAIN, which computes the complex frequency response, or complex gain, of a linear system. We include enough theory to cover the operation of the function and to be able to relate the function's output, the complex gain, to the amplitude gain, power gain, and phase shift of a linear system. For more theory, numerous texts are available on linear digital transfer functions [A–E].

Following the description of SPGAIN, we give some examples of its use to compute the gain and phase shift of some linear systems. Then we introduce the subroutine SPRESP, which is used to compute time-domain responses of linear systems. Examples of the use of SPRESP are also included.

5.2 The Frequency Response of a Linear System

Suppose we know the transfer function, $H(z)$, of a linear digital system. We can then determine its frequency response, or complex gain, by substituting $e^{j\omega}$ for z in $H(z)$. The result is a complex function of the normalized frequency ω, with the units of ω being in radians. In Table 3.1 we saw that the sampling frequency is always 2π radians or $1/T$ hertz, so in general the normalized frequency in radians must be $2\pi T$ times the frequency in hertz, T being the time step or sampling interval in seconds. Thus, if f is the frequency in hertz,

$$\text{Complex gain} = H(e^{j\omega}) = H(e^{j2\pi fT}) \tag{5.1}$$

The complex gain tells us how the linear system affects the amplitude and phase of a signal at any frequency f. Usually we are interested in frequencies only in the range from 0 to half the sampling rate, that is, $0 \le f \le 1/2T$. For example, consider the illustration in Fig. 5.1. Here we see that the output is given by

$$y_k = 2.0(x_k - 0.3y_{k-1}) \tag{5.2}$$

From Chapter 2, we know that the transfer function corresponding with this difference equation is

$$H(z) = \frac{2.0}{1 + 0.6z^{-1}} \tag{5.3}$$

From Equation (5.1), then, we have the complex gain in this example given by

$$H(e^{j\omega}) = \frac{2.0}{1 + 0.6e^{-j\omega}} \tag{5.4}$$

As just indicated, we are interested usually in evaluating (5.4) when ω is in the range from zero to half the sampling rate, that is, $0 \le \omega \le \pi$. As suggested in

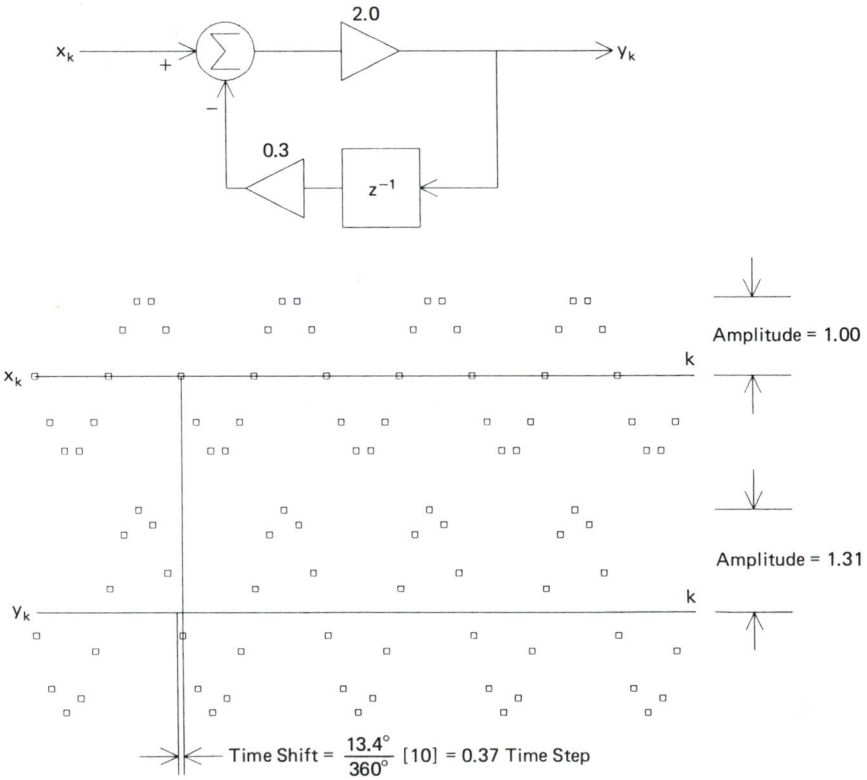

Figure 5.1. Response of a linear system to a sinusoidal input.

Fig. 5.1, suppose the input to the system is a sinusoidal wave having ten samples per cycle, that is, at 0.1 times the sampling rate, or at $\omega_0 = \pi/5$. The complex gain at this frequency is

$$H(e^{j\omega_0}) = \frac{2.0}{1 + 0.6e^{-j\pi/5}}$$

$$= 1.310e^{j0.233} \tag{5.5}$$

Since the system in Fig. 5.1 is linear, we know that an input sinusoid at frequency ω_0 produces an output sinusoid at this same frequency. Then, from (5.5), we know that when $\omega_0 = \pi/5$, the output is 1.310 times the input in amplitude and is shifted ahead in phase by 0.233 rad, or 13.4°. These results are illustrated in the figure.

The *amplitude gain* of a linear system is the magnitude of the complex gain $H(e^{j\omega})$ in (5.1). The *phase shift* is the phase, or polar angle, of $H(e^{j\omega})$. Thus, in (5.5), the amplitude gain is 1.310, and the phase shift is 0.233 rad, or 13.4°. The amplitude gain and the phase are often plotted versus frequency. In Fig. 5.1, for example, the amplitude gain and phase shift at any frequency ω are

$$\text{Amplitude gain} = \left| \frac{2.0}{1 + 0.6e^{-j\omega}} \right|$$

$$= \frac{2.0}{\sqrt{1.36 + 1.2 \cos \omega}} \tag{5.6}$$

$$\text{Phase shift} = \text{phase}\left[\frac{2.0}{1 + 0.6e^{-j\omega}} \right]$$

$$= \tan^{-1}\left(\frac{0.6 \sin \omega}{1 + 0.6 \cos \omega} \right) \tag{5.7}$$

These amplitude-gain and phase-shift functions are plotted in Fig. 5.2. They are plotted versus ω in radians and also versus $\nu = \omega/2\pi$ in hertz-seconds, the latter being a more commonly used frequency measure that puts the sampling rate at $\nu = 1.0$. The *power gain* in decibels is also plotted in Fig. 5.2. The power gain characteristic contains the same information as the amplitude gain characteristic but in different units. The formula for converting (5.6) to power gain is

$$\text{Power gain in dB} = 10 \log_{10}(\text{amplitude gain})^2$$

$$= 20 \log_{10}(\text{amplitude gain})$$

$$= 20 \log_{10}(2.0) - 10 \log_{10}(1.36 + 1.2 \cos \omega) \tag{5.8}$$

To summarize the results of this section, when we know the z-transfer function of a linear system, we have its complex gain given by (5.1), that is, $H(z)$ with $z = e^{j\omega}$. From this, the three functions most usually computed and plotted are

$$\text{Amplitude gain} = |H(e^{j\omega})| \tag{5.9}$$

$$\text{Power gain in decibels} = 10 \log_{10}|H(e^{j\omega})|^2 \tag{5.10}$$

$$\text{Phase shift} = \tan^{-1}\left(\frac{\text{Im}[H(e^{j\omega})]}{\text{Re}[H(e^{j\omega})]}\right) \tag{5.11}$$

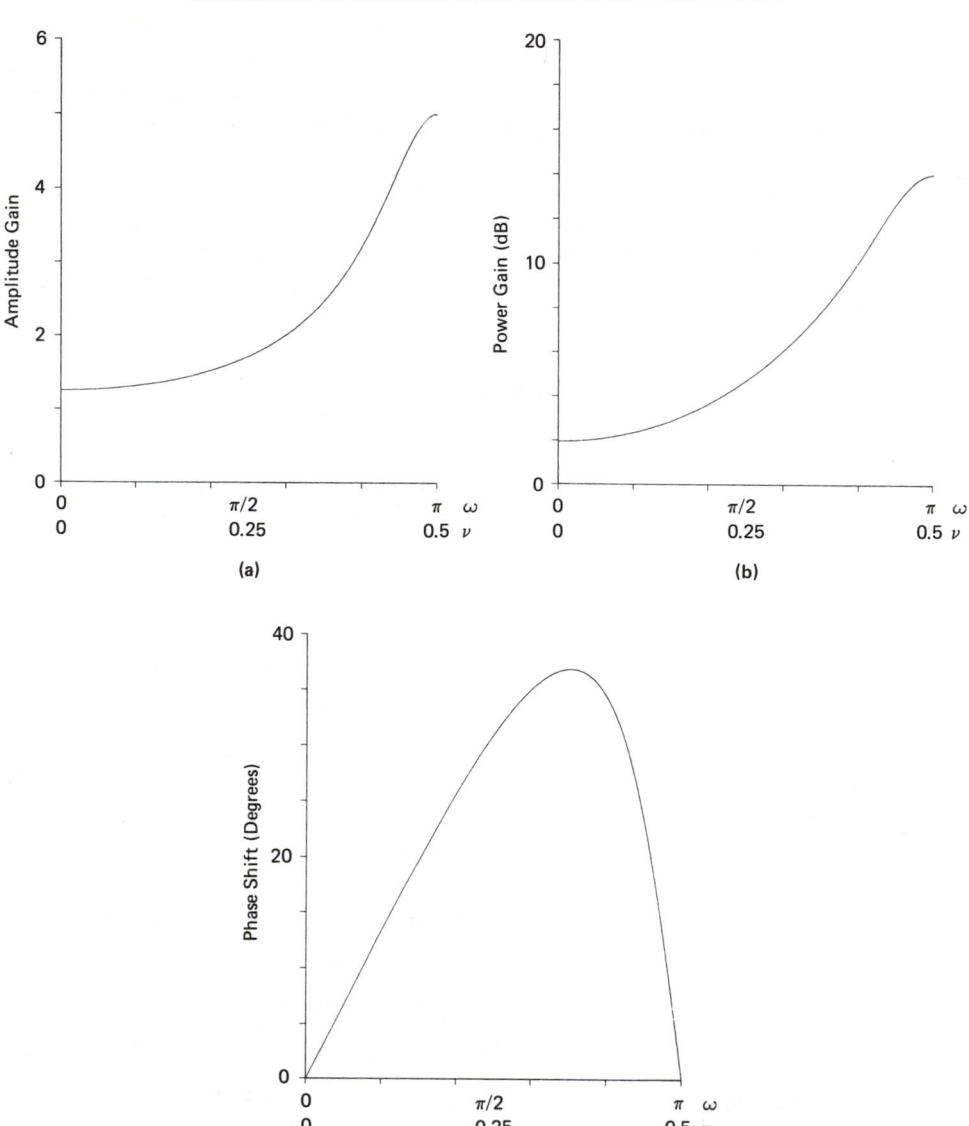

Figure 5.2. Gain and phase characteristics of the linear system in Fig. 5.1.

In all of these formulas the sampling frequency is at $\omega = 2\pi$ radians, which is equivalent to $\nu = 1.0$ hertz-second. In the next section we introduce a simple function that can be used to evaluate the formulas in (5.9) through (5.11). We then demonstrate the use of the function to compute amplitude gain and phase shift. The power gain in decibels, (5.10), is a simple function of the amplitude gain, and we leave its demonstration until later.

5.3 Frequency-Response Routine

The frequency-response routine is called SPGAIN and is listed in Appendix A. It is written as a complex function subprogram and is essentially an implementation of equation (5.1). We demonstrate its use to make gain and phase plots, after discussing the routine and its argument list.

The function specification statement is

COMPLEX FUNCTION SPGAIN (B, A, LB, LA, FREQ)

B = array of numerator coefficients, B(0:LB), as in (5.12)
A = array of denominator coefficients, A(LA), as in (5.12)
LB = last index of B(0:LB)
LA = last index of A(LA)
FREQ = frequency in hertz-seconds at which gain is to be computed

The linear system parameters are contained entirely in the arrays B(0:LB) and A(1:LA), which are both one-dimensional arrays with indices ending at LB and LA, respectively. The numerator coefficients of the transfer function are in B and the denominator coefficients are in A. The transfer function is given by

$$H(z) = \frac{B(0) + B(1)z^{-1} + B(2)z^{-2} + \cdots + B(LB)z^{-LB}}{1.0 + A(1)z^{-1} + A(2)z^{-2} + \cdots + A(LA)z^{-LA}} \qquad (5.12)$$

This notation allows the description of any causal linear system in direct form. For finite impulse response FIR systems with no poles, we have to use LA = 1 and A(1) = 0. This method of specifying an all-zero system allows SPGAIN to be compatible with the filter software in Chapters 6 through 8. The final argument in the calling sequence, FREQ, is the frequency (ν) at which the gain is to be computed, normalized so that the sampling frequency is 1.0. Thus, as earlier, ν is generally in the range $0 \leqslant \nu \leqslant 0.5$. See Table 3.1.

Three simple examples of parameter specifications for SPGAIN are shown in Fig. 5.3. Here the filters are all in direct form. Note that the first example specifies the infinite impulse response (IIR) linear filter in Fig. 5.1 and equation (5.3). The second example shows a causal (FIR) filter, which is specified in D

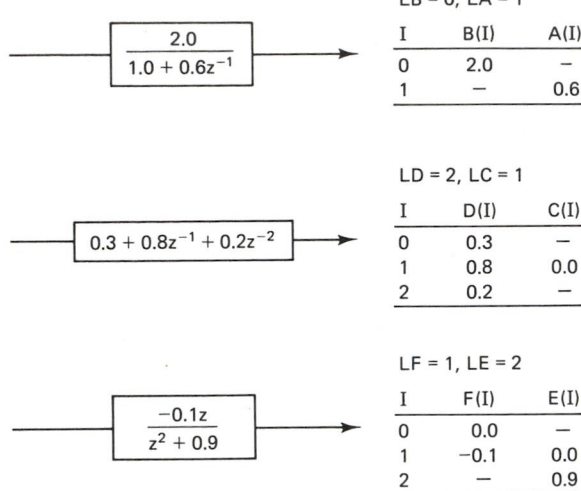

Figure 5.3. Three examples of the parameters in SPGAIN.

and C, with LD = 2 and LC = 1. Note that A, B, and C do not need to be arrays, since each has only one element. In the third example, note first that the system is causal even though its transfer function is in rational form and secondly that the missing coefficients, F(0) and E(1), are both set equal to zero.

To demonstrate the use of SPGAIN, we first compute the amplitude gain of the three systems in Fig. 5.3 at a single frequency FREQ = 0.25, that is, at $\frac{1}{4}$ of the sampling frequency. The gain is computed in the following program, SPA0501. After RUN SPA0501, the program prints the real and imaginary parts of the complex gain in the three systems of Fig. 5.3. The reader can check the results against equation (5.1) with $\omega = 2\pi/4$ or $\pi/2$. For example, in the second case (the FIR system), we have

$$H(e^{j\pi/2}) = 0.3 + 0.8e^{-j\pi/2} + 0.2e^{-j\pi}$$
$$= 0.1 - j0.8$$

(5.13)

```
          PROGRAM SPA0501
C-DEMONSTRATION OF THE USE OF "SPGAIN" AT A SINGLE FREQUENCY.
          DIMENSION D(0:2),E(1:2),F(0:1)
          COMPLEX SPGAIN
          DATA B/2.0/, A/0.6/
          DATA D/0.3,0.8,0.2/, C/0.0/
          DATA F/0.0,-0.1/, E/0.0,0.9/
          PRINT 1,SPGAIN(B,A,0,1,0.25),SPGAIN(D,C,2,1,0.25),
         +        SPGAIN(F,E,1,2,0.25)
        1 FORMAT('       REAL      IMAG'/(2F10.4))
          STOP
          END
$ RUN SPA0501
          REAL      IMAG
          1.4706    0.8824
          0.1000   -0.8000
          0.0000    1.0000
FORTRAN STOP
```

Note how the data in the program is stored in the arrays A through F, just as in Fig. 5.3, and that the arrays A, B, and C, which each have just one element, are not dimensioned. Also note that there are three separate calls to SPGAIN, with the frequency specified as 0.25 in each call. Finally, note especially that SPGAIN must be declared complex in the main program.

5.4 Direct-Form Examples

To provide a more typical example of the use of SPGAIN with a system in direct form, we select the third example in Fig. 5.3. The following program, SPA0502, uses SPGAIN to generate a plot of amplitude gain versus frequency. The program calculates 301 evenly spaced frequency values from 0.0 to 0.5 in the FREQ array and 301 corresponding amplitude gain values in the AMP array. Note that the coefficient arrays, F and E, are the same as in SPA0501. The amplitude gain at each frequency is found in loop 1 of SPA0502 by taking the absolute value of the SPGAIN function at that frequency. The FREQ and AMP arrays are plotted in Fig. 5.4 using subroutine PXY, which is a plot routine described in Appendix B. The reader may wish to substitute his or her own plot routine for PXY. By examining the amplitude gain plot in this example, we see that the system in Fig 5.3 is a resonator with unit peak gain at $\frac{1}{4}$ of the sampling frequency.

```
      PROGRAM SPA0502
C-AMPLITUDE GAIN PLOT FOR H(Z)=-0.1*Z/(Z**2+0.9).
      DIMENSION FREQ(0:300),AMP(0:300),F(0:1),E(1:2)
      COMPLEX SPGAIN
      DATA F/0.0,-0.1/, E/0.0,0.9/
      DO 1 I=0,300
        FREQ(I)=I*0.5/300.0
        AMP(I)=ABS(SPGAIN(F,E,1,2,FREQ(I)))
    1 CONTINUE
      CALL PXY(FREQ,AMP,301,1,0,0,0,2,.1,.1,.9,.9)
      STOP
      END
```

For a second example of the use of SPGAIN with a direct-form system, we construct a plot of the phase shift of this system with $H(z) = -0.1z/(z^2 + 0.9)$. This is done as follows, with SPA0503. The phase shift is calculated in the PHASE array and is plotted in Fig. 5.5. Each phase value is computed by using the Fortran library function ATAN2 to implement equation (5.11). In SPA0503, note that the complex gain is first calculated in H, to avoid using the function

```
      PROGRAM SPA0503
C-PHASE SHIFT PLOT FOR H(Z)=-0.1*Z/(Z**2+0.9).
      DIMENSION FREQ(0:300),PHASE(0:300),F(0:1),E(1:2)
      COMPLEX SPGAIN,H
      DATA F/0.0,-0.1/, E/0.0,0.9/
      DO 1 I=0,300
        FREQ(I)=I*0.5/300.0
        H=SPGAIN(F,E,1,2,FREQ(I))
        PHASE(I)=ATAN2(AIMAG(H),REAL(H))
    1 CONTINUE
      CALL PXY(FREQ,PHASE,301,1,0,0,0,2,.1,.1,.9,.9)
      STOP
      END
```

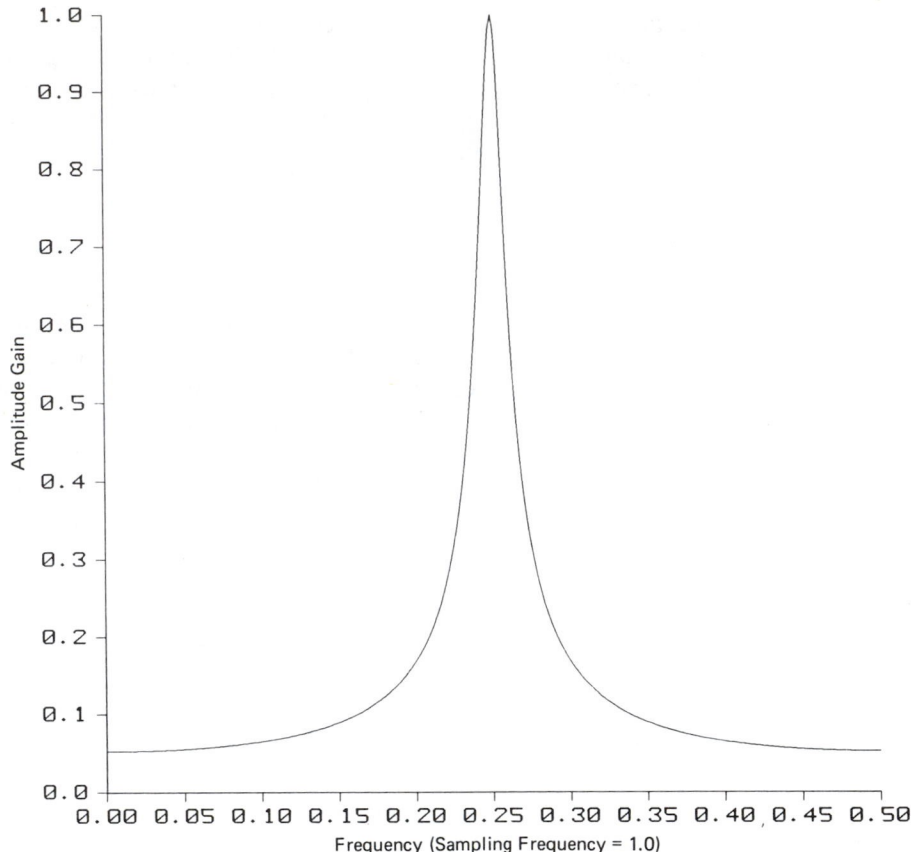

Figure 5.4. Amplitude gain versus frequency computed in SPA0502 for the system in Fig. 5.3 with $H(z) = -0.1z/(z^2 + 0.9)$.

SPGAIN twice in the ATAN2 arguments. In Fig. 5.5, note that the phase shift changes rapidly from π to 0 radians as the frequency passes through resonance at $\frac{1}{4}$ of the sampling frequency.

5.5 Cascade and Parallel Examples

When linear systems are given in cascade or parallel form (Chapter 2, section 2.5), the SPGAIN function is still easy to use without converting the transfer functions to direct form. Here we present two simple examples of amplitude gain computations to illustrate the use of SPGAIN with both cascade and parallel forms. The examples are constructed using the transfer functions

$$H_1(z) = \frac{0.2z}{z - 0.8} \qquad (5.14)$$

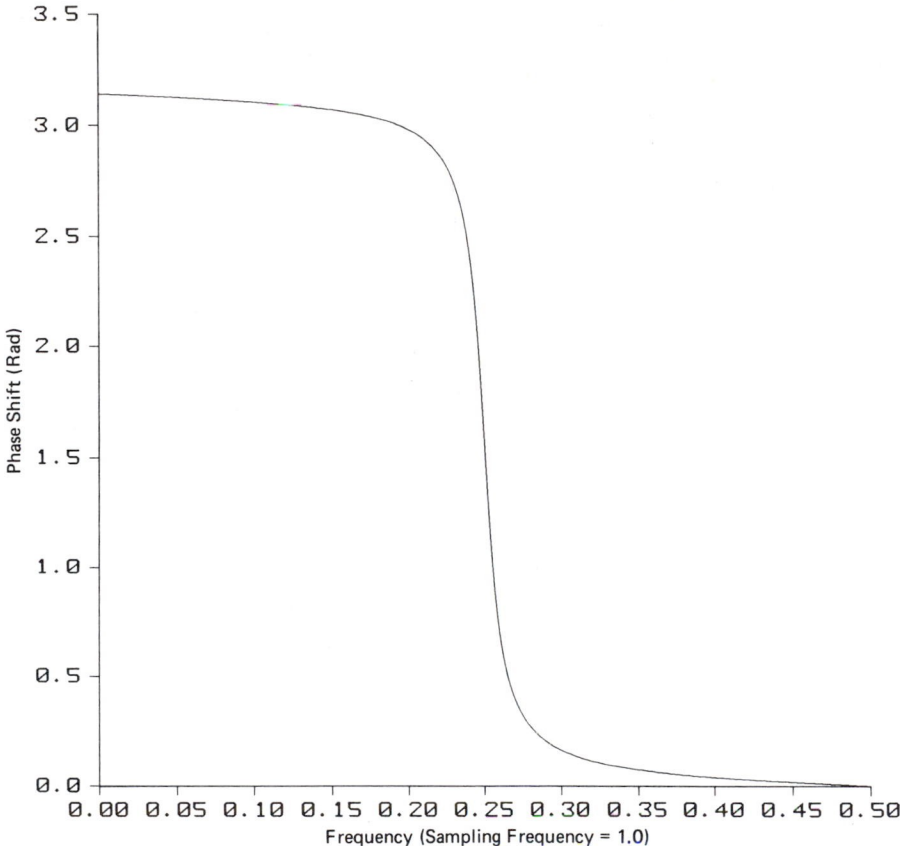

Figure 5.5.   Phase shift versus frequency computed in SPA0503 for the system in Fig. 5.3 with $H(z) = -0.1z/(z^2 + 0.9)$.

$$H_2(z) = \frac{0.05z^2 - 0.1z}{z^2 - 0.95z + 0.9} \tag{5.15}$$

The cascade example evaluates the amplitude gain of the product $H_1 H_2$ and the parallel example evaluates the amplitude gain of the sum $H_1 + H_2$.

For the cascade transfer function $H_1(z)\, H_2(z)$, we simply take the product of the individual gains, as in SPA0504. In loop 1 in the program, note that the

```
        PROGRAM SPA0504
C-AMPLITUDE GAIN PLOT FOR THE CASCADE COMBINATION, H1*H2.
        DIMENSION FREQ(0:300),AMP(0:300),D(0:1),C(1:2)
        COMPLEX SPGAIN
        DATA B/0.2/, A/-0.8/, D/0.05,-0.1/, C/-0.95,0.9/
        DO 1 I=0,300
          FREQ(I)=I*0.5/300.0
          AMP(I)=ABS(SPGAIN(B,A,0,1,FREQ(I))*SPGAIN(D,C,1,2,FREQ(I)))
      1 CONTINUE
        CALL PXY(FREQ,AMP,301,1,0,0,0,2,.1,.1,.9,.9)
        STOP
        END
```

complex gain functions are multiplied together and then the absolute value is taken. Alternately, we could have taken the product of absolute values, that is,

$$|H_1(z)\ H_2(z)| = |H_1(z)||H_2(z)| \tag{5.16}$$

The results of SPA0504, AMP versus FREQ, are plotted in Fig. 5.6 using subroutine PXY. Again, any plot routine could be used.

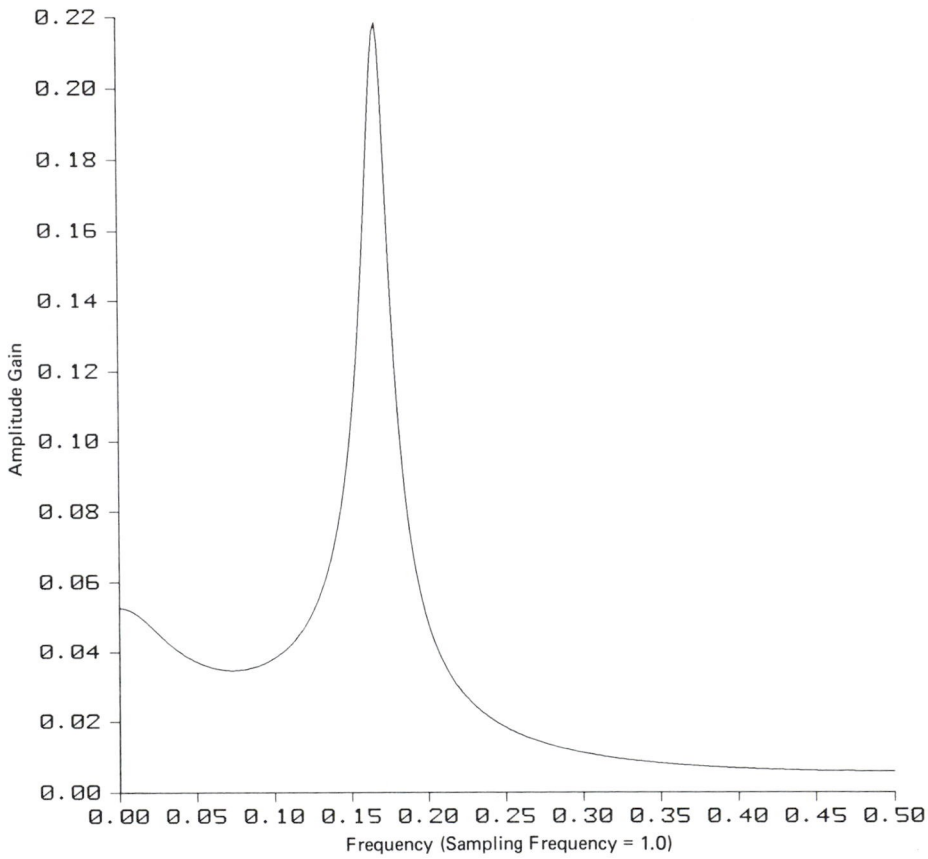

Figure 5.6. Amplitude gain of the cascade combination $H_1(z)\ H_2(z)$ in equations (5.14) and (5.15), computed in SPA0504.

For the parallel transfer function $H_1(z)\ +\ H_2(z)$, we take the absolute value of the sum of the two SPGAIN functions. Note that here we must find the sum first because, unlike (5.16),

$$|H_1(z)\ +\ H_2(z)| \leq |H_1(z)|\ +\ |H_2(z)| \tag{5.17}$$

The amplitude gain of the parallel combination is computed in SPA0505, which is listed next, and plotted in Fig. 5.7. In loop 1, again note that the amplitude gain is computed as the absolute value of the sum of the individual complex gains.

```
      PROGRAM SPA0505
C-AMPLITUDE GAIN PLOT FOR THE PARALLEL COMBINATION, H1+H2.
      DIMENSION FREQ(0:300),AMP(0:300),D(0:1),C(1:2)
      COMPLEX SPGAIN
      DATA B/0.2/, A/-0.8/, D/0.05,-0.1/, C/-0.95,0.9/
      DO 1 I=0,300
        FREQ(I)=I*0.5/300.0
        AMP(I)=ABS(SPGAIN(B,A,0,1,FREQ(I))+SPGAIN(D,C,1,2,FREQ(I)))
    1 CONTINUE
      CALL PXY(FREQ,AMP,301,1,0,0,0,2,.1,.1,.9,.9)
      STOP
      END
```

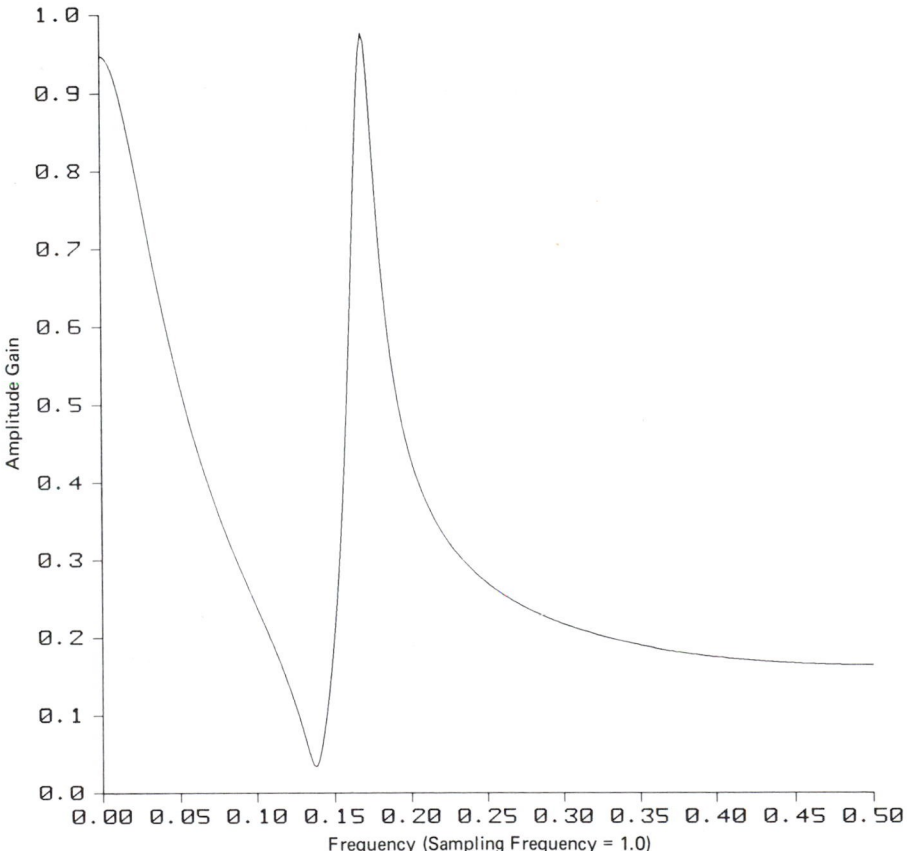

Figure 5.7.   Amplitude gain of the parallel combination $H_1(z) + H_2(z)$ in equations (5.14) and (5.15), computed in SPA0505.

Comparing Figs. 5.6 and 5.7, we can see the effect of combining $H_1(z)$ and $H_2(z)$ in cascade or in parallel. $H_1(z)$ is a simple low-pass function, whereas $H_2(z)$ is a resonator with resonance at $\omega = \pi/3$, or $\nu = 0.167$. These two characteristics can be seen in both plots, combined in different ways.

These examples of cascade and parallel systems can easily be extended to more complicated cases. The systems can always be converted to the direct form before using SPGAIN, but they need not be. (See exercises 5.16 and 5.19.)

For digital filters with large numbers of sections, the SPGAIN function can be used in a loop, with the overall gain initialized to 1 for cascade filters or 0 for parallel filters. For example, see exercises 5.31 and 5.32.

5.6 Noncausal Examples

A special case of the use of SPGAIN occurs when the linear system, or part of the linear system, has a noncausal transfer function. A noncausal linear system is characterized by having an impulse response that begins at a finite interval before the impulse itself occurs. Thus, a noncausal system can always be made causal by appending a suitable delay at its output. For example, $H_3(z)$ and $H_4(z)$ in Fig. 5.8 are both seen to be noncausal, but both overall systems with delays are causal.

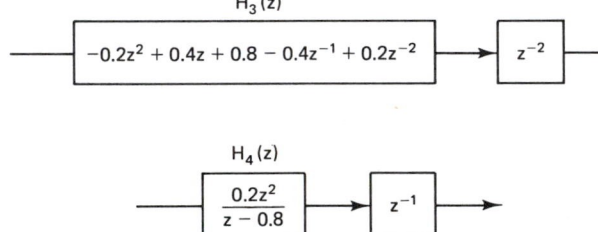

$H_3(z)$

$$-0.2z^2 + 0.4z + 0.8 - 0.4z^{-1} + 0.2z^{-2}$$

$$z^{-2}$$

$H_4(z)$

$$\frac{0.2z^2}{z - 0.8}$$

$$z^{-1}$$

Figure 5.8. Two noncausal transfer functions $H_3(z)$ and $H_4(z)$ made causal by adding delays.

The first property to note about a noncausal system, or any linear system, is that its amplitude gain and power gain are not changed by apppending a delay to the input or output, because the amplitude of $e^{j\omega}$ is always 1. Thus,

$$\text{Amplitude gain of } z^{-n}H(z) = |e^{-j\omega n}H(e^{j\omega})|$$
$$= |e^{-j\omega n}|\,|H(e^{j\omega})|$$
$$= |H(e^{j\omega})| \qquad (5.18)$$

Thus, to use SPGAIN to find the amplitude or power gain of a noncausal system, we first append a delay to make the system causal and then use SPGAIN as described above.

On the other hand, the phase characteristic is changed by appending a delay to $H(z)$. The delay has the effect of subtracting a straight line from the phase characteristic of $H(z)$. That is,

$$\text{Phase shift of } z^{-n}H(z) = \text{phase } [e^{-j\omega n}H(e^{j\omega})]$$
$$= \text{phase } [H(e^{j\omega})] - n\omega \qquad (5.19)$$

Thus, using SPGAIN to obtain the phase shift of a noncausal system involves appending a delay to force the system to be causal, then obtaining the phase shift with SPGAIN, and then adding back the straight line $n\omega$ in (5.19). As an example, in SPA0506 we compute and plot the phase shift of $H_3(z)$ and $H_4(z)$ in Fig. 5.8. The results are plotted in Fig. 5.9. In SPA0506, note first that the

DATA statement defines only the parameters in the array B. The remaining parameters in H_3 and H_4 are specified in the SPGAIN calling sequences. Next, note that the two complex gains are first computed in H3 and H4. Then the phase shifts, PHASE3(I) and PHASE4(I), are computed by adding the $n\omega$ term in (5.19), that is, adding $2\omega = 4\pi\nu$ to PHASE3 and $\omega = 2\pi\nu$ to PHASE4.

```
            PROGRAM SPA0506
      C-PHASE SHIFT PLOTS FOR NONCAUSAL SYSTEMS H3(Z) AND H4(Z).
            DIMENSION FREQ(0:300),PHASE3(0:300),PHASE4(0:300),B(0:4)
            COMPLEX SPGAIN,H3,H4
            DATA B/-0.2,0.4,0.8,-0.4,0.2/
            PI=4.*ATAN(1.)
            DO 1 I=0,300
              FREQ(I)=I*0.5/300.0
              H3=SPGAIN(B,0.,4,1,FREQ(I))
              H4=SPGAIN(0.2,-0.8,0,1,FREQ(I))
              PHASE3(I)=ATAN2(AIMAG(H3),REAL(H3))+4.*PI*FREQ(I)
              PHASE4(I)=ATAN2(AIMAG(H4),REAL(H4))+2.*PI*FREQ(I)
          1 CONTINUE
            CALL SPUNWR(PHASE3,300,1)
            CALL PXY(FREQ,PHASE4,301,1,0,1,0,2,.1,.1,.9,.9)
            CALL PXY(FREQ,PHASE3,301,1,0,3,0,0,0,0,0,0)
            STOP
            END
```

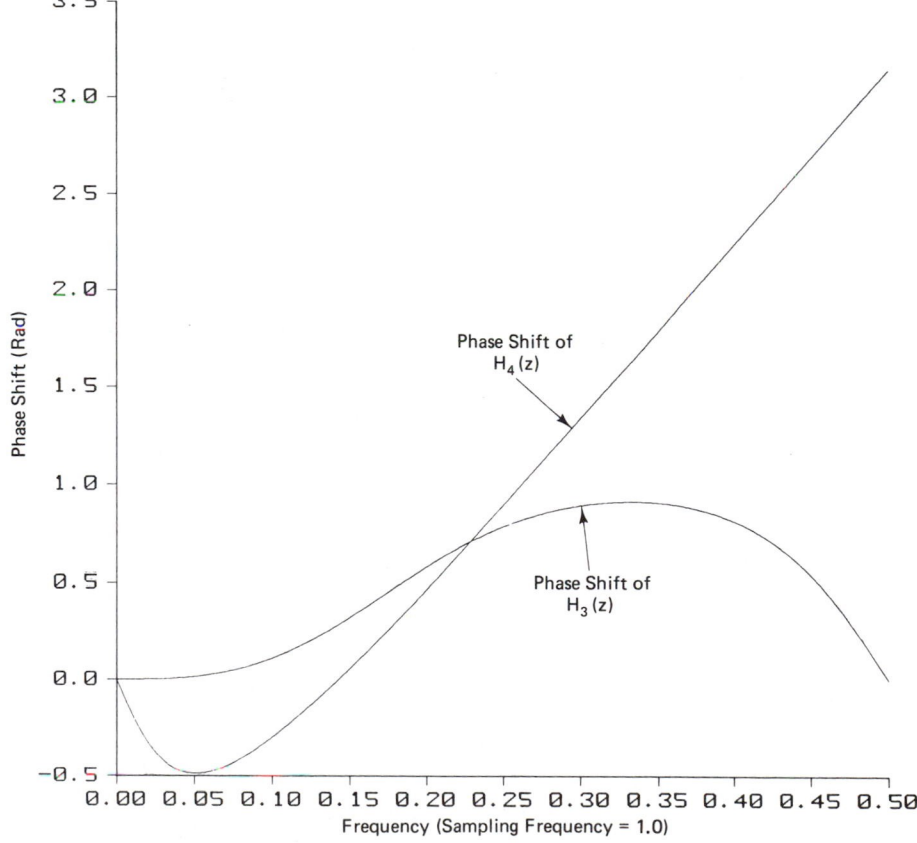

Figure 5.9.   Phase shift of the two noncausal systems, H_3 and H_4, in Fig. 5.8.

Finally, note the phase unwrapping call to SPUNWR just after statement 1, which is needed to avoid a jump of 2π in PHASE3 caused by the ATAN2 function. Phase unwrapping is discussed in Section 14.3. See also exercises 5.6, 5.14, 5.18, and 5.20.

5.7 Time-Domain Response Routine

To conclude this chapter on system response, we describe a time-domain response subroutine called SPRESP. The routine is used to compute the response, or output, of any linear system corresponding to a specified input. The calling sequence for SPRESP is

CALL SPRESP(X,Y,LX,LY,B,A,LB,LA)

$$
\begin{aligned}
X &= \text{input data array, X(0:LX), described later} \\
Y &= \text{output data array, Y(0:LY)} \\
LX &= \text{last index of X(0:LX)} \\
LY &= \text{last index of Y(0:LY)} \\
B &= \text{array of numerator coefficients, B(0:LB), as in (5.22)} \\
A &= \text{array of denominator coefficients, A(LA), as in (5.22)} \\
LB &= \text{last index of B(0:LB)} \\
LA &= \text{last index of A(LA)}
\end{aligned}
$$

The input signal is stored in the array X(0:LX) before calling SPRESP. The corresponding sample set $[x_k]$ is assumed to be 0 before $k = 0$ and to continue with the last specified sample, X(LX). Thus, the input sample set is

$$[x_k] = 0\ 0\ \ldots\ 0\ X(0)\ X(1)\ X(2)\ \ldots\ X(LX)\ X(LX)\ X(LX)\ \ldots \qquad (5.20)$$

The output signal is computed by SPRESP in Y(0:LY). Thus, LY + 1 samples of the output signal are returned. In the computation, SPRESP uses the equivalent of (5.20) to obtain the input samples. Thus, for example, to obtain a unit impulse or unit step response of any length, one could use:

Unit Impulse	Unit Step		
X(0) = 1	X(0) = 1		X(0) = 1
X(1) = 0	LX = 0	or	X(1) = 1
LX = 1			LX = 1

$$(5.21)$$

The linear system $H(z)$ is specified in the arrays B and A just as in equation (5.12), that is,

$$H(z) = \frac{B(0) + B(1)z^{-1} + \cdots + B(LB)z^{-LB}}{1.0 + A(1)z^{-1} + \cdots + A(LA)z^{-LA}} \qquad (5.22)$$

As before, this notation allows us to specify any causal linear system in direct form, including the FIR case with LA = 1 and A(1) = 0.0.

For a simple example, SPA0507 is listed next. This program calls SPRESP to compute the impulse response of $H_2(z)$ in equation (5.15). One hundred one samples of the impulse response are computed by SPA0507 in the array Y, and then the array is plotted using PY (a plot routine described in Appendix B); any other plot routine can be used. Note that the unit impulse function is specified as in (5.21) above. The SPRESP routine automatically extends the function to the right (that is, the last sample = 0.0). The filter $H(z)$ is specified as before in the B and A arrays. Note the last indices, LB = 1 and LA = 2, in this example. The WORK array is a dummy array for use only in PY. The output of SPRESP, Y(K) versus K, is plotted in Fig. 5.10.

To compute the time-domain response of a parallel system, we simply add, sample by sample, the results of applying SPRESP to each section of the system. This simple operation is left to exercises 5.25 and 5.26.

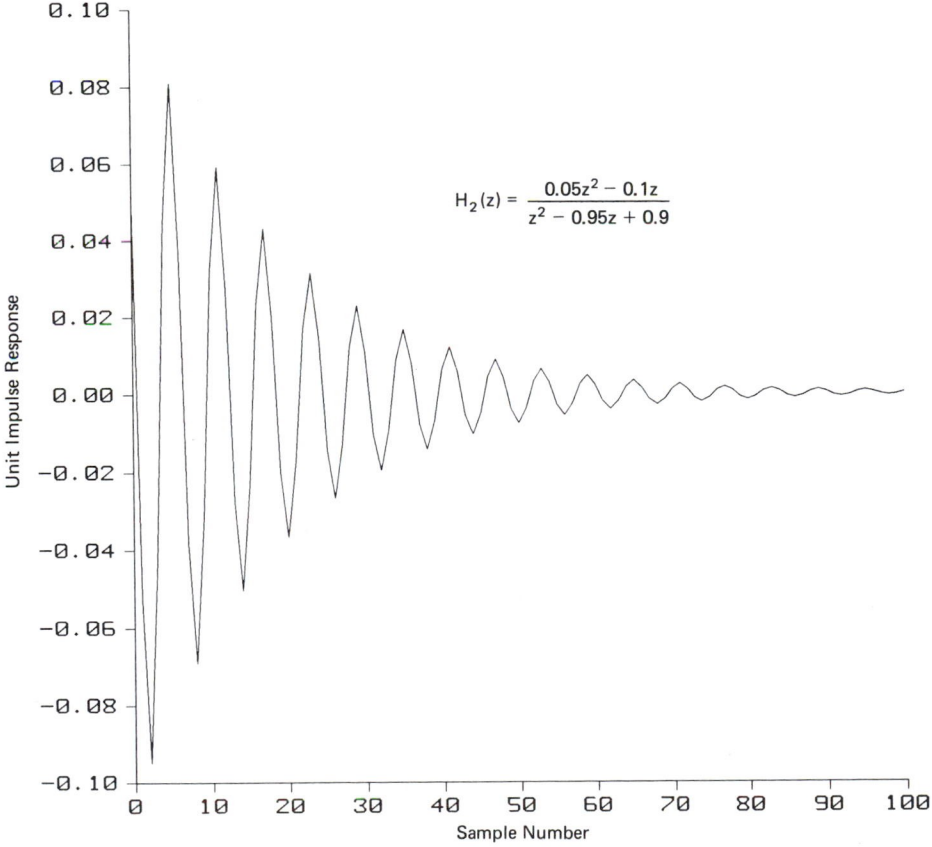

Figure 5.10.   Unit impulse response of $H_2(z) = (0.05z^2 - 0.1z)/(z^2 - 0.95z + 0.9)$, computed in SPA0507 using SPRESP.

```
      PROGRAM SPA0507
C-PLOT IMPULSE RESPONSE OF H(Z)=(0.05*Z**2-0.1*Z)/(Z**2-0.95*Z+0.9).
      DIMENSION X(0:1),Y(0:100),B(0:1),A(1:2),WORK(101)
      DATA X/1.0,0.0/, B/0.05,-0.1/, A/-0.95,0.9/
      CALL SPRESP(X,Y,1,100,B,A,1,2)
      CALL PY(0.,1.,Y,101,1,0,WORK,.1,.1,.9,.9)
      STOP
      END
```

To compute the time-domain response of a system in cascade form, we use the output from the first call to SPRESP as the input to the second call, then use the output of the second call as the input to the third, and so on. An example of a unit step response with two stages, $H_1(z)$ and $H_2(z)$ in (5.14) and (5.15), in cascade is given in SPA0508. Here we see the output (Y1) of the first call to SPRESP being passed to the input of the second call. Then the output of the second call (Y2) is plotted by PY in Fig. 5.11. Comparing the two calling sequences in SPA0508, we note that the first call specifies a final input data index

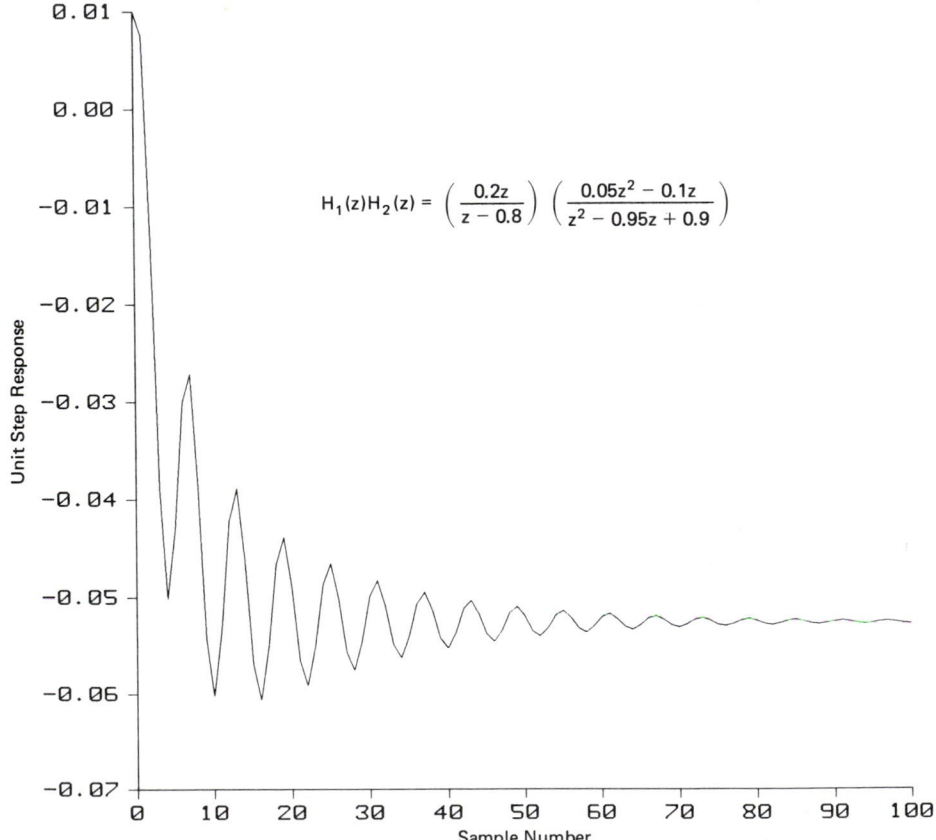

$$H_1(z)H_2(z) = \left(\frac{0.2z}{z - 0.8} \right) \left(\frac{0.05z^2 - 0.1z}{z^2 - 0.95z + 0.9} \right)$$

Figure 5.11. Unit step function response of the cascade combination, $H_1(z)H_2(z)$, in equations (5.14) and (5.15), computed using SPRESP in SPA0508.

```
        PROGRAM SPA0508
C-PLOT THE UNIT STEP RESPONSE OF H1(Z)*H2(Z) IN CASCADE.
        DIMENSION X(0:1),Y1(0:100),Y2(0:100),WORK(101)
        DIMENSION D(0:1),C(1:2)
        DATA X/1.0,1.0/, D/0.05,-0.1/, C/-0.95,0.9/
        CALL SPRESP(X,Y1,1,100,0.2,-0.8,0,1)
        CALL SPRESP(Y1,Y2,100,100,D,C,1,2)
        CALL PY(0.,1.,Y2,101,1,0,WORK,.1,.1,.9,.9)
        STOP
        END
```

LX = 1 and specifies the system $H_1(z)$, whereas the second call specifies a last input data index LX = 100 and specifies $H_2(z)$ by using D and C in the calling sequence.*

For noncausal systems, as in Fig. 5.8 in the preceding section, we simply introduce a bulk delay to make the composite system causal and then apply SPRESP to the composite system. The output of SPRESP can then be shifted to the left, toward decreasing sample numbers, to remove the delay. Note that input signals beginning at different times can also be accommodated in the same way. Noncausal examples may be found in exercises 5.29 and 5.30.

5.8 Summary of Calling Sequences

Two new routines, SPGAIN and SPRESP, were introduced in this chapter. The calling sequences and variable definitions are given in the following statements and list.

COMPLEX FUNCTION SPGAIN (B,A,LB,LA,FREQ)
CALL SPRESP (X,Y,LX,LY,B,A,LB,LA)

Variable	Definition	Input-Output	Remarks
B	Numerator array, B(0:LB)	I	See (5.12) or (5.22)
A	Denominator array, A(LA)	I	See (5.12) or (5.22)
LB	Last index of B	I	
LA	Last index of A	I	
FREQ	Frequency (ν) in Hz-s.	I	Sampling frequency = 1.0
X	Input data array, X(0:LX)	I	See (5.20) and (5.21)
Y	Output data array, Y(0:LY)	O	
LX	Last index of X	I	
LY	Last index of Y	I	Typically, LY $\gg$ LX

As in previous chapters, the symbols I and O indicate whether the variable is an input to the routine or an output computed by the routine.

---

* Alternatively, we could have specified LX = 0 in the first call to SPRESP and used 1.0 in place of X in the calling sequence.

5.9 Exercises

5.1. If we are computing the gain of a digital filter that operates at 2000 samples per second, what is the usual frequency range of interest (a) in terms of hertz and (b) in terms of radians?

5.2. An FIR digital filter has the transfer function $H(z) = 4 + 2z^{-1} + z^{-2}$. Write an expression for the complex gain in terms of frequency, ω, in radians, and in terms of f, in hertz.

5.3. Write an expression for the amplitude gain of the FIR filter in exercise 5.2.

5.4. Write an expression for the phase shift of the FIR filter in exercise 5.2.

5.5. An IIR filter has the transfer function $H(z) = (z + 0.7)/(z - 0.8)$. Write expressions for the complex gain and the amplitude gain as functions of ω.

5.6. Write an expression for the phase shift of the IIR filter in exercise 5.5. Will there be an unwrapping problem when the phase shift is plotted?

5.7. Write an expression for the power gain in decibels of the IIR filter in exercise 5.5.

5.8. Use the SPGAIN routine in a program to find the complex gain and the amplitude gain of the FIR filter in exercise 5.2 at $\frac{1}{8}$ of the sampling frequency.

5.9. Use the SPGAIN routine in a program to compute the complex gain and the phase shift of the IIR filter in exercise 5.5 at 0.1 times the sampling frequency.

5.10. A certain computer smoothing routine averages each sample in a data sequence with each of its neighbors in the following way:
$$y_k = 0.25x_{k-1} + 0.5x_k + 0.25x_{k+1}, \qquad -\infty < k < \infty$$
where $[x_k]$ is the original sequence and $[y_k]$ is the result. (a) Write an expression for the amplitude gain of this averaging operation. (b) What is the phase shift?

5.11. Use the SPGAIN routine in a program to plot the amplitude gain versus frequency for the operation in exercise 5.10 over the range $0 < \omega < \pi$. Use 301 points to produce a smooth plot.

5.12. Use the SPGAIN routine in a program to plot the amplitude gain and phase shift versus frequency for $H(z) = 2.0/(1.0 + 0.6z^{-1})$ in Fig. 5.3.

5.13. Use the SPGAIN routine in a program to plot power gain in decibels versus frequency for $H(z) = 0.3 + 0.8z^{-1} + 0.2z^{-2}$ in Fig. 5.3.

5.14. Do exercise 5.13 with unwrapped phase shift in place of power gain.

5.15. Describe the difference in phase shift between $H(z) = 0.3z^2 + 0.8z + 0.2$ and the filter in exercise 5.13.

5.16. Use the SPGAIN routine in a program to plot amplitude gain versus frequency for the first two systems in Fig. 5.3 in cascade. Then, in a second program, obtain the same result for the system in direct form.

5.17. Use the SPGAIN routine in a program to plot power gain in decibels versus frequency for the three systems in Fig. 5.3 in cascade.

5.18. Use the SPGAIN routine in a program to plot unwrapped phase shift versus frequency for the three systems in Fig. 5.3 in parallel.

5.19. Use the SPGAIN routine in a program to plot the amplitude gains of $H_3(z)$ and $H_4(z)$ in Fig. 5.8 in parallel. Then, in a second program, obtain the same result for the system in direct form.

5.20. Use the SPGAIN routine in a program to plot the unwrapped phase shift of $H_3(z)$ and $H_4(z)$ in parallel in Fig. 5.8.

5.21. A certain low-pass filter uses the following two sections in cascade:

$$H_1(z) = 0.11526(z^2 + 2z + 1)/(z^2 - 1.1113z + 0.5741)$$

$$H_2(z) = 0.08579(z^2 + 2z + 1)/(z^2 - 0.8554z + 0.2097)$$

Use the SPGAIN routine to plot (a) power gain in decibels versus frequency and (b) phase shift versus frequency.

5.22. Using the SPGAIN routine, write your own general-purpose program to read the coefficients of a linear system and plot the amplitude gain or power gain in decibels versus frequency. Allow the user to specify a limited frequency range.

5.23. Use the SPRESP routine in a program to compute and print the entire unit impulse response of $H(z) = 4 + 2z^{-1} + z^{-2}$.

5.24. Use the SPRESP routine in a program to compute and print the first ten samples of the unit impulse response of $H(z) = (z + 0.7)/(z - 0.5)$.

5.25. Use the SPRESP routine in a program to compute and plot the complete unit step function response of the parallel combination, $H_1(z) + H_2(z)$ in equations (5.14) and (5.15).

5.26. Use the SPRESP routine in a program to plot the complete response of the parallel combination of the three systems in Fig. 5.3 to a rectangular pulse of the form $[x_k] = [1\ 1\ 1\ 1\ 0\ 0\ 0\ 0\ 0\ldots]$.

5.27. For the cascade filter in exercise 5.21, use SPRESP to compute and plot the complete impulse response.

5.28. For the cascade filter in exercise 5.21, use SPRESP to compute and plot 100 samples of the response (y_k) to the time series $x_k = 5\sin(2\pi k/16) + \sin(2\pi k/3)$. Plot x_k and y_k on the same graph, and explain the effect of the filter in the light of the answer to exercise 5.21.

5.29. Compute and plot the entire unit step response of $H_4(z) = 0.2z^2/(z - 0.8)$ in Fig. 5.8. Show the abscissa scale clearly.

5.30. Compute and plot the entire impulse response of the cascade combination $H_3(z)\,H_4(z)$ in Fig. 5.8. Show the abscissa scale clearly.

5.31. An *all-pole* filter has eight sections in cascade. Each section has a transfer function of the form $H_n(z) = z^2/(z^2 - 1.9z\cos\theta_n + 0.9025)$, with $\theta_n = n\pi/4$ and $n = 1, 2, \ldots, 8$. Use SPGAIN in a loop, starting with complex gain $= 1.0$, to compute and plot the amplitude gain of the cascade system.

5.32. Use SPGAIN in a loop, starting with complex gain $= 0.0$, to plot the amplitude gain of the eight sections of exercise 5.31 in parallel.

5.33. Use SPRESP in a loop to calculate 200 samples of the impulse response in exercise 5.31 and plot the result.

5.10 References

[A] Gold, B., and C. M. Rader, *Digital Processing of Signals* (New York: McGraw-Hill, 1969), Chapter 2.

[B] Oppenheim, A. V., and R. W. Schafer, *Digital Signal Processing* (Englewood Cliffs, N.J.: Prentice-Hall, 1975), Chapter 2.

[C] Oppenheim, A. V., A. S. Willsky, and I. T. Young, *Signals and Systems* (Englewood Cliffs, N.J.: Prentice-Hall, 1983), Chapter 10.

[D] Ahmed, N., and T. Natarajan, *Discrete-Time Signals and Systems* (Reston, Va.: Reston, 1983), Chapter 5.

[E] Stearns, S. D., *Digital Signal Analysis* (Rochelle Park, N.J.: Hayden, 1975), Chapter 9.

chapter **6**

Digital Filtering Fundamentals

6.1 Introduction

The primary purpose of digital filtering is to alter the spectral information contained in an input signal x_k, thus producing an enhanced output signal y_k. While this can be accomplished in either the time or frequency domain, in the next three chapters we restrict our attention to the time-domain aspects of digital filtering.

Much of the early work in signal processing was done in the analog, or continuous, time domain. While the ultimate goals of digital and analog filtering are the same, the practical aspects vary greatly. In analog filtering we are concerned with active component count and size, termination impedance matching, and lossy reactive elements; in digital filtering we must consider word length, rounding errors, and, in some cases, processing delays.

Since digital filtering has its roots in classical filtering theory, the first digital filters were simulated versions of analog structures. Currently, due to the increase in speed and flexibility of digital systems, most signal processing tasks are being performed in the discrete time domain. While analog filters still provide optimal design criteria, many additional filters for which the analog design is impractical or even impossible can be implemented digitally with relative ease. In Chapters 7 and 8, design routines are provided for both classes of digital filters.

Digital filtering can be performed either off-line using a general-purpose computer or in real time via dedicated hardware. Although numerical precision determined by the available digital word length must be considered in either instance, precision is typically less of a problem with general-purpose computers. For cases where digital processing accuracy is restricted by fixed-point or integer arithmetic, special techniques have been developed for filter design and implementation [F–J]. Since these methods frequently vary according to the specifics of the application, we restrict our discussion to the more general floating-point filter structures and design techniques.

A number of different structures may be used to implement digital filters.

The most common of these, the direct, cascade, and parallel forms, are illustrated in Fig. 6.1. An alternative form, the lattice filter, is discussed later in this chapter. A variety of structures that will not be considered here appear in the literature.

Most digital filters can be implemented using any of the structures just mentioned. Conversion from the direct form to a cascade structure involves factoring the polynomials such that the product of the individual transfer functions in Fig. 6.1(b) is equal to the single transfer function in Fig. 6.1(a). Similarly, conversion to the parallel form requires partial fraction expansion such that the direct-form transfer function is represented by the sum of the transfer functions in Fig. 6.1(c). Conversion to the lattice structure is slightly more complex and is described later in this chapter.

When structural conversions are made, the resulting digital filters are assumed to be equivalent in terms of input-output signal characteristics. This assumption, however, implies that the arithmetic used in both the conversion and implementation is exact. Since, in all practical cases, inexact (finite-precision) arithmetic must be used, the filter structure may significantly affect signal processing results.

The numerical inaccuracies that accumulate during the truncation or rounding necessitated by finite-precision arithmetic as well as those introduced by quantization error due to digitizing an analog signal comprise a separate topic in digital signal processing that is not addressed in detail in this book. In general, the effects are more noticeable in digital systems where the dynamic range is limited by fixed-point arithmetic or a small word size. In these cases, the direct-form implementation, Fig. 6.1(a), tends to provide less accurate results than the other

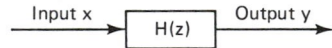

(a) Direct form transfer function.

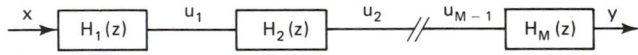

(b) Cascade form transfer function.

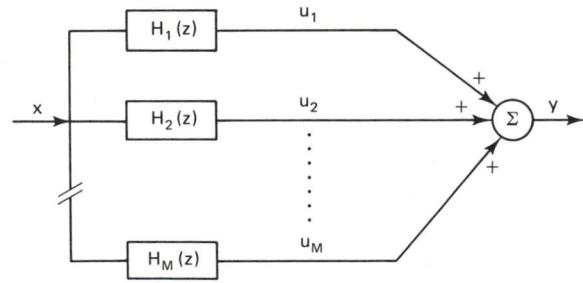

(c) Parallel form transfer function.   **Figure 6.1.** Digital system structures.

structures. There is typically some additional computational cost associated with the use of an alternate form, however, so there are several considerations that affect the selection of a filter structure [A–C].

6.2 Types of Filters

To facilitate discussion of the various types of filters, three basic terms must first be defined. These terms are illustrated pictorially in the context of the normalized low-pass filter in Fig. 6.2. In general, the filter *passband* is defined as the frequency range over which the spectral power of the input signal is passed to the filter output with approximately unity gain. The input spectral power that lies within the filter *stopband* is attenuated to a level that effectively eliminates it from the output signal. The *transition band* is the range of frequencies between the passband and the stopband. In this region, the filter magnitude response typically makes a smooth transition from the passband gain level to that of the stopband. This band has zero width only for the ideal rectangular filter that is not realizable in either the analog or the discrete time domain.

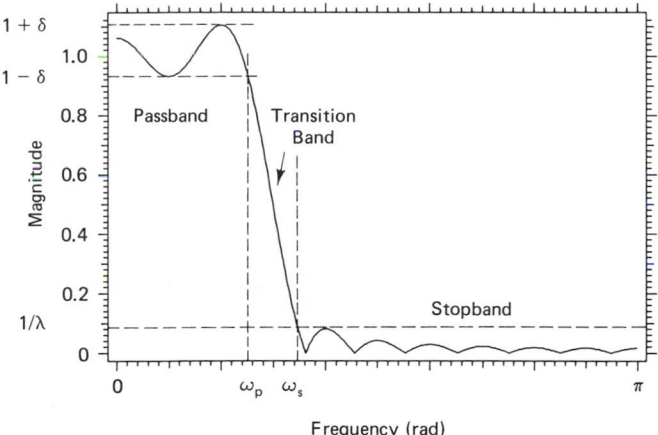

Figure 6.2. Magnitude response of normalized low-pass filter.

Four basic types of filters can now be defined in terms of their frequency response characteristics. The *low-pass filter* passes low-frequency components to the output while attenuating high-frequency components, as illustrated in Fig. 6.2. Conversely, the *high-pass filter,* which is shown in Fig. 6.3, permits high-frequency components to appear at the output while effectively eliminating low-frequency components. The *bandpass filter* rejects both high- and low-frequency components while passing an intermediate range, as illustrated in Fig. 6.4. Note that this filter could, in some cases, be realized as a cascade of a low-pass filter and a high-pass filter whose passbands overlap. The *bandstop filter,* shown in Fig. 6.5, rejects an intermediate band of frequencies while passing high- and low-frequency components. This filter could be implemented using low-pass and high-pass filters with nonoverlapping passbands in the parallel configuration.

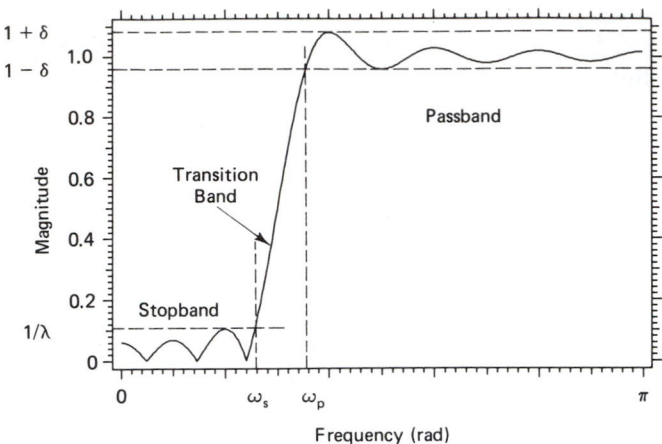

Figure 6.3. Magnitude response of normalized high-pass filter.

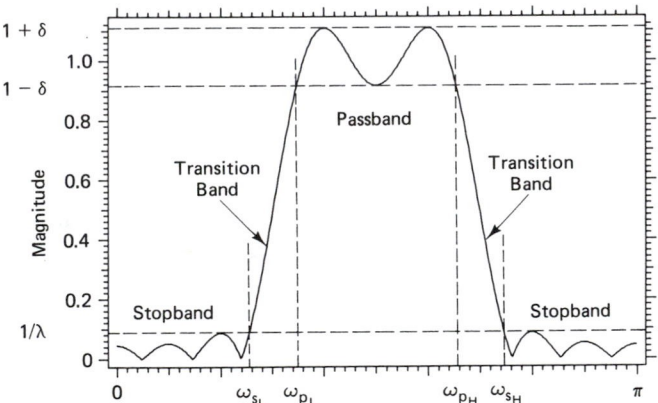

Figure 6.4. Magnitude response of normalized bandpass filter.

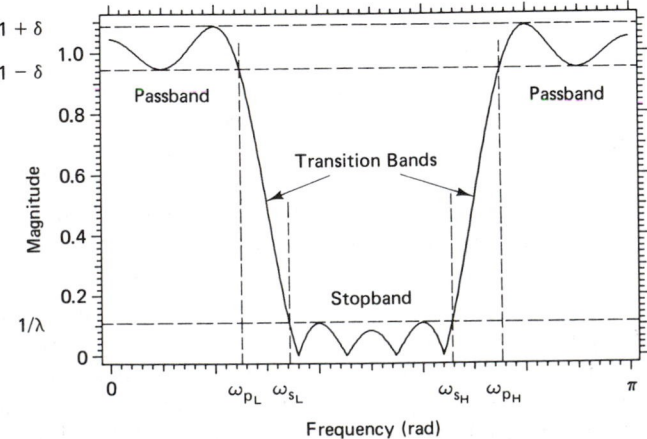

Figure 6.5. Magnitude response of normalized bandstop filter.

The filter response may be specified in terms of the squared magnitude $|H(e^{j\omega})|^2$. Alternatively, the power gain is often defined in decibels as

$$\text{Power gain in decibels} = 10\log_{10}|H(e^{j\omega})|^2 \tag{6.1}$$

Thus, the filter passband with unity magnitude gain corresponds to a power gain of 0 dB. A filter may equivalently be described, as in the following figures, in terms of its amplitude gain or magnitude response characteristic, $|H(e^{j\omega})|$.

The filters illustrated in Figs. 6.2 through 6.5 are normalized examples in the sense that the passbands are specified to have approximately unity gain. In general, this need not be the case. It is, however, a simple matter to adjust the gain of a filter by using a single multiplicative coefficient, for example, $A{\cdot}H(z)$, where A denotes the amplitude gain. Unless otherwise specified, the digital filter-design routines provided in this book produce normalized transfer functions.

Digital filters are further categorized in terms of their impulse responses. In this context, there are infinite impulse response (IIR) and finite impulse response (FIR) digital filters. For each task, the digital filter category is typically determined by weighing the specific requirements of the application against the digital processing capacity available. The primary advantage of IIR filters is that sharp frequency cutoff characteristics are attainable with a relatively low-order structure. This translates to a large savings in processing time and/or hardware complexity. In addition, several familiar analog filters are easily converted to IIR digital structures. On the other hand, one of the most important features of FIR filters is that they can be designed to have exactly linear phase characteristics. Whereas FIR filters typically require many coefficients, implementation via fast convolution will reduce the number of computations required, thus making this filter category more widely applicable. Routines for fast convolution are discussed in Chapter 9.

Each of the four types of filters (that is, low-pass, high-pass, bandpass, bandstop) described can, in general, be realized by either an IIR or an FIR filter. The remainder of this chapter is devoted to the description of algorithms that enable time-domain realization of IIR and FIR filters.

6.3 Direct-Form IIR Digital Filtering

The pole-zero transfer function that describes an Lth-order direct form IIR digital filter is given by a slightly modified version of (5.12):

$$H(z) = \frac{B(z)}{A(z)} = \frac{b_0 + b_1 z^{-1} + \cdots + b_L z^{-L}}{1 + a_1 z^{-1} + \cdots + a_L z^{-L}} \tag{6.2}$$

If the polynomial $B(z) = 1$, $H(z)$ describes an all-pole IIR filter. Conversely, if $A(z) = 1$, an all-zero, or FIR, filter is defined. The FIR filter is discussed in Section 6.6.

A block diagram that illustrates an implementation of the transfer function in (6.2) is provided in Fig. 6.6. Note the feedback of the delayed output values y_{k-n} via the a_n coefficients. It is this feature that generates the infinite impulse

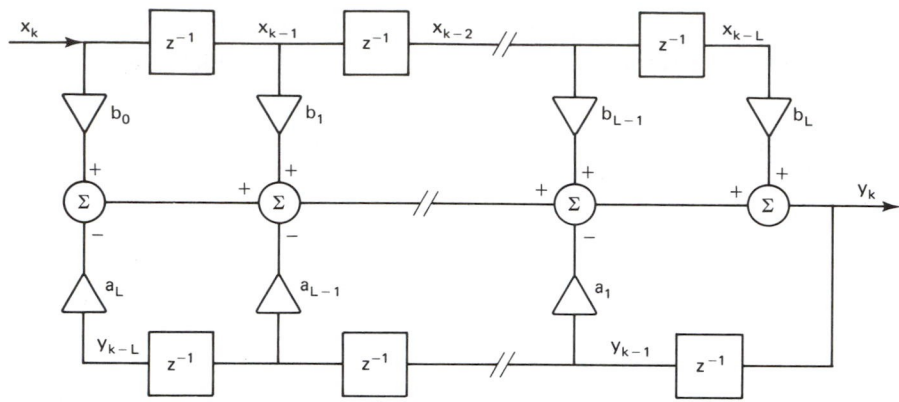

Figure 6.6.   Block diagram of direct-form IIR digital filter.

response, although for all practical purposes the output generally becomes negligible within a finite time period. Figure 6.6 demonstrates a realization of the difference equation

$$y_k = \sum_{n=0}^{L} b_n x_{k-n} - \sum_{n=1}^{L} a_n y_{k-n} \tag{6.3}$$

which corresponds to (6.2). Thus, time-domain IIR digital filtering with a direct-form transfer function involves software implementation of (6.3).

If the filter transfer function is defined in terms of one of the alternate structures described in Section 6.1, there are two options available. The function could, of course, be converted to the direct-form representation of (6.2) so that (6.3) would apply directly. This may, however, be undesirable due to potential numerical inaccuracies. If the transfer function is provided in the cascade or parallel structure, (6.3) can still be used to generate the intermediate signals $[u_m]$, which are then utilized as depicted in Fig. 6.1(b) and (c). Subroutines that facilitate filtering with these structures are described in the next two sections. Implementation of the lattice structure is somewhat more complex and is treated as a separate topic later in this chapter.

The first digital filtering routine to be discussed is called SPFILT and is simply a software implementation of (6.3). This routine is similar to, but slightly more general than, the SPRESP routine in Chapter 5. After defining the calling sequence and its argument list, examples are provided that demonstrate the use of SPFILT.

The subroutine calling sequence is

CALL SPFILT(B,A,LB,LA,X,N,PX,PY,IERROR)

B(0:LB)   = $[b_n]$ filter coefficients from (6.2)
A(LA)   = $[a_n]$ filter coefficients from (6.2)
LB   = order of numerator polynomial in (6.2)

LA	=	order of denominator polynomial in (6.2)
X(0:N−1)	=	data array (input and output)
N	=	number of data samples in X
PX(0:LB)	=	retains past inputs for block mode filtering
PY(LA)	=	retains past outputs for block mode filtering
IERROR	=	0 no errors detected
		1 filter response exceeds 1.E10

The IERROR = 1 condition indicates that the filter output is potentially unbounded. This could result from an invalid input data sequence but more typically is caused by an unstable filter transfer function (i.e., a filter with poles outside the unit circle $|z| = 1$).

The function of arrays PX and PY is to retain past values of the input and output signals, thus enabling the filtering of data sequences whose lengths exceed the size of the input data array. From (6.3) we see that the filter output at time k, y_k, is a function of present and past inputs and past outputs. During the filtering operation in SPFILT, PX and PY are continually updated to the current values necessary for evaluation of (6.3), that is, $PX(0:LB) = [x_k, \ldots, x_{k-LB}]$ and $PY(LA) = [y_{k-1}, y_{k-2}, \ldots, y_{k-LA}]$. Since typically we assume zero initial conditions, array elements should be set to zero prior to initiating the filtering process. Subsequent to this initialization, these arrays should be left alone, since they are updated as necessary by SPFILT. The program steps required for block mode filtering are summarized next. Note that in general the blocks of data could differ in length, since argument N reflects the number of data samples contained in array X.

```
C-SET ZERO INITIAL CONDITIONS
      DATA PX/(LB+1)*0./,PY/LA*0./
C-FILTER DATA IN NBLK BLOCKS
      DO 1 I=1,NBLK
C-GET ITH BLOCK OF INPUT DATA
          CALL INPUT(I,X,N)
C-FILTER DATA
          CALL SPFILT(B,A,LB,LA,X,N,PX,PY,IERROR)
C-SAVE ITH BLOCK OF OUTPUT DATA
          CALL OUTPUT(I,X,N)
      1 CONTINUE
```

Two simple examples that demonstrate the use of SPFILT are provided. These examples are also useful for verifying correct operation of the routine.

The first test program, SPA0601, computes the 10-point inpulse response of a first-order IIR filter with transfer function

$$H(z) = \frac{1. + 0.5z^{-1}}{1. - 0.8z^{-1}} \tag{6.4}$$

(This task could also be performed via the SPRESP routine described in Chapter 5. See exercise 6.19.) In this case, the input data sequence X(0:9) is composed of a single sample of unity magnitude followed by nine zeros. Note that arrays

PX and PY are initialized to zero before SPFILT is called. Since the impulse occurs at $k = 0$ and (6.4) describes a causal filter, all system inputs and outputs for $k < 0$ must be zero. A listing of SPA0601 together with the resulting output data vector is provided.

```
            PROGRAM SPA0601
      C-DEMONSTRATE THE USE OF SPFILT
            DIMENSION X(0:9),B(0:1),PX(0:1),A(1),PY(1)
            DATA X/1.,9*0./,PX/2*0./,PY/0./
            DATA B/1.,.5/,A/-.8/
            DATA LB/1/,LA/1/,N/10/
            CALL SPFILT(B,A,LB,LA,X,N,PX,PY,IERROR)
            IF(IERROR.NE.0)PRINT *,' SPFILT ERROR = ',IERROR
            PRINT 100,(K,X(K),K=0,9)
      100   FORMAT(' SAMPLE    OUTPUT '/(I5,F12.4))
            END
```

```
            $ RUN SPA0601
            SAMPLE    OUTPUT
                0      1.0000
                1      1.3000
                2      1.0400
                3      0.8320
                4      0.6656
                5      0.5325
                6      0.4260
                7      0.3408
                8      0.2726
                9      0.2181
            $
```

The second example in SPA0602 demonstrates the use of SPFILT for block mode filtering of a 100-point data sequence. Realistically, the block mode option would be used primarily in cases where the length of the input data sequence exceeded available array space, but this program provides a simple demonstration of the use of the PX and PY arrays. The input data is again a single impulse at $k = 0$. The filter is now the direct form of a fourth-order Chebyshev low-pass filter that was generated via bilinear transformation [C]. The transfer function is

$$H(z) = \frac{0.001836 + 0.007344z^{-1} + 0.011016z^{-2} + 0.007344z^{-3} + 0.001836z^{-4}}{1.0 - 3.0544z^{-1} + 3.8291z^{-2} - 2.2925z^{-3} + 0.55075z^{-4}}$$

$$(6.5)$$

The data is filtered in 25-point blocks, thus necessitating four calls to SPFILT. Past data arrays PX and PY must again be initialized to zero before the first call, but initial values are provided by SPFILT for subsequent calls. This operation is performed by SPA0602, which is listed next. Since the filtering is again performed in place using the X data array, after each call to SPFILT this array must be emptied of the output data and loaded with the next block of input data (zeros in this case since we are computing the impulse response). In SPA0602 the DATA array is used to save the filter response. The first 10 points of the impulse response are again printed; a plot of the entire response is provided in Fig. 6.7. Note that whereas the $H(z)$ in (6.5) by definition describes an IIR transfer function, the impulse response decays to approximately zero after only 65 data samples.

```
                PROGRAM SPA0602
C-DEMONSTRATE THE USE OF SPFILT FOR BLOCK MODE FILTERING
                DIMENSION X(0:24),B(0:4),A(4),PX(0:4),PY(4)
                DIMENSION DATA(0:99)
                DATA B/.001836,.007344,.011016,.007344,.001836/
                DATA A/-3.0544,3.8291,-2.2925,.55075/
                DATA PX/5*0./,PY/4*0./,X/1.,24*0./
                DATA LB/4/,LA/4/,N/25/
                DO 5 NBLK=0,3
                  CALL SPFILT(B,A,LB,LA,X,N,PX,PY,IERROR)
                  IF(IERROR.NE.0)THEN
                    PRINT *,' SPFILT ERROR = ',IERROR
                    STOP
                  ENDIF
                  DO 4 K=0,24
                    DATA(NBLK*25+K)=X(K)
                    X(K)=0.
        4         CONTINUE
        5       CONTINUE
                CALL MCPLOT(0.,1.,DATA,100,0,0,1,1,2,1,5.,3.6)
                PRINT 100,(K,DATA(K),K=0,9)
      100       FORMAT(' SAMPLE    RESPONSE '/(I6,F12.4))
                END
```

```
              $ RUN SPA0602
              SAMPLE    RESPONSE
                0        0.0018
                1        0.0130
                2        0.0435
                3        0.0950
                4        0.1538
                5        0.1989
                6        0.2123
                7        0.1871
                8        0.1299
                9        0.0574
              $
```

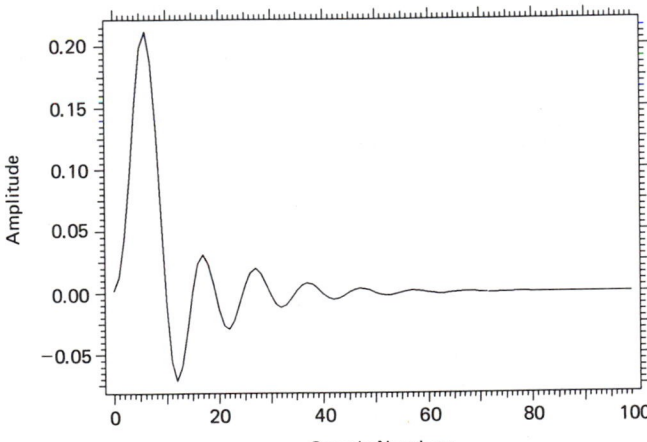

Figure 6.7. SPA0602: Impulse response computed via SPFILT.

In some cases it may not be desirable to perform the filtering operation in place using the input data array. This is particularly true when the transfer function is provided in the parallel form illustrated in Fig. 6.1(c), since the same input sequence $[x_k]$ is required by each of the filter sections. We therefore provide an alternate subroutine in which the filtering is no longer performed in place.

The subroutine calling sequence for SPFLTR is

CALL SPFLTR(B,A,LB,LA,X,N,Y,PX,PY,IERROR)

B(0:LB)	=	$[b_n]$ coefficients from (6.2)
A(LA)	=	$[a_n]$ coefficients from (6.2)
LB	=	order of numerator polynomial in (6.2)
LA	=	order of denominator polynomial in (6.2)
X(0:N − 1)	=	input data array
N	=	number of data samples
Y(0:N − 1)	=	output data array; needs to be initialized
PX(0:LB)	=	retains past inputs for block mode filtering
PY(LA)	=	retains past outputs for block mode filtering
IERROR	=	0 no errors detected
		1 filter response exceeds 1.E10

This subroutine operates identically to SPFILT except that the filtered output data is now returned in array Y(0:N − 1). This array must be initialized appropriately by the calling program, since SPFLTR merely accumulates (adds) values into Y.

Program SPA0603 is provided to demonstrate the use of SPFLTR. In this example, we again use the filter transfer function defined in (6.5) and now compute the 100-point step response. While SPFLTR provides block mode operation, in this case we perform the filtering in a single pass. Thus, we now have an input data array X(0:N − 1) and an output data array Y(0:N − 1), where N = 100 for this example. Note that the X array now contains the input step function and the Y array must be initialized to zero. Upon return from SPFLTR, the X array still contains the input data sequence, and the filter response is returned in Y. A plot of the output data sequence is provided in Fig. 6.8.

```
      PROGRAM SPA0603
C-DEMONSTRATE THE USE OF SPFLTR
      DIMENSION X(0:99),Y(0:99),B(0:4),A(4),PX(0:4),PY(4)
      DATA B/.001836,.007344,.011016,.007344,.001836/
      DATA A/-3.0544,3.8291,-2.2925,.55075/
      DATA PX/5*0./,PY/4*0./,X/100*1./,Y/100*0./
      DATA LB/4/,LA/4/,N/100/
      CALL SPFLTR(B,A,LB,LA,X,N,Y,PX,PY,IERROR)
      IF(IERROR.NE.0)THEN
        PRINT *,' SPFLTR ERROR = ',IERROR
        STOP
      ENDIF
      CALL MCPLOT(0.,1.,Y,100,0,0,1,1,2,1,5.,3.6)
      END
```

Cascade and parallel structures are typically defined in terms of second-order sections. If we assume that the direct-form transfer function in (6.2) contains only real coefficients, we know that all imaginary poles and zeros occur in complex-conjugate pairs. By combining these conjugate pairs to form second-order sections, the equivalent cascade or parallel structure can also be implemented

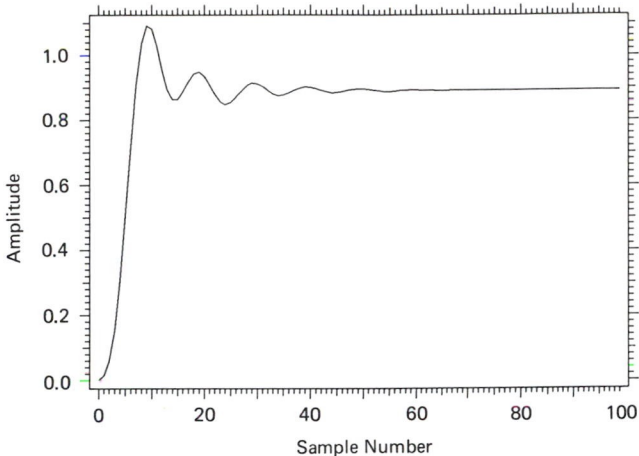

Figure 6.8. SPA0603: Step response computed via SPFLTR.

using only real coefficients. Note that a second-order section could also be used to describe a first-order section by setting the coefficients of z^{-2} to zero.

In this text we consider a more general case by allowing the cascade and parallel structures to be composed from a set of Lth-order sections, each having the form described in (6.2). The only constraints imposed are that all sections must be of equal order and all coefficients are real. The equal-order limitation is easily circumvented by setting unused coefficients equal to zero, as already suggested. This assumption permits the filtering operation for cascade and parallel structures to be accomplished directly via calls to SPFILT or SPFLTR with combination of the results as indicated in Fig. 6.1(b) and (c). To simplify this process, however, we provide alternate routines, which handle the intermediate bookkeeping tasks.

6.4 Cascade-Form IIR Digital Filtering

Here we consider a digital filter defined in the cascade form with transfer function

$$H(z) = H_1(z) \cdot H_2(z) \cdots H_M(z) \tag{6.6}$$

where each $H_m(z)$ is the Lth-order section defined in (6.2). Filtering of an input data sequence contained in array X is accomplished in place via SPCFLT, whose calling sequence and argument list are described next. An example is provided to demonstrate the use of this routine.

The subroutine calling sequence is

CALL SPCFLT(B,A,LS,NS,X,N,PX,PY,IERROR)

B(0:LS,NS)	=	$[b_n]$ coefficients for each section
A(LS,NS)	=	$[a_n]$ coefficients for each section
LS	=	order of sections
NS	=	number of sections in filter

X(0:N − 1)	=	data array (input and output)
N	=	number of data samples in X
PX(0:LS,NS)	=	retains past inputs for each section
PY(LS,NS)	=	retains past outputs for each section
IERROR	=	0 no errors detected;
		1 − NS response of section [IERROR] exceeds 1.E10

In case of an unstable filter section, the IERROR parameter will indicate only the first stage whose output exceeded the value 1.E10. Note that due to the retention of past inputs and outputs in PX and PY, SPCFLT can also be used to filter long data sequences in the block mode.

To demonstrate the use of SPCFLT, we compute the 100-point impulse response of the fourth-order Chebyshev low-pass filter with transfer function

$$H(z) = \frac{0.001836(1 + z^{-1})^4}{(1 - 1.4996z^{-1} + 0.8482z^{-2})(1 - 1.5548z^{-1} + 0.6493z^{-2})} \tag{6.7}$$

Equation (6.7) is the cascade-form equivalent of (6.5). The data is again filtered in 25-point blocks. This task is accomplished by test program SPA0604, which is listed next. The first 10 values of the impulse response are printed to enable verification of the user software. Note that these values are equal to those computed by the direct form filter in SPA0602. A plot of the complete output data array is provided in Fig. 6.9.

```
          PROGRAM SPA0604
C-DEMONSTRATE THE USE OF SPCFLT
          DIMENSION X(0:24),B(0:2,2),A(2,2),PX(0:2,2),PY(2,2),DATA(0:99)
          DATA X/1.,24*0./,PX/6*0./,PY/4*0./,LS/2/,NS/2/,N/25/
          DATA B/.001836,.003672,.001836,1.,2.,1./
          DATA A/-1.4996,.8482,-1.5548,.6493/
          DO 5 NBLK=0,3
            CALL SPCFLT(B,A,LS,NS,X,N,PX,PY,IERROR)
            IF(IERROR.NE.0)THEN
              PRINT *,' SPCFLT ERROR = ',IERROR
              STOP
            ENDIF
            DO 4 K=0,24
              DATA(NBLK*25+K)=X(K)
              X(K)=0.
    4       CONTINUE
    5     CONTINUE
          CALL MCPLOT(0.,1.,DATA,100,0,0,1,1,2,1,5.,3.6)
          PRINT 100,(K,DATA(K),K=0,9)
  100     FORMAT(' SAMPLE    RESPONSE '/(I6,F12.4))
          END
```

```
                    $ RUN SPA0604
                    SAMPLE    RESPONSE
                       0      0.0018
                       1      0.0130
                       2      0.0435
                       3      0.0950
                       4      0.1538
                       5      0.1990
                       6      0.2123
                       7      0.1871
                       8      0.1299
                       9      0.0574
                    $
```

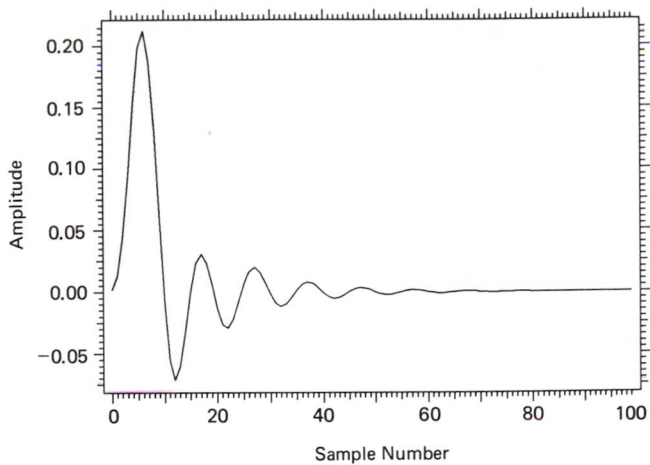

Figure 6.9. SPA0604: Impulse response computed via SPCFLT.

6.5 Parallel-Form IIR Digital Filtering

We now consider filtering via a parallel structure whose transfer function is defined by

$$H(z) = H_1(z) + H_2(z) + \cdots + H_M(z) \tag{6.8}$$

where each $H_m(z)$ is the Lth-order section defined in (6.2). The subroutine is SPPFLT and its calling sequence is

CALL SPPFLT(B,A,LS,NS,X,N,Y,PX,PY,IERROR)

B(0:LS,NS)	= $[b_n]$ coefficients for each section
A(LS,NS)	= $[a_n]$ coefficients for each section
LS	= order of each section
NS	= number of sections in filter
X(0:N−1)	= input data array
N	= number of data samples
Y(0:N−1)	= output data array
PX(0:LS,NS)	= retains past inputs for each section
PY(LS,NS)	= retains past outputs for each section
IERROR	= 0 no errors detected;
	1−NS response from stage [IERROR] exceeds 1.E10

Initialization of the Y array is accomplished internally by SPPFLT. Arrays PX and PY again retain past values of the input and output for each stage, thus enabling block mode filtering.

To demonstrate the use of SPPFLT, we compute the 100-point impulse response of the fourth-order filter with transfer function

$$H(z) = \frac{0.08327 + 0.0239z^{-1}}{1 - 1.5658z^{-1} + 0.6549z^{-2}} - \frac{0.08327 + 0.0246z^{-1}}{1 - 1.4934z^{-1} + 0.8392z^{-2}} \tag{6.9}$$

This transfer function describes a Chebyshev low-pass filter transformed by impulse invariance [C]. The filtering task is performed by test program SPA0605, which is listed next. The first 10 output samples are printed following the program listing, and a plot of the entire output array is provided in Fig. 6.10.

```
            PROGRAM SPA0605
      C-DEMONSTRATE THE USE OF SPPFLT
            DIMENSION X(0:24),B(0:2,2),A(2,2),PX(0:2,2),PY(2,2),Y(0:24)
            DIMENSION DATA(0:99)
            DATA X/1.,24*0./,PX/6*0./,PY/4*0./,LS/2/,NS/2/,N/25/
            DATA B/.08327,.0239,0.,-.08327,-.0246,0./
            DATA A/-1.5658,.6549,-1.4934,.8392/
            DO 5 NBLK=0,3
              CALL SPPFLT(B,A,LS,NS,X,N,Y,PX,PY,IERROR)
              IF(IERROR.NE.0)THEN
                PRINT *,' SPPFLT ERROR = ',IERROR
                STOP
              ENDIF
              DO 4 K=0,24
                DATA(NBLK*25+K)=Y(K)
                X(K)=0.
      4       CONTINUE
      5     CONTINUE
            CALL MCPLOT(0.,1.,DATA,100,0,0,1,1,2,1,5.,3.6)
            PRINT 100,(K,DATA(K),K=0,9)
      100   FORMAT(' SAMPLE    RESPONSE'/(I6,F12.4))
            END
```

```
            $ RUN SPA0605
            SAMPLE    RESPONSE
               0      0.0000
               1      0.0053
               2      0.0345
               3      0.0890
               4      0.1523
               5      0.2010
               6      0.2162
               7      0.1913
               8      0.1336
               9      0.0606
            $
```

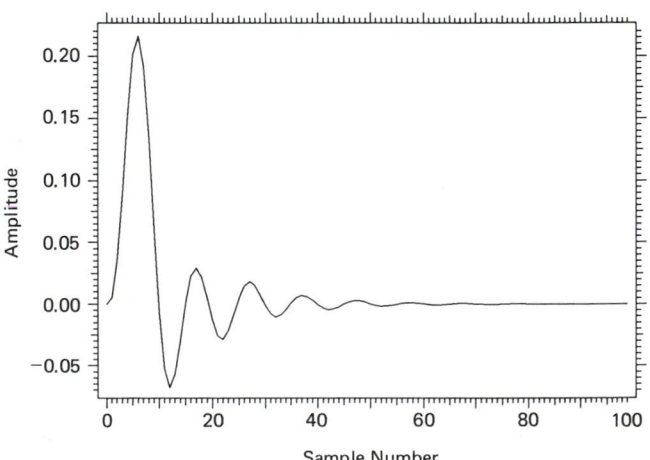

Figure 6.10. SPA0605: Impulse response computed via SPPFLT.

Note that although the sample values do not exactly match those obtained from the direct and cascade forms in SPA0602 and SPA0604, respectively, the output plots are very similar. The discrepancy results from the type of trans-formation (i.e., bilinear versus impulse invariant) as well as from the parallel structure. IIR filter design is discussed further in Chapter 7.

We have described four subroutines, SPFILT, SPFLTR, SPCFLT, and SPPFLT, that perform IIR digital filtering. We next consider FIR filtering and demonstrate the use of these same routines in that context.

6.6 FIR Digital Filters

The all-zero transfer function that describes an FIR digital filter is typically defined in direct-form representation as

$$H(z) = B(z) = b_0 + b_1 z^{-1} + \cdots + b_L z^{-L} \tag{6.10}$$

Note that (6.10) is equivalent to (6.2) with $A(z) = 1$. A block diagram that illustrates the implementation of the transfer function in (6.10) is provided in Fig. 6.11.

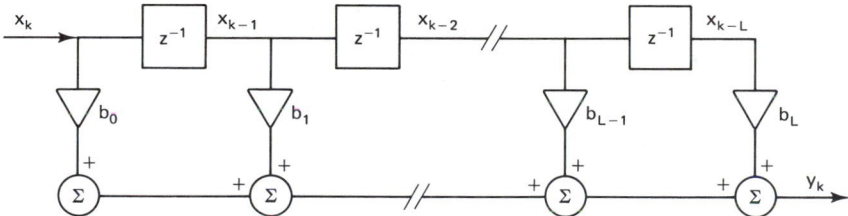

Figure 6.11.   Block diagram of direct-form FIR digital filter.

The difference equation corresponding to the FIR direct-form implementation in Fig. 6.11, similar to (6.3), is given by

$$y_k = \sum_{n=0}^{L} b_n x_{k-n} \tag{6.11}$$

This equation describes an FIR filter, since for an input sequence consisting of a unit impulse at $k = 0$, the filter output y_k is equal to the filter coefficient b_k for $k = 0, 1, \ldots, L$ and is equal to zero thereafter. Thus, the filter coefficients $[b_n]$ also define the impulse response of an FIR filter. In this context, (6.11) defines the filter output y_k as the discrete convolution of the input data sequence $[x_k]$ with the filter impulse response $[b_k]$. The subject of convolution is discussed in more detail in Chapter 9.

The FIR filter can also be implemented in a cascade structure, where the transfer function from (6.10) is factored such that

$$H(z) = B_1(z) \cdot B_2(z) \cdots B_M(z) \tag{6.12}$$

where each $B_m(z)$ is now the section of length $L + 1$ defined in (6.10).

The parallel structure described in the previous section is less useful for the FIR filter, since the length of the impulse response is restricted to L by the factorization. Additional delay elements could be inserted to circumvent this problem, but the topic will not be considered further here.

As stated previously, (6.10) is equivalent to the IIR transfer function in (6.2) when $A(z) = 1$. This suggests that the subroutines described in Sections 6.3 through 6.5 are also applicable to FIR filtering. The examples that follow demonstrate this fact.

We first consider the simple task of using subroutine SPFILT to compute a 10-point impulse response of the FIR filter described by

$$H(z) = 1 + 4z^{-1} + 6z^{-2} + 4z^{-3} + z^{-4} \qquad (6.13)$$

From the previous discussion, we would expect the filter outputs to match the $[b_n]$ coefficients in (6.13) for the first five samples, after which the filter output should be zero. This is demonstrated by test program SPA0606, which is given together with its output data. Note especially that arguments A, LA, and PY are initialized as LA = 1, and A(1) = PY(1) = 0.0 in order to specify the FIR filter.

```
              PROGRAM SPA0606
        C-FIR FILTERING WITH SUBROUTINE SPFILT
              DIMENSION X(0:9),B(0:4),PX(0:4),A(1),PY(1)
              DATA A/0./,PY/0./,X/1.,9*0./,PX/5*0./
              DATA B/1.,4.,6.,4.,1./,LB/4/,LA/1/,N/10/
              CALL SPFILT(B,A,LB,LA,X,N,PX,PY,IERROR)
              IF(IERROR.NE.0)PRINT *,' SPFILT ERROR = ',IERROR
              PRINT 100,(K,X(K),K=0,9)
        100   FORMAT(' SAMPLE    RESPONSE '/(I6,F12.4))
              END
```

```
              $ RUN SPA0606
              SAMPLE    RESPONSE
                 0      1.0000
                 1      4.0000
                 2      6.0000
                 3      4.0000
                 4      1.0000
                 5      0.0000
                 6      0.0000
                 7      0.0000
                 8      0.0000
                 9      0.0000
              $
```

Test program SPA0607 performs the task just described via a call to SPCFLT. A cascade equivalent of the transfer function in (6.13) is

$$H(z) = (1 + 2z^{-1} + z^{-2})^2 \qquad (6.14)$$

A listing of SPA0607 and the 10-point impulse response are provided next. Note that since the dimensions of A and PY are determined by LS and NS in SPCFLT, for each array LS × NS values must be initialized to zero in order to obtain the desired FIR response.

```
              PROGRAM SPA0607
        C-FIR FILTERING WITH SUBROUTINE SPCFLT
              DIMENSION X(0:9),B(0:2,2),PX(0:2,2),A(2,2),PY(2,2)
              DATA A/4*0./,PY/4*0./,X/1.,9*0./,PX/6*0./
              DATA B/1.,2.,1.,1.,2.,1./,LS/2/,NS/2/,N/10/
              CALL SPCFLT(B,A,LS,NS,X,N,PX,PY,IERROR)
              IF(IERROR.NE.0)PRINT *,' SPCFLT ERROR = ',IERROR
              PRINT 100,(K,X(K),K=0,9)
        100   FORMAT(' SAMPLE    RESPONSE '/(I6,F12.4))
              END
```

```
$ RUN SPA0607
SAMPLE      RESPONSE
    0        1.0000
    1        4.0000
    2        6.0000
    3        4.0000
    4        1.0000
    5        0.0000
    6        0.0000
    7        0.0000
    8        0.0000
    9        0.0000
$
```

Other more interesting examples of digital filtering with SPFILT, SPFLTR, SPCFLT, and SPPFLT are provided in Chapters 7 and 8. There we consider the complete task from filter specification and design through the actual filtering operation.

6.7 Digital Filtering via an IIR Lattice Structure

Recent work in the area of digital filter synthesis suggests that lattice structures can play an important role in the solution of problems arising from the use of a finite word length [D]. In filtering applications where coefficient sensitivity is a primary concern, the lattice structure may be preferable to the direct, cascade, or parallel form; however, its implementation nearly doubles the computational cost.

Because the emphasis of this presentation is on the utilization of the filter structure, a detailed derivation of the lattice form is not included here. The block diagram in Fig. 6.12 illustrates the symmetrical IIR lattice filter on which this discussion is based.

Since the coefficient sets $[\kappa_n]$ and $[\nu_n]$ are derived from, but not equal to, the $[b_n]$ and $[a_n]$ of the direct-form transfer function in (6.2), we first provide a conversion routine that computes the lattice coefficients. The algorithm was developed by Gray and Markel; a more complete description appears in reference [D].

The subroutine calling sequence is

CALL SPLTCF(B,A,L,KAPPA,NU,WORK,IERROR)

B(0:L)	=	$[b_n]$ coefficients from (6.2)
A(L)	=	$[a_n]$ coefficients from (6.2)
L	=	order of filter
KAPPA(0:L − 1)	=	$[\kappa_n]$ coefficients − REAL array
NU(0:L)	=	$[\nu_n]$ coefficients − REAL array
WORK(0:L,2)	=	work array used internally
IERROR	=	0 no errors detected
		1 unstable transfer function

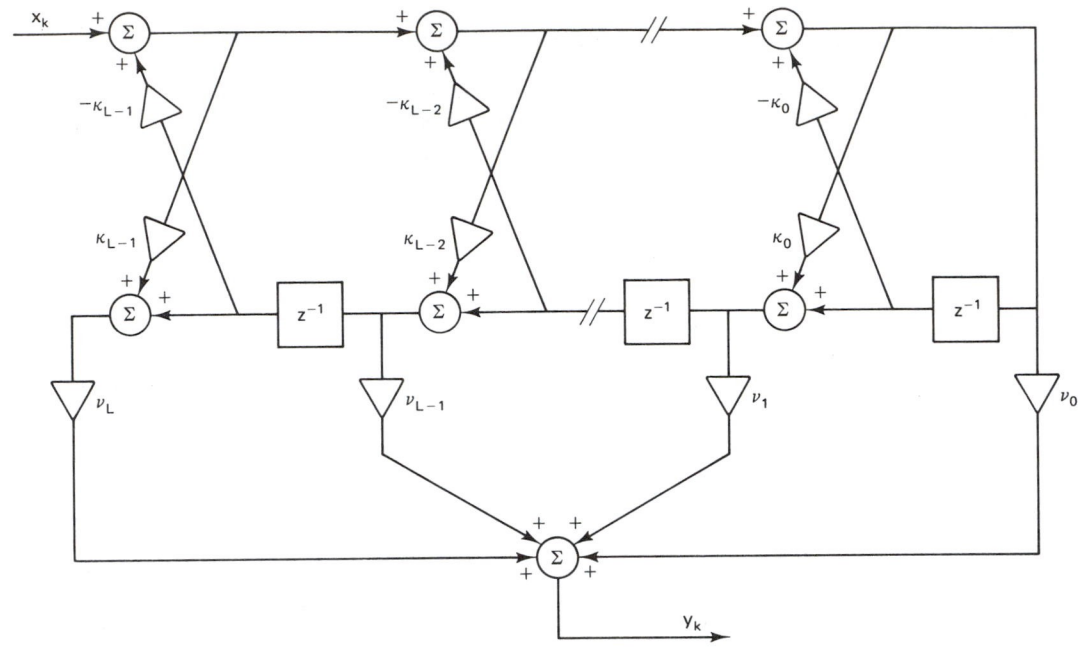

Figure 6.12. Symmetric IIR lattice structure with order = L.

The IERROR = 1 return results when $|\kappa| \geq 1$ for some computed coefficient κ_n. This indicates that the transfer function has poles on or outside the unit circle and is therefore unstable.

Before discussing an example that demonstrates the use of SPLTCF, we describe a digital filtering routine, SPLFLT, which utilizes the computed coefficients. The calling sequence is

CALL SPLFLT(KAPPA,NU,L,X,N,PAST,IERROR)

KAPPA(0:L − 1)	=	$[\kappa_n]$ coefficients − REAL array
NU(0:L)	=	$[\nu_n]$ coefficients − REAL array
L	=	order of filter
X(0:N − 1)	=	data array (input and output)
N	=	number of data samples in X
PAST(0:L)	=	retains past values for block mode filtering
IERROR	=	0 no errors detected
		1 filter response exceeds 1.E10

A return of IERROR = 1 should in this case indicate a problem with the input data sequence if the lattice coefficients were computed by SPLTCF with no errors detected.

Test program SPA0608 demonstrates the use of both SPLTCF and SPLFLT.

The problem is once again that of computing the 100-point impulse response of the fourth-order transfer function in (6.5) by using four 25-point data blocks. Direct-form transfer function coefficients are defined by arrays B(0:4) and A(4). Subroutine SPLTCF is used to convert these values to the lattice coefficients KAPPA(0:3) and NU(0:4). The input impulse sequence is then filtered via four calls to SPLFLT. Note that whereas the past value array, PAST(0:4), is initialized to zero prior to the first call, subsequent values are provided by SPLFLT. A listing of SPA0608 is provided together with a printout of the lattice coefficients and the first 10 output data values. A plot of the complete output data array is provided in Fig. 6.13.

```
            PROGRAM SPA0608
   C-IIR FILTERING WITH SPLFLT - LATTICE STRUCTURE
            DIMENSION X(0:24),B(0:4),A(4),PAST(0:4),WORK(0:4,2),DATA(0:99)
            REAL KAPPA(0:3),NU(0:4)
            DATA B/.001836,.007344,.011016,.007344,.001836/
            DATA A/-3.0544,3.8291,-2.2925,.55075/
            DATA PAST/5*0./,X/1.,24*0./,L/4/,N/25/
            CALL SPLTCF(B,A,L,KAPPA,NU,WORK,IERROR)
            IF(IERROR.NE.0)THEN
              PRINT *,' UNSTABLE TRANSFER FUNCTION H(Z)'
              STOP
            ENDIF
            DO 5 NBLK=0,3
              CALL SPLFLT(KAPPA,NU,L,X,N,PAST,IERROR)
              IF(IERROR.NE.0)THEN
                PRINT *,' SPLFLT ERROR = ',IERROR
                STOP
              ENDIF
              DO 4 K=0,24
                DATA(NBLK*25+K)=X(K)
                X(K)=0.
      4       CONTINUE
      5     CONTINUE
            CALL MCPLOT(0.,1.,DATA,100,0,0,1,1,2,1,5.,3.6)
            PRINT 100,KAPPA,NU
            PRINT 200,(K,DATA(K),K=0,9)
    100     FORMAT(' KAPPA COEF = ',4F10.4,/,' NU COEF =      ',5F10.4)
    200     FORMAT(' SAMPLE     RESPONSE '/(I6,F12.4))
            END


    $ RUN SPA0608
    KAPPA COEF =     -0.9112     0.9293    -0.8760     0.5508
    NU COEF =         0.0186     0.0451     0.0373     0.0130     0.0018
    SAMPLE     RESPONSE
         0       0.0018
         1       0.0130
         2       0.0435
         3       0.0950
         4       0.1538
         5       0.1989
         6       0.2123
         7       0.1871
         8       0.1299
         9       0.0574
    $
```

The subroutines introduced in this section can be used to convert any stable IIR transfer function from the form of (6.2) to the equivalent lattice structure, and then to filter an input data sequence. They are especially useful in applications where the lattice structure provides numerical advantages, since IIR filter design routines do not, in general, provide lattice filter coefficients.

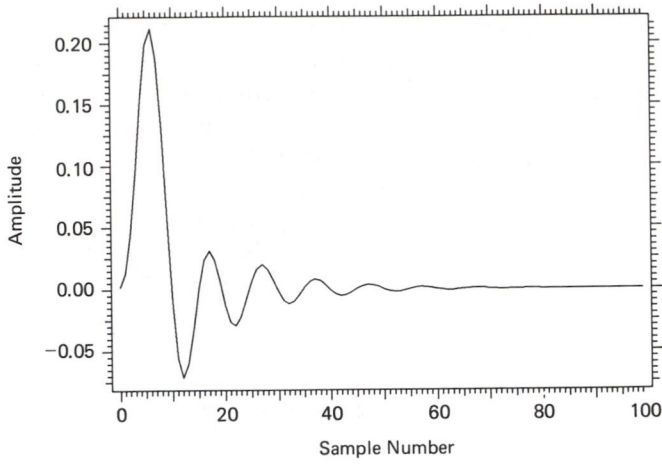

Figure 6.13. SPA0608: Impulse response computed via SPLFLT.

6.8 Summary of Calling Sequences

In this chapter we have described subroutines that enable digital filtering of a data sequence $[x_k]$ via a filter transfer function, which is defined in direct, cascade, parallel, or lattice form. A more useful digital filtering package is formed when these routines are coupled with the filter design routines in Chapters 7 and 8. The calling sequences and argument lists for the Chapter 6 subroutines are summarized as follows.

CALL SPFILT(B,A,LB,LA,X,N,PX,PY,IERROR)
CALL SPFLTR(B,A,LB,LA,X,N,Y,PX,PY,IERROR)
CALL SPCFLT(B,A,LS,NS,X,N,PX,PY,IERROR)
CALL SPPFLT(B,A,LS,NS,X,N,Y,PX,PY,IERROR)

Variable	Definition	Routines			
		SPFILT	SPFLTR	SPCFLT	SPPFLT
B(0:LB)	$[b_n]$ from (6.2)	I	I	—	—
A(LA)	$[a_n]$ from (6.2)	I	I	—	—
B(0:LS,NS)	$[b_n]$ for each section	—	—	I	I
A(LS,NS)	$[a_n]$ for each section	—	—	I	I
LB	Degree of numerator	I	I	—	—
LA	Degree of denominator	I	I	—	—
LS	Order of each section	—	—	I	I
NS	Number of sections	—	—	I	I
X(0:N − 1)	Data array	I/O	I	I/O	I
Y(0:N − 1)	Output data array	—	O	—	O
N	Number of samples in X,Y	I	I	I	I
PX(0:LB)	Retains past inputs	I/O	I/O	—	—
PY(LA)	Retains past outputs	I/O	I/O	—	—
PX(0:LS,NS)	Retains past inputs	—	—	I/O	I/O
PY(LS,NS)	Retains past outputs	—	—	I/O	I/O
IERROR	Return error indicator	O	O	O	O

CALL SPLFLT(KAPPA,NU,L,X,N,PAST,IERROR)
CALL SPLTCF(B,A,L,KAPPA,NU,WORK,IERROR)

Variable	Definition	Routines	
		SPLTCF	SPLFLT
B(0:L)	$[b_n]$ from (6.2)	I	—
A(L)	$[a_n]$ from (6.2)	I	—
L	Order of filter	I	I
KAPPA(0:L−1)	$[\kappa_n]$ coefficients − REAL array	O	I
NU(0:L)	$[\nu_n]$ coefficients − REAL array	O	I
X(0:N−1)	Data array	—	I/O
N	Number of samples in X	—	I
PAST(0:L)	Retains past values	—	I/O
WORK(0:L,2)	Work array	X	—
IERROR	Error indicator	O	O

In this summary, I indicates that the variable is an input to the routine, O denotes a value returned by the routine, and — indicates that the argument is not used. Work space that must be allocated but not necessarily initialized by the calling program is denoted by an X.

6.9 Exercises

6.1. Using 25-point data blocks, use SPFILT to generate the 100-point impulse response of the digital filter with

$$H(z) = \frac{0.2871 - 0.4466z^{-1}}{1.0 - 1.2971z^{-1} + 0.6949z^{-2}}$$

Plot the output data array.

6.2. Work exercise 6.1 using SPFLTR.

6.3. Work exercise 6.1 using SPLTCF and SPLFLT.

6.4. Generate a 1000-point input data sequence as

$$x_k = \sin(0.3k) + \sin(0.9k)$$

Using SPFILT with 250-point data blocks, filter $[x_k]$ with

$$H_1(z) = \frac{0.0931 - 0.0975z^{-1}}{1.0 - 1.8175z^{-1} + 0.9025z^{-2}}$$

Plot the output data array.

6.5. Work exercise 6.4 with

$$H_2(z) = \frac{0.0606 - 0.0975z^{-1}}{1.0 - 1.1826z^{-1} + 0.9025z^{-2}}$$

6.6. Using $H_1(z)$ and $H_2(z)$ from exercises 6.4 and 6.5 in parallel, filter the input data sequence $[x_k]$ defined in exercise 6.4 using SPPFLT. Plot both the input and output data arrays. Explain the results.

6.7. Repeat exercise 6.6 using SPCFLT to filter the data.

6.8. Use SPFLTR to compute the 10-point impulse response of the FIR filter with coefficients $b_n = \sin(0.3n)$ for $0 \leqslant n \leqslant 10$. List $[b_n]$ and $[y_k]$.

6.9. Use SPFILT to generate the 15-point step response of the filter in exercise 6.8.

6.10. Use SPCFLT to generate the 100-point impulse response of

$$H(z) = \frac{0.02426(1 + z^{-1})^4}{(1 - 1.0416z^{-1} + 0.4019z^{-2})(1 - 0.5561z^{-1} + 0.7647z^{-2})}$$

Use 40-point data blocks. Plot the output data array.

6.11. Calculate the direct-form equivalent of the transfer function in exercise 6.10. Repeat the problem using SPFILT.

6.12. Work exercise 6.10 using SPLTCF and SPLFLT.

6.13. Compute the parallel equivalent of the transfer function in exercise 6.10. Repeat the exercise using SPPFLT.

6.14. Generate a 1000-point input data sequence as $x_k = \sin(0.3k) + 0.2r_k$, where r_k is a sample of zero-mean white noise with unit power density. (Use the SPRAND function from Chapter 4 with ISEED = 123.) Filter the data using SPFLTR with $H_1(z)$ from exercise 6.4. Plot the input and output data arrays for $0 \leqslant k < 200$.

6.15. Repeat exercise 6.14 using SPLTCF and SPLFLT.

6.16. Generate a 1000-point input data sequence as

$$x_k = \sin(0.3k) + \sin(0.9k) + 0.3r_k$$

where r_k is the random sequence described in exercise 6.14. Using $H_1(z)$ and $H_2(z)$ from exercises 6.4 and 6.5, filter the data with SPPFLT. Plot the input and output data arrays for $0 \leqslant k < 200$.

6.17. Repeat exercise 6.16 using SPCFLT. Explain the results.

6.18. Repeat exercise 6.10, computing the step response rather than the impulse response.

6.19. Verify the results of SPA0601 by using the SPRESP routine in Chapter 5 in place of SPFILT.

6.20. Verify the results of SPA0604 by using SPFILT in place of SPCFLT.

6.21. Verify the results of SPA0605 by using SPFLTR in place of SPPFLT.

6.10 References

[A] Stearns, S. D., *Digital Signal Analysis* (Rochelle Park, N.J.: Hayden, 1975).

[B] Ahmed, N., and T. Natarajan, *Discrete-Time Signals and Systems* (Englewood Cliffs, N.J.: Prentice-Hall, Inc., 1983).

[C] Oppenheim, A. V., and R. W. Schafer, *Digital Signal Processing* (Englewood Cliffs, N.J.: Prentice-Hall, Inc., 1975).

[D] Gray, A. H., and J. D. Markel, "Digital Lattice and Ladder Filter Synthesis," *IEEE Trans. on Audio and Electroacoustics,* AU-21, no. 6 (December 1973).

[E] Tretter, S. A., *Introduction to Discrete-Time Signal Processing* (New York: John Wiley, 1976).

[F] Mitra, S. K., and Y. Neuvo, "Canonic Ladder Realizations of IIR Digital Filters," *Proc. of the IEEE,* 70, no. 7 (July 1982).

[G] Mitra, S. K., and J. Szczupak, ''Minimization of Coefficient Sensitivities of Digital Filter Structures,'' *Proc. of the IEEE,* 67, no. 1 (January 1979).

[H] Chang, Tien-Lin, ''Comparison of Round-Off Noise Variances in Several Low Round-Off Noise Digital Filter Structures,'' *Proc. of the IEEE,* 68, no. 1 (January 1980).

[I] Mitra, S. K., and Gong-Tao Yan, ''Modified Coupled-Form Digital Filter Structures,'' *Proc. of the IEEE,* 70, no. 7 (July 1982).

[J] Lim, Y. C., ''On the Synthesis of IIR Digital Filters Derived from Single Channel AR Lattice Network,'' *IEEE Trans. on ASSP,* ASSP-32, no. 4 (August 1984).

chapter 7

IIR Filter Design

7.1 Introduction

In the previous chapter we discussed routines that filter a set of data samples $[x_k]$ using the digital filter coefficients $[b_n]$ and $[a_n]$ and generate an output data sequence $[y_k]$ according to the following relationship:

$$y_k = \sum_{n=0}^{L} b_n x_{k-n} - \sum_{n=1}^{L} a_n y_{k-n} \tag{7.1}$$

We now consider filter design, which is the process of determining values for the filter coefficients. By assuming that at least one of the a_n coefficients in (7.1) is nonzero, we focus the discussion in this chapter on IIR filter design. The topic of FIR filter design is addressed in Chapter 8.

There are two hard constraints that must be satisfied by any practical IIR filter design. The system must be both realizable and stable. Realizability in real time implies causality or, equivalently, $h_n = 0$ for $n < 0$, where $[h_n]$ defines the impulse response of the IIR filter. A stable system is one whose impulse response satisfies the condition [B]

$$\sum_{n=0}^{\infty} |h_n| < \infty \tag{7.2}$$

The direct-form transfer function that corresponds to the difference equation in (7.1) can be expressed as in (6.2) in Chapter 6:

$$H(z) = \frac{B(z)}{A(z)} = \frac{b_0 + b_1 z^{-1} + \cdots + b_L z^{-L}}{1 + a_1 z^{-1} + \cdots + a_L z^{-L}} \tag{7.3}$$

Stability requires the zeros of $A(z)$ (i.e., poles of $H(z)$) to be located inside the unit circle on the z-plane [A–D].

It is interesting to note that a highly desirable feature of digital filters, linear phase response, is not generally attainable with an IIR structure. This is due to the fact that linear phase response requires that the filter transfer function satisfy the relationship $H(z) = H(z^{-1})$. This implies that for every pole inside the unit circle at $z = z_1$, there must be a pole outside the unit circle at $z = 1/z_1$. Since an unstable filter design would result, alternate means for deriving an acceptable phase response must be explored. Phase equalization via an all-pass filter is one technique that has been suggested [C]. Alternatively, exactly linear phase response may be attained via a combination of multiple filtering passes and time reversal [C]. Both of these approaches, which are illustrated in block diagram form in Fig. 7.1, increase the computational burden of the filtering process. It is, therefore, more common to design a filter with acceptable phase response as well as good magnitude response in the filter passband [S–U].

The first step in digital filter design involves a specification of the desired properties of the system. This step typically determines the necessary filter order. The requirements may be stated in terms of frequency-response characteristics, an equivalent analog design, or some other criteria. In IIR filter design, the system is often specified in terms of magnitude tolerances in the passband and stopband, with the only phase constraints being those imposed by stability and causality. Since this step in the filter-design process is highly dependent upon the specific application, it will not be addressed in detail in this text.

In this chapter we focus on the second step in the design process, which is the approximation of the desired filter specifications using a causal discrete-time system. This step often requires several passes as trade-offs are made between design specification tolerances and the available processing power. This is especially true when the filter is to be implemented in real time or using fixed-point or integer arithmetic. In general, the coefficients used in the filter implementation are not exact. Thus, the actual poles and zeros differ from the desired poles and zeros. If these quantization errors are large, the resulting digital system may not satisfy the design specifications. In fact, coefficient quantization errors could result in an unstable design as poles move outside the unit circle. The actual effect of coefficient quantization is highly dependent upon the structure used to implement the filter [A–D]. A variety of design techniques has been

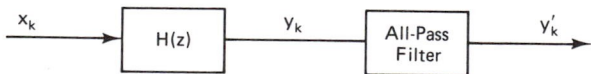

(a) Use of all-pass filter. Magnitude response is constant; phase shift can be adjusted to compensate for nonlinear shift of H(z).

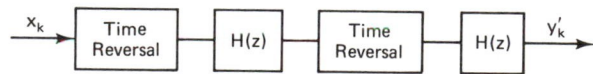

(b) Use of time reversal and refiltering. Equivalent transfer function from x to y′ is [H(z)H(z⁻¹)].

Figure 7.1. Techniques for phase compensation in IIR filters.

developed in an attempt to circumvent the problem of limited computing capacity. The references at the end of this chapter suggest a number of sources for additional reading in this area.

Historically, many digital filter-design techniques have evolved from classical filtering theory. This approach to the design problem has a number of advantages. It allows the engineer to make use of the optimization procedures that have been developed for analog filter design. In many cases, closed-form design formulas are available, which yield a simple digital implementation. For engineers who are more familiar with classical filter theory, this has the added benefit of easing the transition into the world of digital signal processing. Thus, we begin our discussion of IIR filter design with techniques for conversion of analog structures. In later sections, we provide routines that may be used to design analog Butterworth, Chebyshev, and Bessel filters for subsequent conversion to digital IIR filters.

7.2 Filter Design via Analog Conversion

In this section we discuss techniques that enable the translation of an analog filter transfer function to a digital transfer function with comparable frequency-response characteristics. Analog transfer functions can, in general, be described as

$$H(s) = \frac{D(s)}{C(s)} = \frac{d_L s^L + d_{L-1} s^{L-1} + \cdots + d_0}{c_L s^L + c_{L-1} s^{L-1} + \cdots + c_0} \tag{7.4}$$

The frequency response of $H(s)$ is determined by the s-plane locations of the roots of $D(s)$ and $C(s)$, where the s-plane is as illustrated in Fig. 7.2. The s-plane stability requirement states that the zeros of $C(s)$ (i.e., poles of $H(s)$) must lie in the left half-plane (LHP). This fact, when coupled with the z-plane stability requirement, suggests that any viable mapping technique must translate stable s-plane poles to z-plane locations inside the unit circle. In addition, mapping of the $j\Omega$-axis in the s-plane onto the unit circle in the z-plane enables preservation

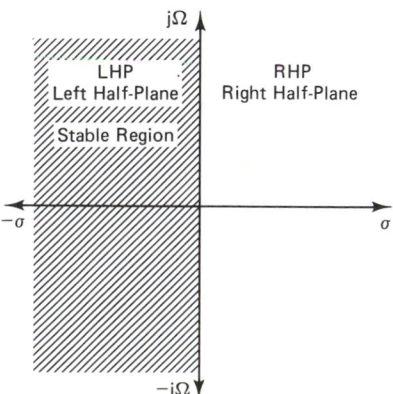

Figure 7.2. The s-plane associated with analog systems.

of the frequency-selective characteristics of the continuous system. (We are using Ω to denote analog frequency in radians per second. See Table 3.1.)

One transformation technique that has been suggested involves replacement of the differentials in the analog system differential equation by finite differences [C]. Although the use of first-order forward or backward differences yields a simple digital structure, neither will adequately preserve frequency response characteristics.

The matched z-transform is another technique that has been described briefly in the literature [C]. This mapping has the property that an s-plane pole or zero at $s = \sigma + j\Omega$ is translated to a z-plane pole or zero at $e^{(\sigma + j\Omega)T}$, where T is the sample interval in seconds. Since the analog transfer function must be in factored form and aliased digital transfer functions often result, this technique has not gained widespread usage.

The two methods that are most commonly used are the impulse invariant transformation and the bilinear transformation. The impulse invariant technique preserves the impulse response of the analog filter by specifying the digital filter impulse response to be a sampled version of the continuous-time response [A, C]. As a result, the digital filter frequency response is an aliased version of the corresponding analog filter response. In order to obtain an acceptable digital filter, the analog filter response must be absolutely bandlimited to the Nyquist range $(-\pi/T) < \Omega < (\pi/T)$. Since the impulse invariant transformation is somewhat limited in applicability for our purposes, we do not provide details here.

Probably the most widely used technique—and certainly the most appropriate for this presentation—is the bilinear transformation. While the basics are described herein, interested readers should refer to a general text on digital signal processing for further details [A–D].

A primary advantage of the bilinear transformation is that it provides a one-to-one mapping of poles and zeros from the continuous time s-plane to the discrete time z-plane. The key features of this method are [A]:

- Points in the left half of the s-plane are mapped to points inside the unit circle.
- Points in the right half of the s-plane are mapped to points outside the unit circle.
- The positive portion $(0 \rightarrow \infty)$ of the s-plane $j\Omega$-axis is mapped onto the $0 \rightarrow \pi$ range of the z-plane unit circle.
- The negative portion $(0 \rightarrow -\infty)$ of the s-plane $j\Omega$-axis is mapped onto the $0 \rightarrow -\pi$ range of the z-plane unit circle.

Since the entire $j\Omega$-axis on the s-plane is uniquely mapped onto the unit circle in the z-plane, the bilinear transformation has a compression effect, known as *frequency warping,* on the frequency-response characteristics. This problem is alleviated by prewarping critical frequencies and using frequency scaling. These steps are described later in this section.

The bilinear transformation is defined by the following substitution:

$$s = \frac{z - 1}{z + 1} \tag{7.5}$$

To examine the effect of this mapping, we consider points at $s = j\Omega_0$ on the s-plane and $z = e^{j\omega_0}$ on the z-plane. (As in previous chapters, we are using ω to denote the normalized digital frequency in radians. See Table 3.1, for example.) The value Ω_0 can be interpreted as an analog frequency, which corresponds to the normalized frequency ω_0 of the desired digital filter. By substituting these values for s and z in (7.5), the following relationship, which is illustrated in Fig. 7.3, can be derived:

$$\Omega_0 = \tan\left(\frac{\omega_0}{2}\right) = \tan(\pi f_0 T) \tag{7.6}$$

The nonlinear character of the bilinear transformation can be seen by considering the mappings listed below.

s-Plane (imaginary axis)	z-Plane (unit circle)
$\Omega = 0$	$\omega = 0$
$0 < \Omega < 1$	$0 < \omega < \dfrac{\pi}{2}$
$\Omega \geqslant 1$	$\dfrac{\pi}{2} \leqslant \omega \leqslant \pi$
$\Omega = \infty$	$\omega = \pi$

Corresponding values of the analog frequency Ω_0 and the normalized digital frequency ω_0, as determined by (7.6), are designated by matching symbols in Fig. 7.3. This figure illustrates the warping effect of this transformation as the analog frequency range $1 \leqslant \Omega \leqslant \infty$ is compressed onto an arc on the unit circle that is equal in length to the unit circle segment corresponding to the analog range $0 \leqslant \Omega \leqslant 1$. This warping of the frequency axis may, for some types of filters, result in some degradation of the magnitude and/or phase characteristics. The advantage is that the aliasing problems that arise in other techniques for analog-to-digital (A/D) transfer function conversion are eliminated [B].

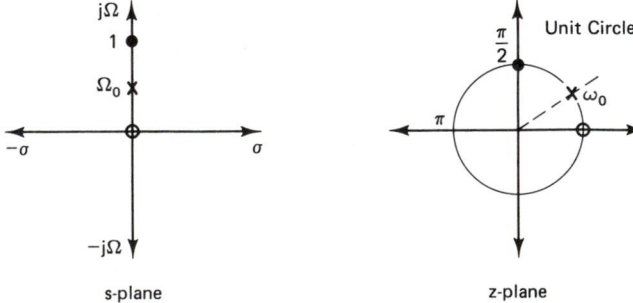

s-plane   z-plane

Figure 7.3. Bilinear transformation mapping.

Because of the nonlinear frequency compression of the bilinear transformation in (7.5), we begin the design process by defining a normalized low-pass analog transfer function $H(s)$ having a bandwidth of 1 rad/s. We describe how $H(s)$ is converted to a digital $H(z)$ with an arbitrary cutoff frequency. Finally, transformation techniques are described that enable filter type conversion to yield high-pass, bandpass, or bandstop filters.

To facilitate our discussion, we first introduce the concept of frequency scaling in the analog domain [A]. Given an $H(s)$ with specific magnitude and phase characteristics, we can define a new analog transfer function $\hat{H}(s)$ where

$$\hat{H}(s) = H(s)\big|_{s=s/\alpha} = H(s/\alpha) \qquad \alpha \text{ real, positive} \tag{7.7}$$

It can be shown that the magnitude and phase responses of $H(s)$ and $\hat{H}(s)$ have identical shapes, but corresponding points along the frequency axis are scaled by the factor α.

As an example, consider an $H(s)$ that describes a low-pass filter with a bandwidth of 1 rad/s. In order to extend the bandwidth to 100 rad/s, we simply evaluate $\hat{H}(s)$ from (7.7) with $\alpha = 100$.

With the idea of frequency scaling in mind, we are ready to outline the steps required for low-pass digital filter design via the bilinear transformation. These steps are summarized next:

1. Pick $H(s)$: Low-pass analog transfer function with cutoff frequency at 1 rad/s.

2. Compute Ω_c via (7.6), where ω_c is the normalized cutoff frequency in radians of the desired digital filter:

$$\Omega_c = \tan\left(\frac{\omega_c}{2}\right) = \tan(\pi f_c T)$$

3. Frequency scale as in (7.7) with $\alpha = \Omega_c$:

$$\hat{H}(s) = H(s)\big|_{s=s/\Omega_c} = H\left(\frac{s}{\Omega_c}\right)$$

4. Find $H(z)$ by making the substitution in (7.5):

$$H(z) = \hat{H}(s)\big|_{s=(z-1)/(z+1)}$$

Our first subroutine SPBILN performs the substitution in Step 4 of this design process. The calling sequence is

CALL SPBILN(D,C,LN,B,A,WORK,IERROR)

D(0:LN)	= analog filter coefficients of $D(s)$ in (7.4)
C(0:LN)	= analog filter coefficients of $C(s)$ in (7.4)
LN	= upper dimension of coefficient arrays
B(0:LN)	= $B(z)$ coefficients in (7.3)—returned
A(LN)	= $A(z)$ coefficients in (7.3)—returned

WORK(0:LN,0:LN) = work array used internally
IERROR   = 0 no errors detected
                    1 all-zero analog transfer function
                    2 invalid transfer function

Subroutine SPBILN computes the filter order L internally by examining the D and C arrays to find the highest-order nonzero coefficient. If L is less than the array size LN, unused coefficients are set to zero. The IERROR = 2 return is generated if, during normalization of the digital transfer function to yield the $y(k)$ coefficient $a_0 = 1$, this computed coefficient is found to be zero. This result implies that either the filter is noncausal or there is some pure delay factor which should be extracted.

To demonstrate the use of SPBILN, consider the following task. Given the second-order normalized low-pass Butterworth filter

$$H(s) = \frac{1}{s^2 + \sqrt{2}s + 1} \tag{7.8}$$

use the bilinear transformation to design a digital filter with a cutoff frequency at 1 kHz and sample interval equal to 0.0001 s. Recall from equation (5.1) that the normalized cutoff frequency is computed as $\omega_c = 2\pi f_c T$, where f_c is the cutoff frequency in hertz and T is the sample interval in seconds. Following Step 2 outlined previously, we have:

$$\Omega_c = \tan\left(\frac{2\pi(1000)(0.0001)}{2}\right) = 0.3249$$

$$\hat{H}(s) = \frac{1}{\left(\dfrac{s}{\Omega_c}\right)^2 + \sqrt{2}\left(\dfrac{s}{\Omega_c}\right) + 1} = \frac{0.1056}{s^2 + 0.4595s + 0.1056}$$

We now call SPBILN with LN = 2 and arrays D and C with the elements defined in $\hat{H}(s)$. This is accomplished by program SPA0701, which is listed next, together with its output.

```
        PROGRAM SPA0701
C-TRANSFORMATION VIA THE BILINEAR TRANSFORM
        DIMENSION B(0:2),A(2),D(0:2),C(0:2),WORK(0:2,0:2)
        DATA D/0.1056,0.0,0.0/
        DATA C/0.1056,0.4595,1.0/
        DATA LN/2/
        CALL SPBILN(D,C,LN,B,A,WORK,IERROR)
        IF(IERROR.NE.0)PRINT *,' SPBILN ERROR =',IERROR
        PRINT 100,B,A
100     FORMAT(' B(Z) COEFFICIENTS:',3F10.4,/,
      +        ' A(Z) COEFFICIENTS:',2F10.4)
        END

    $ RUN SPA0701
    B(Z) COEFFICIENTS:     0.0675     0.1349     0.0675
    A(Z) COEFFICIENTS:    -1.1429     0.4128
    $
```

Before considering additional examples and routines, we define a set of frequency substitutions that enable us to design high-pass, bandpass, and bandstop filters. As before, we begin the design process with a normalized low-pass analog transfer function (i.e., $\Omega_c = 1$ rad/s). The frequency substitutions are summarized below [A]. The values Ω_L (low edge) and Ω_H (high edge) define the analog band edges for the bandpass and bandstop filters in radians per second. As defined next, W is the analog bandwidth and Ω_0 is the center frequency on a logarithmic scale.

Desired Filter	Substitution

High-pass

$$s \longleftarrow \frac{1}{s}$$

Bandpass

$$s \longleftarrow \frac{s^2 + \Omega_0^2}{Ws}$$

Bandstop

$$s \longleftarrow \frac{1}{s}$$
$$s \longleftarrow \frac{s^2 + \Omega_0^2}{Ws}$$

where

$$W = \Omega_H - \Omega_L$$
$$\Omega_0 = \sqrt{\Omega_L \Omega_H} \tag{7.9}$$

Note that conversion to the bandstop characteristic actually involves two substitutions. The transfer function is first converted to a high-pass filter; then the bandpass substitution is made. Note also that both the bandpass and bandstop conversions double the filter order.

The subroutine SPFBLT described next uses these substitutions together with the SPBILN routine and produces a more complete IIR filter design module. Prewarping, frequency scaling, and frequency substitutions are all provided internally. The calling sequence is

CALL SPFBLT(D,C,LN,IBAND,FLN,FHN,B,A,WORK,IERROR)

D(0:LN)	= $D(s)$ coefficients of analog filter in (7.4)
C(0:LN)	= $C(s)$ coefficients of analog filter in (7.4)
LN	= upper dimension of coefficient arrays
IBAND	= desired digital filter type:
	1—low-pass 2—high-pass
	3—bandpass 4—bandstop
FLN	= normalized cutoff in hertz-seconds (low)
FHN	= normalized cutoff in hertz-seconds (high)
B(0:LN)	= $B(z)$ coefficients in (7.3)—returned

A(LN)	$= A(z)$ coefficients in (7.3)—returned
WORK(0:LN,0:LN)	= work array used internally
IERROR	= 0 no errors detected
	1 all-zero transfer function
	2 SPBILN error: invalid transfer function
	3 filter order exceeds array size
	4 invalid filter type (IBAND < 1 or > 4)
	5 invalid cutoff specifications

The filter order L is again computed internally. The IERROR $= 3$ return will occur during bandpass or bandstop filter design if LN is less than two times the order of the analog filter described by arrays D and C. The normalized cutoff frequencies are computed as FLN $= f_L T$ and FHN $= f_H T$, where f_L and f_H are the cutoff frequencies in hertz and T is the sample interval in seconds. Note that FHN is used only for bandpass and bandstop filter design. The IERROR $= 5$ return indicates that either FLN or FHN (if used) is outside the range $0.0 <$ FLN, FHN < 0.5 Hz-s, or FLN $\geq$ FHN for bandpass or bandstop filter design. The use of this routine is demonstrated in examples that follow.

For each of the four cases discussed next, we begin with the second-order normalized low-pass Butterworth filter defined in (7.8).

The first example repeats the task performed by SPA0701, but now the prewarping and frequency scaling are done internally. The desired filter is a second-order low-pass digital filter with a cutoff frequency at 1 kHz and sample interval $T = 0.0001$ s. Program SPA0702 uses this information together with the analog filter coefficients from (7.8) to generate the digital filter coefficients. The computed $H(z)$ coefficients are printed below the program listing, and a plot of the magnitude response of $H(z)$ is provided in Fig. 7.4. The SPGAIN function

```
                PROGRAM SPA0702
      C-LOWPASS ANALOG TO LOWPASS DIGITAL SCALED IN FREQUENCY
                DIMENSION D(0:2),C(0:2),B(0:2),A(2),WORK(0:2,0:2),AMP(0:500)
                COMPLEX SPGAIN
                DATA D/1.,0.,0./,C/1.,1.41421,1./
                DATA IBAND/1/,FL/1000./,FH/0./,T/.0001/,LN/2/
                FLN=FL*T
                FHN=FH*T
                CALL SPFBLT(D,C,LN,IBAND,FLN,FHN,B,A,WORK,IERROR)
                IF(IERROR.NE.0)THEN
                  PRINT *,' SPFBLT ERROR = ',IERROR
                  STOP
                ENDIF
                DO 1 I=0,500
                  AMP(I)=ABS(SPGAIN(B,A,LN,LN,I*.5/500.))
            1   CONTINUE
                CALL MCPLOT(0.,.01,AMP,501,0,0,1,1,4,2,5.,3.6)
                PRINT 100,B,A
          100   FORMAT(' B(Z) COEFFICIENTS:',3F10.4,/,
                +        ' A(Z) COEFFICIENTS:',2F10.4)
                END

              $ RUN SPA0702
              B(Z) COEFFICIENTS:     0.0675    0.1349    0.0675
              A(Z) COEFFICIENTS:    -1.1430    0.4128
              $
```

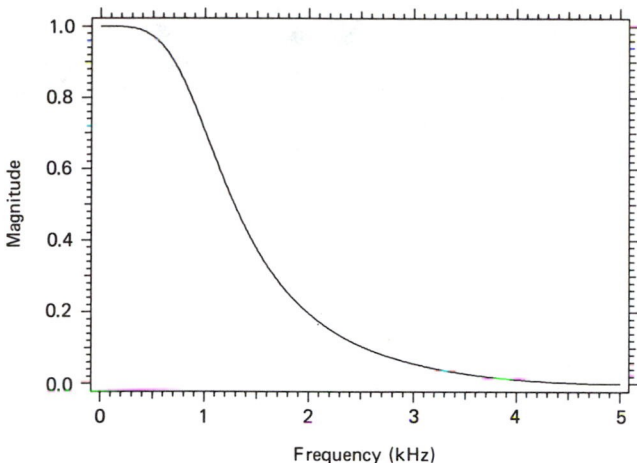

Figure 7.4. SPA0702: Magnitude response of low-pass digital Butterworth filter; f_L = 1 kHz; T = 0.0001 s.

from Chapter 5 was used to compute this response. The plot in Figure 7.4 exhibits the expected magnitude response, that is, the gain is approximately 0.7 at the 1-kHz cutoff point and is 0 at the 5-kHz Nyquist frequency.

The goal of the next example is to design a second-order high-pass filter. The band edge is now 3.5 kHz and the sample interval is again T = 0.0001 s. The design is performed by program SPA0703. With the exception of the initialization ITYPE=2 and the change in f_L, SPA0703 is identical to SPA0702. Note that the magnitude response in Fig. 7.5 again exhibits a gain of 0.7 at the 3.5-kHz cutoff, but the characteristic now describes a high-pass filter.

```
         PROGRAM SPA0703
C-LOWPASS ANALOG TO HIGHPASS DIGITAL SCALED IN FREQUENCY
         DIMENSION D(0:2),C(0:2),B(0:2),A(2),WORK(0:2,0:2),AMP(0:500)
         COMPLEX SPGAIN
         DATA D/1.,0.,0./,C/1.,1.41421,1./
         DATA IBAND/2/,FL/3500./,FH/0./,T/.0001/,LN/2/
         FLN=FL*T
         FHN=FH*T
         CALL SPFBLT(D,C,LN,IBAND,FLN,FHN,B,A,WORK,IERROR)
         IF(IERROR.NE.0)THEN
           PRINT *,' SPFBLT ERROR = ',IERROR
           STOP
         ENDIF
         DO 1 I=0,500
           AMP(I)=ABS(SPGAIN(B,A,LN,LN,I*.5/500.))
       1 CONTINUE
         CALL MCPLOT(0.,.01,AMP,501,0,0,1,1,4,2,5.,3.6)
         PRINT 100,B,A
     100 FORMAT(' B(Z) COEFFICIENTS:',3F10.4,/,
        +        ' A(Z) COEFFICIENTS:',2F10.4)
         END

         $ RUN SPA0703
         B(Z) COEFFICIENTS:    0.1311    -0.2622    0.1311
         A(Z) COEFFICIENTS:    0.7478    0.2722
         $
```

A digital bandpass filter can be designed by making a few changes in test

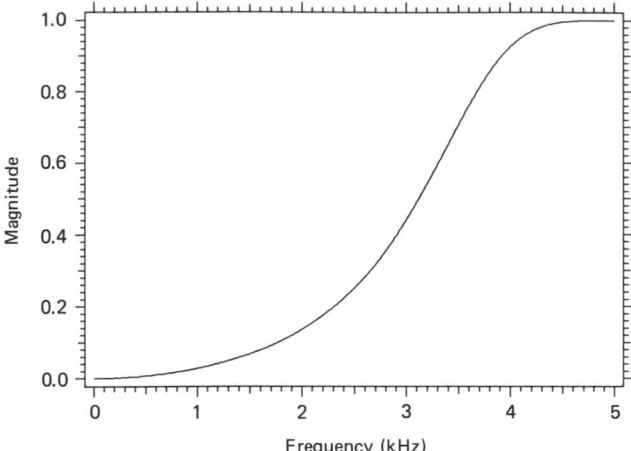

Figure 7.5. SPA0703: Magnitude response of high-pass digital Butterworth filter; $f_L = 3.5$ kHz; $T = 0.0001$ s.

program SPA0702. We must now dimension all coefficient arrays such that LN ≥ 4, since we are beginning with a second-order analog filter. In addition, two band-edge frequencies must be specified. Program SPA0704 designs a fourth-order digital bandpass filter with cutoff frequencies $f_L = 1500$ and $f_H = 3500$ Hz and sample interval $T = 0.0001$ s. Once again, only minor changes from SPA0702 occur in the initialization and format statements. Digital filter coefficients are shown and a plot of the amplitude gain is provided in Fig. 7.6. Note that the magnitude response is now approximately 0.7 at both cutoff frequencies (1.5 kHz and 3.5 kHz).

```
            PROGRAM SPA0704
    C-LOWPASS ANALOG TO BANDPASS DIGITAL SCALED IN FREQUENCY
            DIMENSION D(0:4),C(0:4),B(0:4),A(4),WORK(0:4,0:4),AMP(0:500)
            COMPLEX SPGAIN
            DATA D/1.,0.,0.,0.,0./,C/1.,1.41421,1.,0.,0./
            DATA IBAND/3/,FL/1500./,FH/3500./,T/.0001/,LN/4/
            FLN=FL*T
            FHN=FH*T
            CALL SPFBLT(D,C,LN,IBAND,FLN,FHN,B,A,WORK,IERROR)
            IF(IERROR.NE.0)THEN
              PRINT *,' SPFBLT ERROR = ',IERROR
              STOP
            ENDIF
            DO 1 I=0,500
              AMP(I)=ABS(SPGAIN(B,A,LN,LN,I*.5/500.))
        1   CONTINUE
            CALL MCPLOT(0.,.01,AMP,501,0,0,1,1,4,2,5.,3.6)
            PRINT 100,B,A
    100     FORMAT(' B(Z) COEFFICIENTS:',5F10.4,/,
        +          ' A(Z) COEFFICIENTS:',4F10.4)
            END


    $ RUN SPA0704
    B(Z) COEFFICIENTS:    0.2066    0.0000   -0.4131    0.0000    0.2066
    A(Z) COEFFICIENTS:    0.0000    0.3695    0.0000    0.1958
    $
```

The final example of this section demonstrates bandstop filter design via the SPFBLT routine. Once again, the cutoff frequencies are 1500 and 3500 Hz,

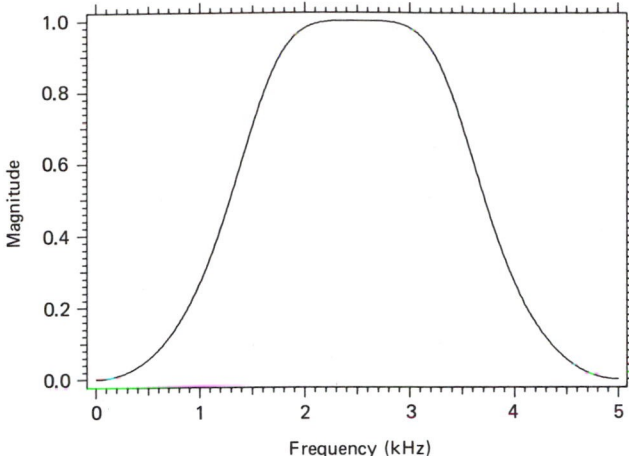

Figure 7.6. SPA0704: Magnitude response of bandpass digital Butterworth filter; $f_L = 1.5$ kHz; $f_H = 3.5$ kHz; $T = 0.0001$ s.

and the sample interval is $T = 0.0001$ s. The only change required from program SPA0704 is in the initialization IBAND=4. This program is designated SPA0705 and is listed together with its output. In Fig. 7.7, we see that the magnitude response at both cutoff frequencies is once again approximately 0.7, as expected.

```
          PROGRAM SPA0705
C-LOWPASS ANALOG TO BANDSTOP DIGITAL SCALED IN FREQUENCY
          DIMENSION D(0:4),C(0:4),B(0:4),A(4),WORK(0:4,0:4),AMP(0:500)
          COMPLEX SPGAIN
          DATA D/1.,0.,0.,0.,0./,C/1.,1.41421,1.,0.,0./
          DATA IBAND/4/,FL/1500./,FH/3500./,T/.0001/,LN/4/
          FLN=FL*T
          FHN=FH*T
          CALL SPFBLT(D,C,LN,IBAND,FLN,FHN,B,A,WORK,IERROR)
          IF(IERROR.NE.0)THEN
            PRINT *,' SPFBLT ERROR = ',IERROR
            STOP
          ENDIF
          DO 1 I=0,500
            AMP(I)=ABS(SPGAIN(B,A,LN,LN,I*.5/500.))
        1 CONTINUE
          CALL MCPLOT(0.,.01,AMP,501,0,0,1,1,4,2,5.,3.6)
          PRINT 100,B,A
      100 FORMAT(' B(Z) COEFFICIENTS:',5F10.4,/,
         +       ' A(Z) COEFFICIENTS:',4F10.4)
          END

$ RUN SPA0705
B(Z) COEFFICIENTS:      0.3913    0.0000    0.7827    0.0000    0.3913
A(Z) COEFFICIENTS:      0.0000    0.3695    0.0000    0.1958
$
```

In this section we have introduced two general routines for IIR filter design. The basic building block is a normalized low-pass analog filter $H(s)$. We now consider techniques for specifying $H(s)$.

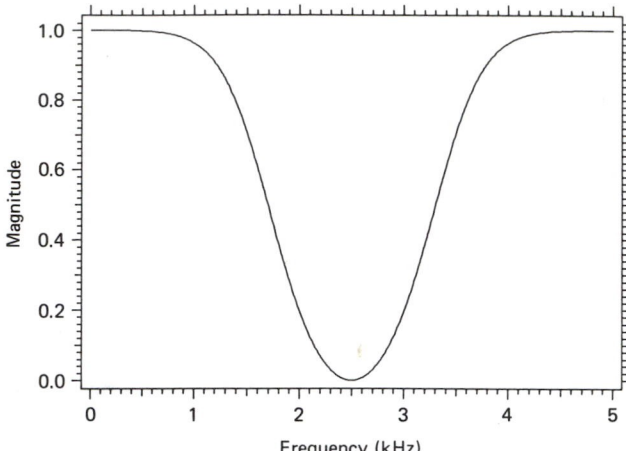

Figure 7.7. SPA0705: Magnitude response of bandstop digital Butterworth filter; $f_L = 1.5$ kHz; $f_H = 3.5$ kHz; $T = 0.0001$ s.

7.3 Butterworth Filter Design

In order to make use of the filter-design routines described in the previous section, here we discuss the design of a normalized low-pass analog Butterworth filter. We begin by describing the characteristics of this filter in terms of its analog frequency response.

The Lth-order analog Butterworth magnitude response is described by

$$|P_L(\Omega)| = \frac{1}{\sqrt{1 + \Omega^{2L}}}, \qquad L \geq 1 \qquad (7.10)$$

This function has a number of desirable properties. The maximum value occurs at $\Omega = 0$, where $|P_L(0)| = 1$. In addition, $|P_L(1)| = 0.7071$ for all L; thus, the bandwidth of the low-pass filters represented by (7.10) is 1 rad/s. As $L \to \infty$, the magnitude response approaches that of the ideal rectangular filter. If the function in (7.10) is evaluated at $\Omega = 0$, it can be shown that the first $(2L - 1)$ derivatives are zero. As a result, the Butterworth magnitude response is said to be maximally flat in the passband. The plots in Fig. 7.8 illustrate the shape of this function for three different filter orders.

The s-plane poles associated with the Butterworth response function in (7.10) are given by

$$s_k = \exp\left(\frac{j(2k + L - 1)\pi}{2L}\right), \qquad 1 \leq k \leq 2L \qquad (7.11)$$

These poles are located on a unit circle in the s-plane. Figure 7.9 illustrates the pole locations for functions of order $L = 3$ and $L = 4$. Note the symmetry in the pole locations about both the real and imaginary axes. Butterworth poles never fall on the imaginary axis, but there are always poles on the real axis when L is odd. The pole separation around the unit circle is π/L radians.

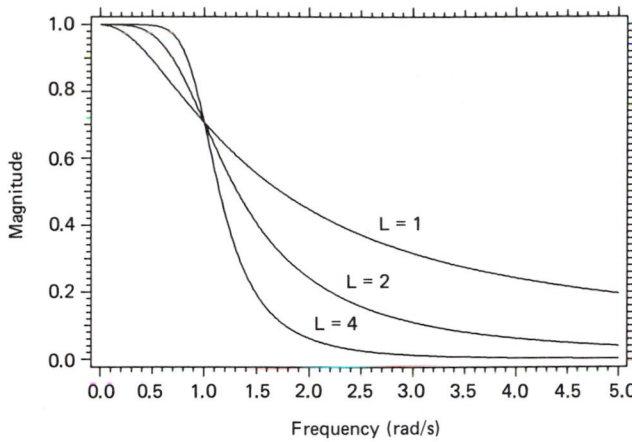

Figure 7.8. Low-pass Butterworth analog filter magnitude response; cutoff at $\Omega_c = 1$ rad/s.

To form a stable Lth-order Butterworth filter, we use only the poles in the left half of the s-plane. These are the first L poles defined in (7.11), that is, s_k for $1 \leq k \leq L$. The related transfer function is

$$P_L(s) = \frac{1}{(s - s_1)(s - s_2) \cdots (s - s_L)} \tag{7.12}$$

Note that this function is defined in terms of its pole locations rather than in the direct form of $H(s)$ in (7.4). Rather than fully expanding the denominator polynomial in (7.12), we design Butterworth filters in a cascade structure. This is a simple task due to the s-plane pole placement.

We first consider filters of even order, that is, $L = 2, 4, 6 \ldots$. In this case, all poles occur in complex conjugate pairs of the form

$$s_k, s_{L+1-k}, \qquad 1 \leq k \leq \frac{L}{2} \tag{7.13}$$

Combining (7.13) with the pole definitions in (7.11), we can write the transfer

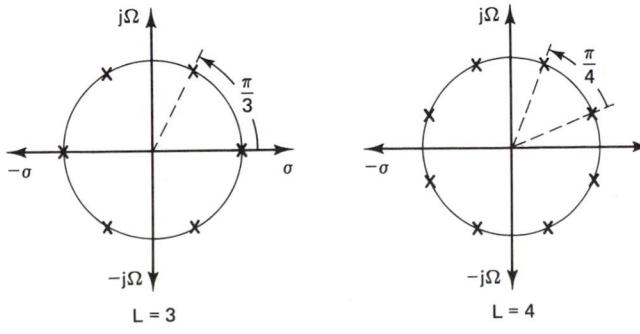

Figure 7.9. s-plane pole locations for Butterworth filters; the poles of $H(s)$ are those in the left half-plane.

function for a Butterworth second-order section as

$$P_k(s) = \frac{1}{(s - s_k)(s - s_{L+1-k})}$$

$$= \frac{1}{s^2 - 2s \cos\left(\dfrac{2k + L - 1}{2L}\pi\right) + 1} \tag{7.14}$$

The analog transfer function for a normalized low-pass Butterworth filter can now be written as

$$P_L(s) = \prod_{k=1}^{L/2} P_k(s), \qquad L \text{ even} \tag{7.15}$$

where $P_k(s)$ is as defined in (7.14). The only modification necessary to design an odd order filter is the addition of a pole at $s = -1$. Thus, we can write

$$P_L(s) = \frac{1}{s + 1} \prod_{k=1}^{(L-1)/2} P_k(s), \qquad L \text{ odd} \tag{7.16}$$

where $P_k(s)$ is again the second-order section in (7.14). Using these transfer functions, together with the subroutines described in the previous section, digital Butterworth filters of any type may be designed.

Before considering examples of Butterworth filter design, we introduce a simple routine, SPBWCF, which provides the analog filter coefficients in the form needed by the SPFBLT subroutine. The calling sequence is

CALL SPBWCF(L,K,LN,D,C,IERROR)

L	= order of normalized low-pass analog filter
K	= section number
LN	= upper dimension of coefficient arrays
D(0:LN)	= $D(s)$ coefficients for Kth section—returned
C(0:LN)	= $C(s)$ coefficients for Kth section—returned
IERROR	= 0 no errors detected
	1 invalid filter order (L ≤ 0)
	2 invalid section number $\left(K \leq 0 \text{ or } K > \dfrac{L + 1}{2}\right)$

This routine generates the analog filter coefficients for the Kth second-order section of an Lth-order Butterworth filter. For valid filter design, we must have $1 \leq K \leq L/2$ for even L and $1 \leq K \leq (L + 1)/2$ for odd L. If the filter order is odd, the final section is the first-order pole at $s = -1$, whereas all other sections are as defined in (7.14). Argument LN allows the coefficient arrays to be larger than second order if desired. Unused coefficients are set to zero.

A sixth-order low-pass Butterworth filter with cutoff frequency $f_L = 1.0$ kHz and sample interval $T = 0.0001$ s is designed by program SPA0706. The filter is configured in the form of three second-order sections; SPFBLT is used to transform each section individually. Note that the majority of this test program is devoted to listing coefficients and generating the amplitude plot. The actual filter design is accomplished by the three sets of calls to SPBWCF and SPFBLT. A plot of the magnitude response of the resulting digital filter is provided in Fig. 7.10. Note that the gain of the filter is approximately 0.7 at the 1-kHz cutoff frequency, as predicted from the Butterworth characteristic.

```
            PROGRAM SPA0706
C-DESIGN 6TH ORDER LOWPASS BUTTERWORTH FILTER
            DIMENSION AMP(0:500),WORK(0:2,0:2),D(0:2),C(0:2),B(0:2),A(2)
            COMPLEX SPGAIN
            DATA IBAND/1/,FL/1000./,FH/0./,T/.0001/,L/6/,LN/2/,AMP/501*1./
            FLN=FL*T
            FHN=FH*T
            DO 5 K=1,3
              CALL SPBWCF(L,K,LN,D,C,IERROR)
                IF(IERROR.NE.0)THEN
                  PRINT *,' SPBWCF ERROR = ',IERROR
                  STOP
                ENDIF
              CALL SPFBLT(D,C,LN,IBAND,FLN,FHN,B,A,WORK,IERROR)
                IF(IERROR.NE.0)THEN
                  PRINT *,' SPFBLT ERROR = ',IERROR
                  STOP
                ENDIF
              DO 3 I=0,500
                AMP(I)=AMP(I)*ABS(SPGAIN(B,A,LN,LN,I*.5/500.))
    3         CONTINUE
              PRINT 100,K,B,A
    5       CONTINUE
            PAUSE
            CALL MCPLOT(0.,.01,AMP,501,0,0,1,1,4,2,5.,3.6)
  100       FORMAT(' SECTION ',I2,' COEFFICIENTS = ',3F8.4,5X,2F8.4)
            END

$ RUN SPA0706
SECTION  1 COEFFICIENTS =    0.0829  0.1658  0.0829      -1.4044  0.7359
SECTION  2 COEFFICIENTS =    0.0675  0.1349  0.0675      -1.1430  0.4128
SECTION  3 COEFFICIENTS =    0.0609  0.1218  0.0609      -1.0321  0.2757
FORTRAN PAUSE
$
```

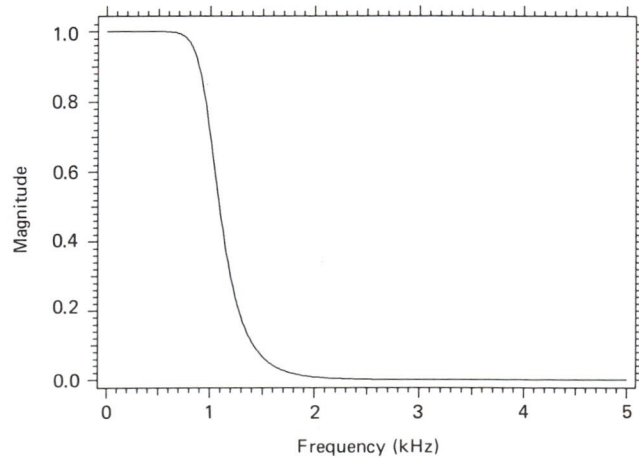

Figure 7.10. SPA0706: Magnitude response of sixth-order low-pass digital Butterworth filter; $f_L = 1.0$ kHz; $T = 0.0001$ s.

In the next example, we wish to design an eighth-order bandpass digital Butterworth filter with band-edge frequencies f_L = 1.5 kHz and f_H = 3.5 kHz and sample interval T = 0.0001 s. This task is accomplished by program SPA0707, which is listed. Note that very few changes were required from program SPA0706. Since the bandpass transformation doubles the filter order, the coefficient array dimensions must be adjusted by changing LN from 2 to 4. Two fourth-order filter sections are generated. Recall that in SPBWCF, low-pass analog filter coefficients are being generated, so the corresponding filter order is L/2. The

```
                PROGRAM SPA0707
        C-DESIGN 8TH ORDER BANDPASS BUTTERWORTH FILTER
                DIMENSION AMP(0:500),WORK(0:4,0:4),D(0:4),C(0:4),B(0:4),A(4)
                COMPLEX SPGAIN
                DATA IBAND/3/,FL/1500./,FH/3500./,T/.0001/,L/8/,LN/4/
                DATA AMP/501*1./
                FLN=FL*T
                FHN=FH*T
                DO 5 K=1,2
                  CALL SPBWCF(L/2,K,LN,D,C,IERROR)
                    IF(IERROR.NE.0)THEN
                      PRINT *,' SPBWCF ERROR =',IERROR
                      STOP
                    ENDIF
                  CALL SPFBLT(D,C,LN,IBAND,FLN,FHN,B,A,WORK,IERROR)
                    IF(IERROR.NE.0)THEN
                      PRINT *,' SPFBLT ERROR =',IERROR
                      STOP
                    ENDIF
                  DO 3 I=0,500
                    AMP(I)=AMP(I)*ABS(SPGAIN(B,A,LN,LN,I*.5/500.))
            3       CONTINUE
                  PRINT 100,K,B,A
            5   CONTINUE
                PAUSE
                CALL MCPLOT(0.,.01,AMP,501,0,0,1,1,4,2,5.,3.6)
          100   FORMAT(' SECTION ',I2,' COEFFICIENTS = ',5F9.4,/,27X,4F9.4)
                END

        $ RUN SPA0707
        SECTION  1 COEFFICIENTS =    0.2533    0.0000   -0.5066    0.0000    0.2533
                                     0.0000    0.4531    0.0000    0.4663
        SECTION  2 COEFFICIENTS =    0.1839    0.0000   -0.3678    0.0000    0.1839
                                     0.0000    0.3290    0.0000    0.0646

        FORTRAN PAUSE
        $
```

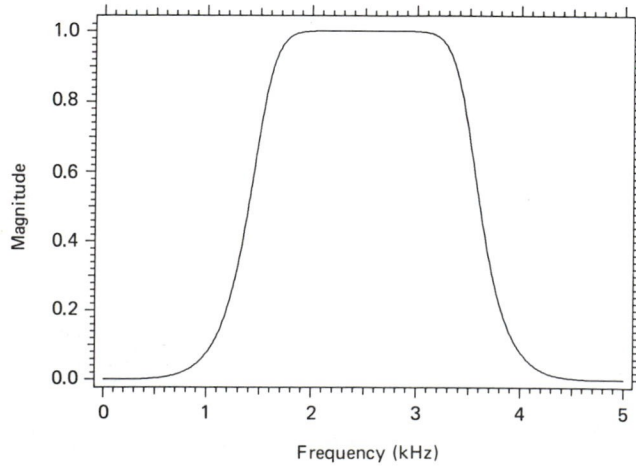

Figure 7.11. SPA0707: Magnitude response of eighth-order bandpass digital Butterworth filter; f_L = 1.5 kHz; f_H = 3.5 kHz; T = 0.0001 s.

results from SPA0707 are listed next, and a magnitude response plot is shown in Fig. 7.11.

The goal of SPA0708 is to design a fifth-order high-pass Butterworth filter with cutoff $f_L = 2.5$ kHz and sample interval $T = 0.0001$ s. This necessitates three sets of calls to SPBWCF and SPFBLT to generate and transform the filter coefficients. Once again, the key differences from SPA0706 occur in the initialization. The corresponding digital filter coefficients and magnitude response are also shown (Fig. 7.12).

```
            PROGRAM SPA0708
C-DESIGN 5TH ORDER HIGHPASS BUTTERWORTH FILTER
            DIMENSION AMP(0:500),WORK(0:2,0:2),D(0:2),C(0:2),B(0:2),A(2)
            COMPLEX SPGAIN
            DATA IBAND/2/,FL/2500./,FH/0./,T/.0001/,L/5/,LN/2/,AMP/501*1./
            FLN=FL*T
            FHN=FH*T
            DO 5 K=1,3
               CALL SPBWCF(L,K,LN,D,C,IERROR)
                  IF(IERROR.NE.0)THEN
                     PRINT *,' SPBWCF ERROR =',IERROR
                     STOP
                  ENDIF
               CALL SPFBLT(D,C,LN,IBAND,FLN,FHN,B,A,WORK,IERROR)
                  IF(IERROR.NE.0)THEN
                     PRINT *,' SPFBLT ERROR =',IERROR
                     STOP
                  ENDIF
               DO 3 I=0,500
                  AMP(I)=AMP(I)*ABS(SPGAIN(B,A,LN,LN,I*.5/500.))
    3          CONTINUE
               PRINT 100,K,B,A
    5       CONTINUE
            PAUSE
            CALL MCPLOT(0.,.01,AMP,501,0,0,1,1,4,2,5.,3.6)
  100       FORMAT(' SECTION ',I2,' COEFFICIENTS = ',3F8.4,5X,2F8.4)
            END


$ RUN SPA0708
SECTION  1 COEFFICIENTS =    0.3820 -0.7639  0.3820      0.0000  0.5279
SECTION  2 COEFFICIENTS =    0.2764 -0.5528  0.2764      0.0000  0.1056
SECTION  3 COEFFICIENTS =    0.5000 -0.5000  0.0000      0.0000  0.0000
FORTRAN PAUSE
$
```

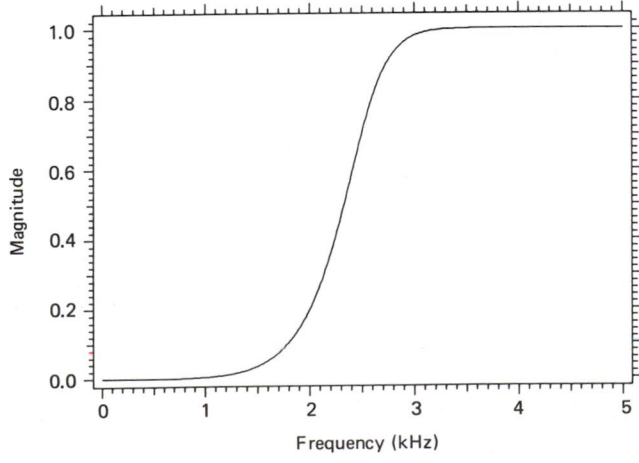

Figure 7.12. SPA0708: Magnitude response of fifth-order high-pass digital Butterworth filter; $f_L = 2.5$ kHz; $T = 0.0001$ s.

In this discussion, we have assumed that the filter order L is given. In some applications, it is more desirable to treat L as a variable and specify the filter in terms of the magnitude-response characteristics. The simplest criterion for Butterworth filter-order determination is stated in terms of stopband attenuation. With reference to the normalized low-pass analog filter illustrated in Fig. 6.2, the necessary order can be determined by the relationship

$$L \geqslant \frac{\log_{10}(\lambda^2 - 1)}{2 \log_{10}\Omega_s} \tag{7.17}$$

where the magnitude response of the filter is $1/\lambda$ at the specified stopband frequency Ω_s. In (7.17) we assume that the related cutoff frequency is $\Omega_c = 1$ rad/s, so we must specify $\Omega_s > 1$. To find the corresponding digital filter relationship, we must compensate for the frequency compression and scaling that occurs in the bilinear transformation. Thus, we would first select some digital filter stopband frequency in hertz, f_s, and the desired attenuation parameter, λ. The normalized analog frequency Ω_s can then be evaluated as

$$\Omega_s = \frac{\tan(2\pi f_s T/2)}{\tan(2\pi f_c T/2)} \tag{7.18}$$

where f_s is the stopband frequency, f_c is the filter cutoff (both in hertz), and T is the sample interval in seconds. The required filter order is found by substituting Ω_s and λ into (7.17).

For example, suppose we wish to design a low-pass digital Butterworth filter that has a magnitude gain of 0.1 at twice the cutoff frequency. The digital filter cutoff is 1 kHz and the sample interval is 0.0001 s. From (7.18) we find that the corresponding normalized analog filter stopband frequency is $\Omega_s = 2.24$ rad/s. By evaluating (7.17) with $\lambda = 10$ and $\Omega_s = 2.24$, we obtain $L \geqslant 2.86$. Thus, a third-order filter would be required to satisfy the amplitude specifications.

While determination of the necessary filter order works in a similar manner for high-pass filter design, it should be noted in the case of bandpass or bandstop design that the actual digital filter order is twice the value of the L computed from (7.17) and (7.18). This results from the fact that we transform the Lth-order low-pass analog filter into a digital filter of order $2L$.

7.4 Chebyshev Filter Design

As in the previous section, we begin our discussion of Chebyshev filter design with a description of the properties of the normalized low-pass analog filter. Rather than the monotonic frequency response exhibited by Butterworth filters, Chebyshev filters are characterized by a response that oscillates between specified limits over a prescribed band of frequencies.

The Chebyshev Type I filter is an all-pole structure, which has an equiripple passband response and a monotonic stopband. This filter is said to be *optimal* in the sense that no other Lth-order all-pole filter has equal or better performance in both the passband and the stopband. The magnitude response of an Lth-order Type I Chebyshev filter is described by

$$|H_{\mathrm{I}}(\Omega)| = \frac{1}{\sqrt{1 + \varepsilon^2\, V_L^2(\Omega)}} \tag{7.19}$$

where $V_L(\Omega)$ is the Lth-order Chebyshev polynomial

$$V_L(\Omega) = \begin{cases} \cos(L\cos^{-1}\Omega) & \Omega \leqslant 1 \\ \cosh(L\cosh^{-1}\Omega) & \Omega \geqslant 1 \end{cases} \tag{7.20}$$

and ε is a parameter that regulates the passband ripple [C]. The passband amplitude gain of this filter oscillates between 1 and $1/\sqrt{(1 + \varepsilon^2)}$. The gain at the normalized cutoff frequency of 1 rad/s is always equal to $1/\sqrt{(1 + \varepsilon^2)}$, whereas the gain at $\Omega = 0$ is 1 for odd L and $1/\sqrt{(1 + \varepsilon^2)}$ for even L. The plots in Fig. 7.13 illustrate the shape of the function in (7.19) for a second-order filter with three different values of ε. It is apparent that there is a trade-off between the degree of passband ripple and the sharpness of the frequency roll-off characteristic.

The Chebyshev Type II filter has zeros as well as poles and provides a monotonic passband response with an equiripple stopband characteristic. The magnitude response of the Type II filter is given by

$$|H_{\mathrm{II}}(\Omega)| = \frac{1}{\sqrt{1 + \varepsilon^2\, [V_L(\Omega_s)/V_L(\Omega_s/\Omega)]^2}} \tag{7.21}$$

where $V_L(\Omega)$ is again the Lth-order Chebyshev polynomial in (7.20), Ω_s specifies the low end of the stopband ($\Omega_s > 1$ rad/s), and ε determines the gain at the normalized cutoff frequency, $\Omega_c = 1$ rad/s. The plots in Fig. 7.14 illustrate the shape of the function in (7.21) for a second-order filter with $\Omega_s = 2$ rad/s and three different values of ε. Once again, small values of ε provide better passband characteristics, but the penalty is an increase in the stopband amplitude gain.

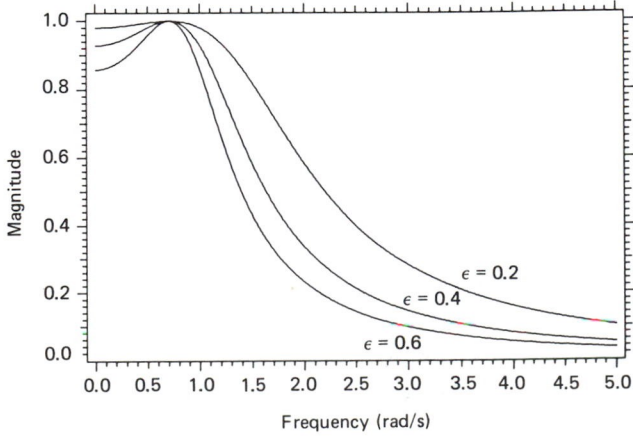

Figure 7.13. Low-pass Chebyshev Type I analog filter magnitude response; cutoff at $\Omega_C = 1$ rad/s.

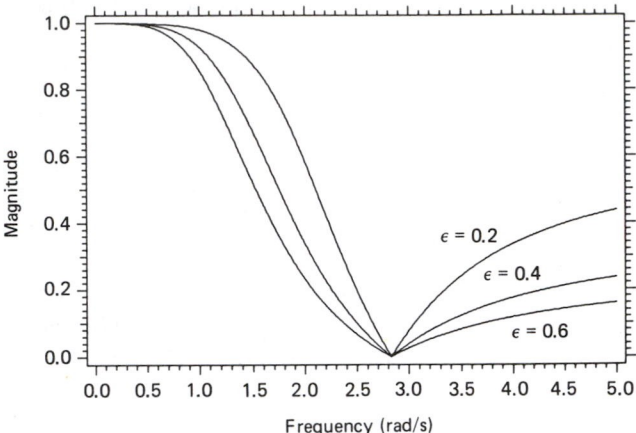

Figure 7.14. Low-pass Chebyshev Type II analog filter magnitude response; cutoff at $\Omega_C = 1$ rad/s.

An alternate parameter used to specify Chebyshev filters is λ, which regulates the stopband attenuation. For the Type II low-pass analog filter, the stopband magnitude response oscillates between $1/\lambda$ and 0 for $\Omega \geq \Omega_s$. The filter specification parameters are related by the following equation [D]:

$$L \geq \frac{\cosh^{-1}(\sqrt{(\lambda^2 - 1)}/\varepsilon)}{\cosh^{-1}(\Omega_s)} \tag{7.22}$$

The order determination in (7.22) is stated in terms of the normalized low-pass analog filter characteristics. Thus, to find the corresponding digital filter order, we again use the procedure outlined in the section on Butterworth filter design. Once the digital filter parameters have been specified, the analog frequency Ω_s is evaluated using (7.18). Equation (7.22) is then used to find the minimum required filter order.

As in the previous section, analog Chebyshev filters are specified in terms of their s-plane pole locations. The analog transfer function of an Lth-order Type I Chebyshev filter can be written

$$H_1(s) = A\frac{s_1 s_2 \cdots s_L}{(s - s_1) \cdots (s - s_L)} \tag{7.23}$$

where A is a scale factor defined as follows:

$$A = -1 \qquad L \text{ odd}$$

$$A = \frac{1}{\sqrt{1 + \varepsilon^2}} \qquad L \text{ even}$$

and ε is the passband ripple parameter defined previously. The poles of this filter lie on an ellipse in the s-plane; once again we select only the poles that lie in the stable region [C]. The stable poles $[s_k]$ are given by

$$s_k = \sigma_k + j\Omega_k \qquad\qquad\qquad k = 1, 2, \dots L$$

$$\sigma_k = -\sinh\varphi \, \sin\left(\frac{2k-1}{2L}\pi\right) \qquad\qquad \Omega_k = \cosh\varphi \, \cos\left(\frac{2k-1}{2L}\pi\right)$$

$$\sinh\varphi = \frac{\gamma - \gamma^{-1}}{2} \qquad\qquad\qquad \cosh\varphi = \frac{\gamma + \gamma^{-1}}{2}$$

$$\gamma = \left(\frac{1 + \sqrt{1 + \varepsilon^2}}{\varepsilon}\right)^{1/L} \tag{7.24}$$

Using the information in (7.23) and (7.24) and combining complex conjugate poles, we can realize the Chebyshev Type I filter as a cascade of second-order sections. To facilitate this process, subroutine SPCHBI uses (7.24) to generate the analog filter coefficients for the Kth section of an Lth-order Chebyshev Type I filter. The scale factor $1/\sqrt{(1 + \varepsilon^2)}$ is incorporated into the first section for even L. For odd L, the final section $(K = (L + 1)/2)$ is of first order and the numerator coefficient is modified to account for the -1 scale factor in (7.23). The calling sequence is

CALL SPCHBI(L,K,LN,EP,D,C,IERROR)

L	=	order of normalized low-pass analog filter
K	=	section number
LN	=	upper dimension of coefficient arrays
EP	=	passband ripple parameter, ε
D(0:LN)	=	$D(s)$ coefficients for Kth section—returned
C(0:LN)	=	$C(s)$ coefficients for Kth section—returned
IERROR	=	0 no errors detected

$\qquad\qquad\qquad$ 1 invalid filter order (L $\leq$ 0)

$\qquad\qquad\qquad$ 2 invalid section number $\left(K \leq 0 \text{ or } K > \dfrac{L+1}{2}\right)$

$\qquad\qquad\qquad$ 3 invalid ripple parameter (EP $\leq$ 0)

Program SPA0709 demonstrates the use of SPCHBI in the design of a sixth-order low-pass filter with a cutoff frequency at 1 kHz and sample interval $T = 0.0001$ s. To enable direct comparison with the Butterworth filter designed in SPA0706, we select the ripple parameter to be $\varepsilon = 1.0$ so that the magnitude response of the Chebyshev filter will match that of the Butterworth filter at the 1-kHz cutoff. With the exception of the design routine initialization and call, SPA0709 is identical to SPA0706. A listing of the program is provided, together with a description of the corresponding digital filter characteristics. The magnitude response is shown in Fig. 7.15.

```
            PROGRAM SPA0709
C-DESIGN 6TH ORDER LOWPASS CHEBYSHEV TYPE I FILTER
            DIMENSION AMP(0:500),WORK(0:2,0:2),D(0:2),C(0:2),B(0:2),A(2)
            COMPLEX SPGAIN
            DATA IBAND/1/,FL/1000./,FH/0./,T/.0001/,L/6/,LN/2/,AMP/501*1./
            DATA EP/1./
            FLN=FL*T
            FHN=FH*T
            DO 5 K=1,3
              CALL SPCHBI(L,K,LN,EP,D,C,IERROR)
                IF(IERROR.NE.0)THEN
                  PRINT *,' SPCHBI ERROR = ',IERROR
                  STOP
                ENDIF
              CALL SPFBLT(D,C,LN,IBAND,FLN,FHN,B,A,WORK,IERROR)
                IF(IERROR.NE.0)THEN
                  PRINT *,' SPFBLT ERROR = ',IERROR
                  STOP
                ENDIF
              DO 3 I=0,500
                AMP(I)=AMP(I)*ABS(SPGAIN(B,A,LN,LN,I*.5/500.))
    3         CONTINUE
              PRINT 100,K,B,A
    5       CONTINUE
            PAUSE
            CALL MCPLOT(0.,.01,AMP,501,0,0,1,1,4,2,5.,3.6)
  100       FORMAT(' SECTION ',I2,' COEFFICIENTS = ',3F8.4,5X,2F8.4)
            END

$ RUN SPA0709
SECTION  1 COEFFICIENTS =    0.0633   0.1266   0.0633     -1.5977  0.9559
SECTION  2 COEFFICIENTS =    0.0491   0.0981   0.0491     -1.6831  0.8793
SECTION  3 COEFFICIENTS =    0.0085   0.0170   0.0085     -1.7980  0.8320
FORTRAN PAUSE
$
```

Comparing the Butterworth magnitude response in Fig. 7.10 to that of the Chebyshev Type I filter in Fig. 7.15, we note that the characteristics match at the 1 kHz cutoff frequency. With ripple in its passband response, the Chebyshev filter yields a sharper roll-off characteristic than that of the Butterworth filter.

Chebyshev Type II filters may be designed in a similar fashion. These filters have an analog transfer function with both poles and zeros that can be expressed

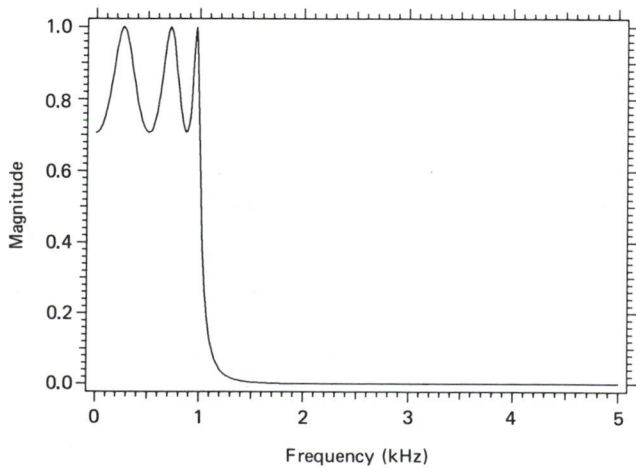

Figure 7.15. SPA0709: Magnitude response of sixth-order low-pass digital Chebyshev Type I filter; $f_L = 1.0$ kHz; $T = 0.0001$ s.

as

$$H_{II}(s) = \frac{s_1 s_2 \dots s_L (s - t_1) (s - t_2) \dots (s - t_L)}{t_1 t_2 \dots t_L (s - s_1) (s - s_2) \dots (s - s_L)} \tag{7.25}$$

The zeros of $H_{II}(s)$ are located on the imaginary axis and are defined by

$$t_k = j \frac{\Omega_s}{\cos [\pi(2k - 1)/(2L)]}, \qquad k = 1, 2, \dots, L \tag{7.26}$$

where Ω_s specifies the low end of the stopband ($\Omega_s > 1$ rad/s). For odd L, the zero associated with $k = (L + 1)/2$ lies at $s = \infty$. The pole locations are specified by

$$s_k = \sigma_k + j\Omega_k, \qquad\qquad k = 1, 2, \dots, L$$

$$\sigma_k = \frac{\Omega_s \alpha_k}{\alpha_k^2 + \beta_k^2} \qquad\qquad \Omega_k = \frac{-\Omega_s \beta_k}{\alpha_k^2 + \beta_k^2}$$

$$\alpha_k = -\sinh \varphi \sin [\pi(2k - 1)/(2L)] \qquad \beta_k = \cosh \varphi \cos [\pi(2k - 1)/(2L)]$$

$$\sinh \varphi = \frac{\gamma - \gamma^{-1}}{2} \qquad\qquad \cosh \varphi = \frac{\gamma + \gamma^{-1}}{2}$$

$$\gamma = (\lambda + \sqrt{\lambda^2 - 1})^{1/L} \tag{7.27}$$

where Ω_s is as defined previously and λ is the parameter related to stopband attenuation. The amplitude gain of the Chebyshev Type II filter is equal to $1/\lambda$ at $\Omega = \Omega_s$ and varies between $1/\lambda$ and 0 for $\Omega > \Omega_s$.

Once again utilizing second-order sections, poles and zeros are grouped in complex conjugate pairs to form the cascade-form transfer function. Subroutine SPCBII uses (7.27) to generate the analog filter coefficients for the Kth section of the Lth-order transfer function. As with SPCHBI and SPBWCF, if L is odd, the final section will contain the first-order real pole. The calling sequence is

CALL SPCBII(L,K,LN,SB,ATTN,D,C,IERROR)

L   = order of normalized low-pass analog filter
K   = section number
LN   = upper dimension of coefficient arrays
SB   = low edge of stopband, Ω_s [SB > 1 rad/sec; See (7.18)]
ATTN   = stopband attenuation parameter, λ
D(0:LN) = $D(s)$ coefficients for Kth section—returned
C(0:LN) = $C(s)$ coefficients for Kth section—returned
IERROR   = 0 no errors detected
           1 invalid filter order (L $\leq$ 0)
           2 invalid section number, $\left(K \leq 0 \text{ or } K > \dfrac{L + 1}{2} \right)$
           3 invalid stopband frequency (SB $\leq$ 1 rad/s)
           4 invalid attenuation parameter (ATTN $\leq$ 0)

In SPA0710 we again design a sixth-order low-pass filter with the cutoff at 1 kHz and a sample interval of 0.0001 s. In this case, we want to match the stopband response of the Butterworth filter in SPA0706 at 1.5 kHz. To find the normalized analog frequency, we evaluate (7.18) and obtain $\Omega_s = 1.57$ rad/s. From (7.10), we find that the corresponding Butterworth filter gain at Ω_s is 0.0666. This yields an attenuation parameter of approximately 15. Once again, this task can be accomplished by changing only the initialization and design routine call in program SPA0706. The resulting program, SPA0710, is listed, together with its output. A plot of the magnitude response of the resulting digital filter is shown in Fig. 7.16. From this plot, it is apparent that the gain at 1.5 kHz is approximately 0.066, as specified, and that the response oscillates between this level and 0 throughout the filter stopband. In this example, there is very little attenuation at the 1-kHz filter cutoff.

```
                PROGRAM SPA0710
        C-DESIGN 6TH ORDER LOWPASS CHEBYSHEV TYPE II FILTER
                DIMENSION AMP(0:500),WORK(0:2,0:2),D(0:2),C(0:2),B(0:2),A(2)
                COMPLEX SPGAIN
                DATA IBAND/1/,FL/1000./,FH/0./,T/.0001/,L/6/,LN/2/,AMP/501*1./
                DATA SB/1.57/,ATTN/15./
                FLN=FL*T
                FHN=FH*T
                DO 5 K=1,3
                  CALL SPCBII(L,K,LN,SB,ATTN,D,C,IERROR)
                    IF(IERROR.NE.0)THEN
                      PRINT *,' SPCBII ERROR = ',IERROR
                      STOP
                    ENDIF
                  CALL SPFBLT(D,C,LN,IBAND,FLN,FHN,B,A,WORK,IERROR)
                    IF(IERROR.NE.0)THEN
                      PRINT *,' SPFBLT ERROR = ',IERROR
                      STOP
                    ENDIF
                  DO 3 I=0,500
                    AMP(I)=AMP(I)*ABS(SPGAIN(B,A,LN,LN,I*.5/500.))
            3     CONTINUE
                  PRINT 100,K,B,A
            5   CONTINUE
                PAUSE
                CALL MCPLOT(0.,.01,AMP,501,0,0,1,1,4,2,5.,3.6)
          100   FORMAT(' SECTION ',I2,' COEFFICIENTS = ',3F8.4,5X,2F8.4)
                END

        $ RUN SPA0710
        SECTION  1 COEFFICIENTS =    0.6986 -0.7878  0.6986     -1.2058  0.8153
        SECTION  2 COEFFICIENTS =    0.4910 -0.3097  0.4910     -0.7709  0.4432
        SECTION  3 COEFFICIENTS =    0.2570  0.3036  0.2570     -0.2573  0.0749
        FORTRAN PAUSE
        $
```

In summary, two simple subroutines were introduced in this section. These routines generate the coefficients for second-order sections of the normalized low-pass analog Chebyshev filter (Type I or II) and may be used in conjunction with the SPFBLT routine described in Section 7.2 to design digital Chebyshev filters.

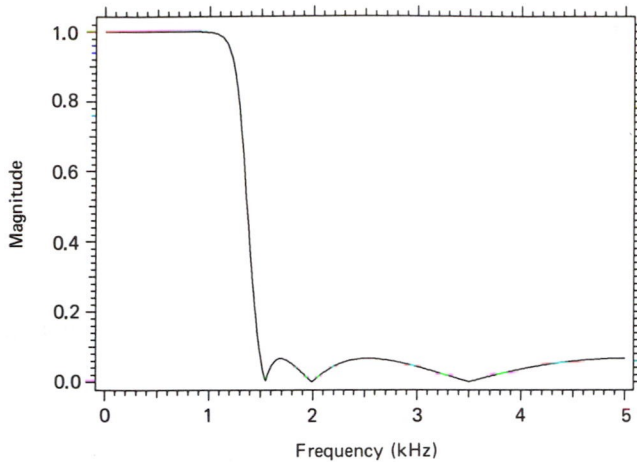

Figure 7.16. SPA0710: Magnitude response of sixth-order low-pass digital Chebyshev Type II filter; $f_L = 1$ kHz; $T = 0.0001$ s.

7.5 Bessel Filter Design

Unlike the Butterworth and Chebyshev analog filters, Bessel filters are most easily specified in terms of direct-form transfer functions rather than in terms of second-order sections. The analog Bessel filter is an all-pole structure whose denominator polynomial is defined by the Bessel function. The Lth-order low-pass Bessel transfer function can be written

$$H(s) = \frac{d_0}{B_L(s)} \tag{7.28}$$

where the Bessel function, $B_L(s)$, can be defined recursively as

$$B_L(s) = (2L - 1)B_{L-1}(s) + s^2 B_{L-2}(s)$$

with initial conditions

$$\begin{aligned} B_0(s) &= 1 \\ B_1(s) &= s + 1 \end{aligned} \tag{7.29}$$

An alternate derivation of the Bessel function yields the required $C(s)$ coefficients, where $H(s) = D(s)/C(s)$, as in (7.4). In this form we have [C]

$$B_L(s) = \sum_{k=0}^{L} c_k s^k = C(s)$$

where

$$c_k = \frac{(2L - k)!}{2^{L-k} k! (L - k)!}, \qquad k = 0, 1, ..., L \tag{7.30}$$

In (7.28), the numerator coefficient d_0 is a normalizing constant that is generally selected to yield a gain equal to 1 at $s = 0$. Thus, we have

$$d_0 = \frac{(2L)!}{2^L L!} = c_0 \tag{7.31}$$

Combining these results, we express the Lth-order Bessel transfer function in the form of (7.4):

$$H(s) = \frac{c_0}{c_L s^L + \cdots + c_0} \tag{7.32}$$

where the c_k coefficients are defined by (7.30).

The analog Bessel filter described above has a number of useful properties. One of its key characteristics is nearly linear phase response (i.e., constant group delay) in the filter passband. Unfortunately, due to the nonlinear warping of the bilinear transformation, this property translates only approximately into the digital Bessel filter. Properties that are preserved through the digital transformation include a step response with extremely low overshoot, and a magnitude response which tends to be Gaussian in shape for high-order filters [C].

The primary difficulty in Bessel filter design is that the passband definition varies with the order of the filter. This is illustrated in Fig. 7.17, where the analog magnitude response of $H(s)$ in (7.32) is plotted for three different orders (values of L). If, as in the previous sections, we define the edge of the passband to be at $\Omega_C = 1$ rad/s, the magnitude response at that point varies from approximately 0.97 for $L = 8$ to 0.84 for $L = 2$. Using this interpretation, we see that the passband characteristic improves as the order increases, but this is achieved at the expense of the stopband attenuation. For higher orders, the curves in Fig. 7.17 become more Gaussian in shape. Thus, it is apparent that some care must be exercised in designing Bessel filters.

This problem can be alleviated somewhat by using the frequency scaling techniques described in Section 7.2. Recall that the transformation subroutine SPFBLT, which is used to convert the normalized analog transfer function into the desired digital $H(z)$, assumes that the magnitude response of $H(s)$ at $\Omega = 1$ rad/s is the desired response of $H(z)$ at its cutoff frequency. For a given filter order L, the transfer function defined by (7.30) and (7.32) could be prescaled to formulate a new $\hat{H}(s)$ whose response at 1 rad/s matches that of the original

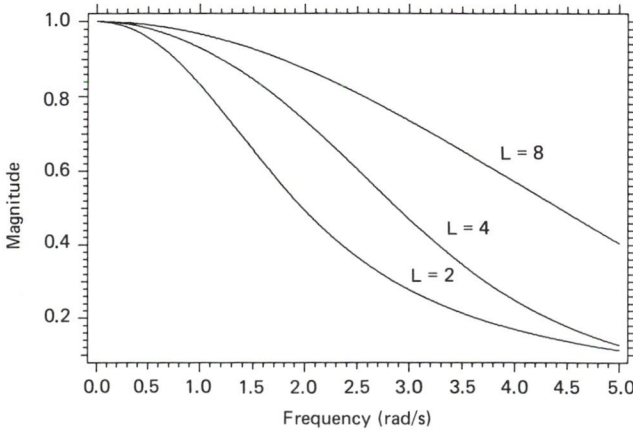

Figure 7.17. Low-pass Bessel analog filter magnitude response; cutoff at $\Omega_C = 1$ rad/s.

$H(s)$ at some arbitrary frequency. This prescaling will maintain the shape of the magnitude and phase curves while scaling the frequency axis.

As an example, suppose a fourth-order low-pass Bessel filter is to be designed, but the desired magnitude response at the cutoff frequency is equal to that of $H(s)$ at $\Omega = 2$ rad/s. This could be accomplished by defining a new transfer function $\hat{H}(s) = H(s/.5)$. The response of $\hat{H}(s)$ at $s = 1$ is now equal to the response $H(1/0.5) = H(2)$. This prescaling is illustrated in Fig. 7.18, where we show the magnitude response of $H(s)$ together with that of $\hat{H}(s)$. By using SPFBLT to convert $\hat{H}(s)$, the prescaled analog transfer function, we obtain the desired response in the digital transfer function $H(z)$.

Design routine SPBSSL first generates the normalized Bessel filter coefficients described by (7.30) and (7.32) and then performs frequency scaling if requested. The calling sequence is

CALL SPBSSL(L,PSCL,LN,D,C,IERROR)

L	=	order of low-pass analog filter
PSCL	=	prescaling specification
LN	=	upper dimension of coefficient arrays
D(0:LN)	=	$D(s)$ coefficients—returned
C(0:LN)	=	$C(s)$ coefficients—returned
IERROR	=	0 no errors detected
		1 invalid filter order ($L \leq 0$)
		2 invalid prescaling parameter (PSCL ≤ 0)

If frequency scaling is desired, PSCL is set to the analog frequency in radians per second at which the magnitude response of the normalized Bessel filter in (7.32) matches that of the desired digital filter at its cutoff frequency. In this case, the response of the filter $\hat{H}(s)$ at $\Omega = 1$ rad/s matches that of the normalized Bessel filter at PSCL radians per second. Thus the scaled coefficients, which

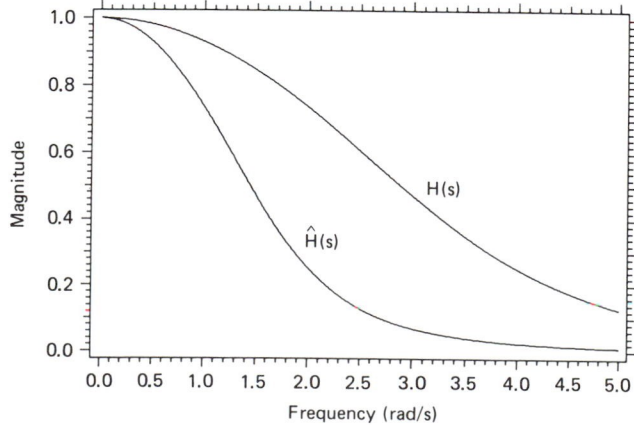

Figure 7.18.   Frequency scaling of the Bessel response.

are returned in arrays D and C, are in the form required for conversion to $H(z)$ by subroutine SPFBLT. No prescaling is performed if PSCL = 1.

In SPA0711, the first example of digital Bessel filter design, we assume that the response characteristics are acceptable, so no frequency scaling is performed. The program designs a sixth-order low-pass digital filter with cutoff frequency $f_L = 1$ kHz and sample interval $T = 0.0001$ s. This task requires a single call to SPBSSL to obtain the $H(s)$ coefficients and a single call to SPFBLT to convert $H(s)$ to the desired digital $H(z)$. Following the program listing, the $H(z)$ coefficients are shown. A plot of the magnitude response of the digital filter is provided in Fig. 7.19.

```
              PROGRAM SPA0711
      C-DESIGN 6TH ORDER LOWPASS BESSEL FILTER
              DIMENSION AMP(0:500),WORK(0:6,0:6),D(0:6),C(0:6),B(0:6),A(6)
              COMPLEX SPGAIN
              DATA IBAND/1/,FL/1000./,FH/0./,T/.0001/,L/6/,LN/6/,PSCL/1./
              FLN=FL*T
              FHN=FH*T
              CALL SPBSSL(L,PSCL,LN,D,C,IERROR)
              IF(IERROR.NE.0)THEN
                PRINT *,' SPBSSL ERROR= ',IERROR
                STOP
              ENDIF
              CALL SPFBLT(D,C,LN,IBAND,FLN,FHN,B,A,WORK,IERROR)
              IF(IERROR.NE.0)THEN
                PRINT *,' SPFBLT ERROR =',IERROR
                STOP
              ENDIF
              DO 3 I=0,500
                AMP(I)=ABS(SPGAIN(B,A,LN,LN,I*.5/500.))
            3 CONTINUE
              CALL MCPLOT(0.,.01,AMP,501,0,0,1,1,4,2,5.,3.6)
              PRINT 100,B,A
          100 FORMAT(' B(Z) COEF = ',7F8.4,/,' A(Z) COEF = ',6F8.4)
              END

      $ RUN SPA0711
      B(Z) COEF =    0.0696   0.4176   1.0439   1.3919   1.0439   0.4176   0.0696
      A(Z) COEF =    1.4319   1.2308   0.5841   0.1762   0.0290   0.0021
      $
```

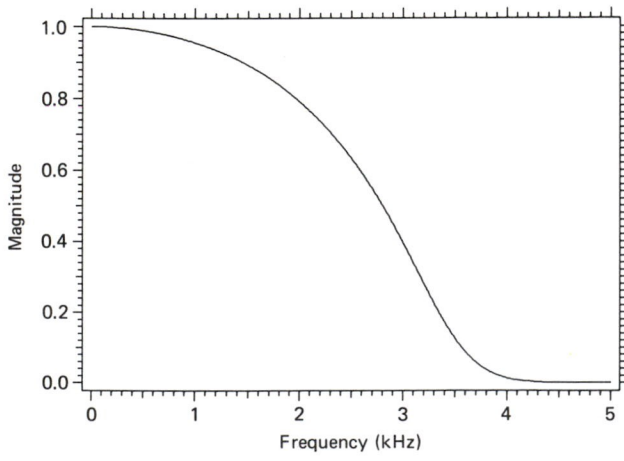

Figure 7.19. SPA0711: Magnitude response of sixth-order low-pass digital Bessel filter; $f_L = 1$ kHz; $T = 0.0001$ s; PSCL = 1.

The task accomplished by test program SPA0712 is the design of an eighth-order bandpass Bessel filter with band edges at 1.5 and 3.5 kHz and a sample interval of $T = 0.0001$ s. In this case, the desired magnitude response is specified to match that of the Butterworth filter at the cutoff frequencies. (Recall from Section 7.3 that the gain of a Butterworth filter at its cutoff frequency is always 0.7071.) To find the necessary value for PSCL, we use the curve in Fig. 7.17 for $L = 4$ since the order will be doubled by the bandpass transformation. The normalized Bessel magnitude response is approximately equal to 0.7071 at 2.1 rad/s. Thus, we use SPBSSL to design a fourth-order analog filter with scale factor PSCL = 2.1. Program SPA0712 is given. A plot of the filter magnitude response is provided in Fig. 7.20.

```
          PROGRAM SPA0712
C-DESIGN 8TH ORDER BANDPASS BESSEL FILTER
          DIMENSION AMP(0:500),WORK(0:8,0:8),D(0:8),C(0:8),B(0:8),A(8)
          COMPLEX SPGAIN
          DATA IBAND/3/,FL/1500./,FH/3500./,T/.0001/,L/8/,LN/8/,PSCL/2.1/
          FLN=FL*T
          FHN=FH*T
          CALL SPBSSL(L/2,PSCL,LN,D,C,IERROR)
          IF(IERROR.NE.0)THEN
            PRINT *,' SPBSSL ERROR=',IERROR
            STOP
          ENDIF
          CALL SPFBLT(D,C,LN,IBAND,FLN,FHN,B,A,WORK,IERROR)
          IF(IERROR.NE.0)THEN
            PRINT *,' SPFBLT ERROR = ',IERROR
            STOP
          ENDIF
          DO 3 I=0,500
            AMP(I)=ABS(SPGAIN(B,A,LN,LN,I*.5/500.))
    3     CONTINUE
          CALL MCPLOT(0.,.01,AMP,501,0,0,1,1,4,2,5.,3.6)
          PRINT 100,B,A
  100     FORMAT(' B(Z) COEF:',9F7.3,/,' A(Z) COEF:',8F7.3)
          END

$ RUN SPA0712
B(Z) COEF =  0.096  0.000 -0.383  0.000  0.575  0.000 -0.383  0.000  0.096
A(Z) COEF =  0.000 -0.242  0.000  0.271  0.000 -0.015  0.000  0.005
$
```

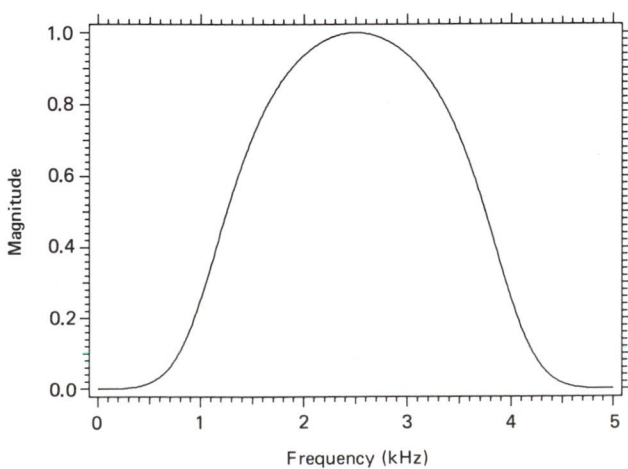

Figure 7.20. SPA0712: Magnitude response of eighth-order bandpass digital Bessel filter; $f_L = 1.5$ kHz; $f_H = 3.5$ kHz; $T = 0.0001$ s; PSCL = 2.1.

Note that in Fig. 7.20 the filter response at the band edges is approximately 0.7 as specified, but the passband gain is less uniform than that of the 8th-order digital Butterworth filter in Fig. 7.11.

7.6 A General IIR Design Routine

In this section we introduce a general IIR design routine, SPIIRD. This routine uses most of the routines described previously in this chapter. Its principal advantage is that it is simple to use; its application does not require an understanding of the details of previous routines, including frequency transformations, the bilinear transformation, or filter design. Its principal disadvantage is that it does not allow some of the flexibility and designs that one could attain through independent use of the lower-order routines described previously.

The SPIIRD routine allows one to design any one of three types of filters with any one of four frequency characteristics:

Filter type options	Frequency characteristics
1) Chebyshev Type I	1) Low-pass
2) Chebyshev Type II	2) High-pass
3) Butterworth	3) Bandpass
	4) Bandstop

The filters are always designed in cascade form, which is the primary reason that the Bessel filter was not included in the options shown. Low-pass and high-pass filters are designed with two-pole sections; bandpass and bandstop filters are designed with four-pole sections. The user specifies the number of sections together with other characteristics of the filter.

To develop a simple rule for choosing the number of filter sections, NS, we consider the power gain of a low-pass analog Butterworth filter with cutoff at $\Omega = 1$, which, as stated in (7.10), is

$$|P_L(\Omega)|^2 = \frac{1}{1 + \Omega^{2L}} \tag{7.33}$$

Here, L is the number of filter poles and is therefore twice the number of 2-pole filter sections. When Ω is large compared with 1, the power gain in decibels is

$$dB = 10 \log_{10}|P_L(\Omega)|^2 \approx -20L \log_{10} \Omega; \qquad \Omega \gg 1 \tag{7.34}$$

Recalling that L is twice the number (NS) of low-pass sections and also that the analog frequency, Ω, in accordance with (7.6), corresponds with $\tan(\pi f T)$, where f is the digital filter frequency in hertz, we rewrite (7.34):

$$dB \approx -10(NS)\log_{10}[\tan(\pi f T)] \tag{7.35}$$

Thus, if we choose two frequencies f_1 and f_2, with $f_2 > f_1$, in the stopband of a digital Butterworth filter and require that the gain change by ΔdB from f_1 to f_2, then our simple rule for choosing *NS*, the number of sections, is, from (7.35),

$$NS > \frac{0.1|\Delta\text{dB}|}{\log_{10}\left[\dfrac{\tan(\pi f_2 T)}{\tan(\pi f_1 T}\right]} \tag{7.36}$$

For example, if we wish a 20-dB drop from 0.2 to 0.4 Hz-s, we find from (7.36) that the filter must have at least *NS* = 4 sections. Note that this rule was derived via the Butterworth response characteristic and is therefore less accurate when applied to Chebyshev filter design. In general, (7.36) is somewhat more stringent than is needed for Chebyshev filters.

The calling sequence for the IIR design routine SPIIRD is

CALL SPIIRD(IFILT,IBAND,NS,LS,F1,F2,F3,F4,DB,B,A,IERROR)

IFILT	= type of filter: 1—Chebyshev Type I
	2—Chebyshev Type II
	3—Butterworth
IBAND	= type of characteristic: 1—Low-pass
	2—High-pass
	3—Bandpass
	4—Bandstop
NS	= number of filter sections; see (7.36)
LS	= order of each section: 2 (IBAND = 1,2)
	4 (IBAND = 3,4)
F1–F4	= cutoff and stopband frequencies in Hz-s; see Fig. 7.21
DB	= stopband rejection; see Fig. 7.21 (must be > 3.0 dB)
B(0:LS,NS)	= numerator coefficients in each section
A(LS,NS)	= denominator coefficients in each section
IERROR	= return error indicator:
	0 No errors detected
	1–5 See SPFBLT error list
	6 IFILT or IBAND out of range
	7 F1–F4 not in order or out of the range 0.0–0.5 Hz-s
	11+ Subtract 10 and see SPCHBI, SPCBII, or SPBWCF error list.

Note that the digital filter coefficients returned by this routine are in the exact format needed for use with the cascade-form filtering routine SPCFLT (Chapter 6).

A complete set of examples of SPIIRD designs is shown in Fig. 7.21. All cases have NS = 2 filter sections. All parameters are shown next to the corresponding example. Note carefully the use of frequencies F1–F4 in each case. Also note that some of the filter parameters are not always used and that the

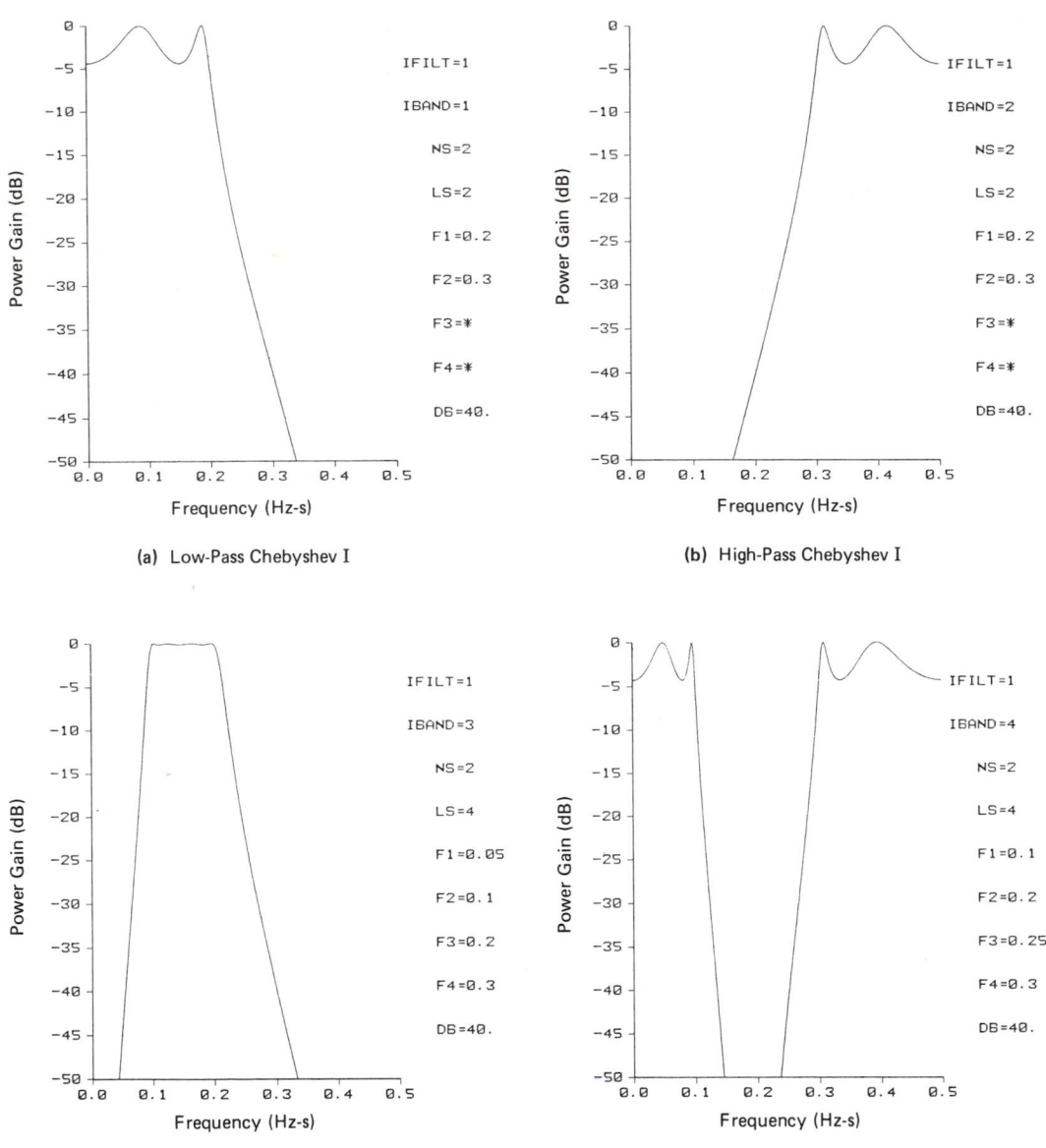

Figure 7.21(a) through (d).   Chebyshev Type I examples.

specifications given by the parameter values are always met or surpassed. For example, in the low-pass Chebyshev Type I design in Fig. 7.21(a), frequencies F3 and F4 are not used, the passband ends at F1 = 0.2 Hz-s, and the gain is down 40 dB at 0.3 Hz-s. An example where one of the specifications is surpassed is in Fig. 7.21(d) as well as the other bandstop filters, where the low cutoff

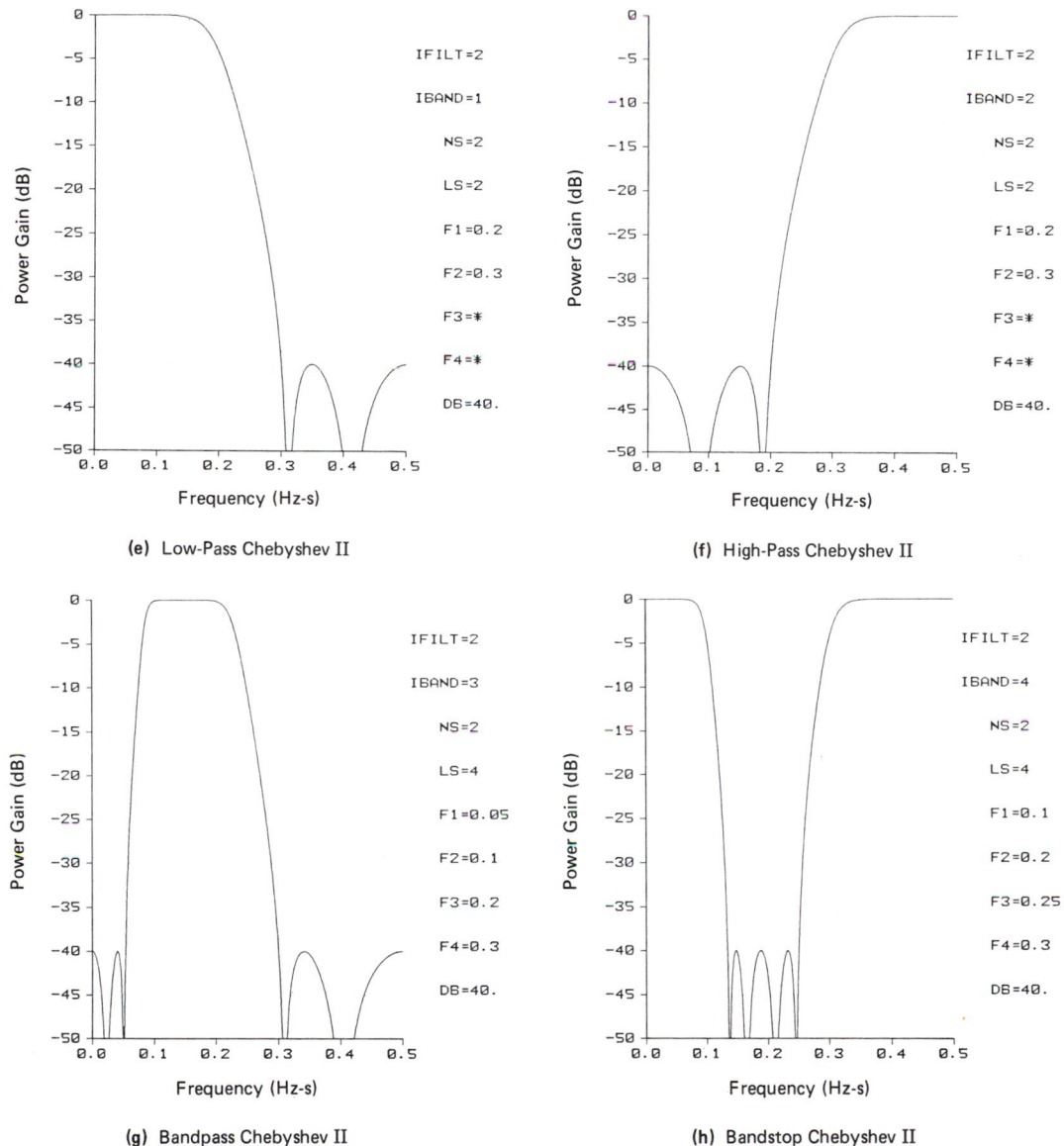

(e) Low-Pass Chebyshev II

(f) High-Pass Chebyshev II

(g) Bandpass Chebyshev II

(h) Bandstop Chebyshev II

Figure 7.21(e) through (h). Chebyshev Type II examples.

exceeds the specification. This type of result occurs with bandpass and bandstop filters in general. In order to meet one cutoff specification, the other may be exceeded.

The Chebyshev Type I design in Fig. 7.21(a)–(d) always has passband ripple. Depending upon the specifications, the ripple may be negligible as in Fig. 7.21(c). It can always be reduced by increasing NS, the number of filter sections.

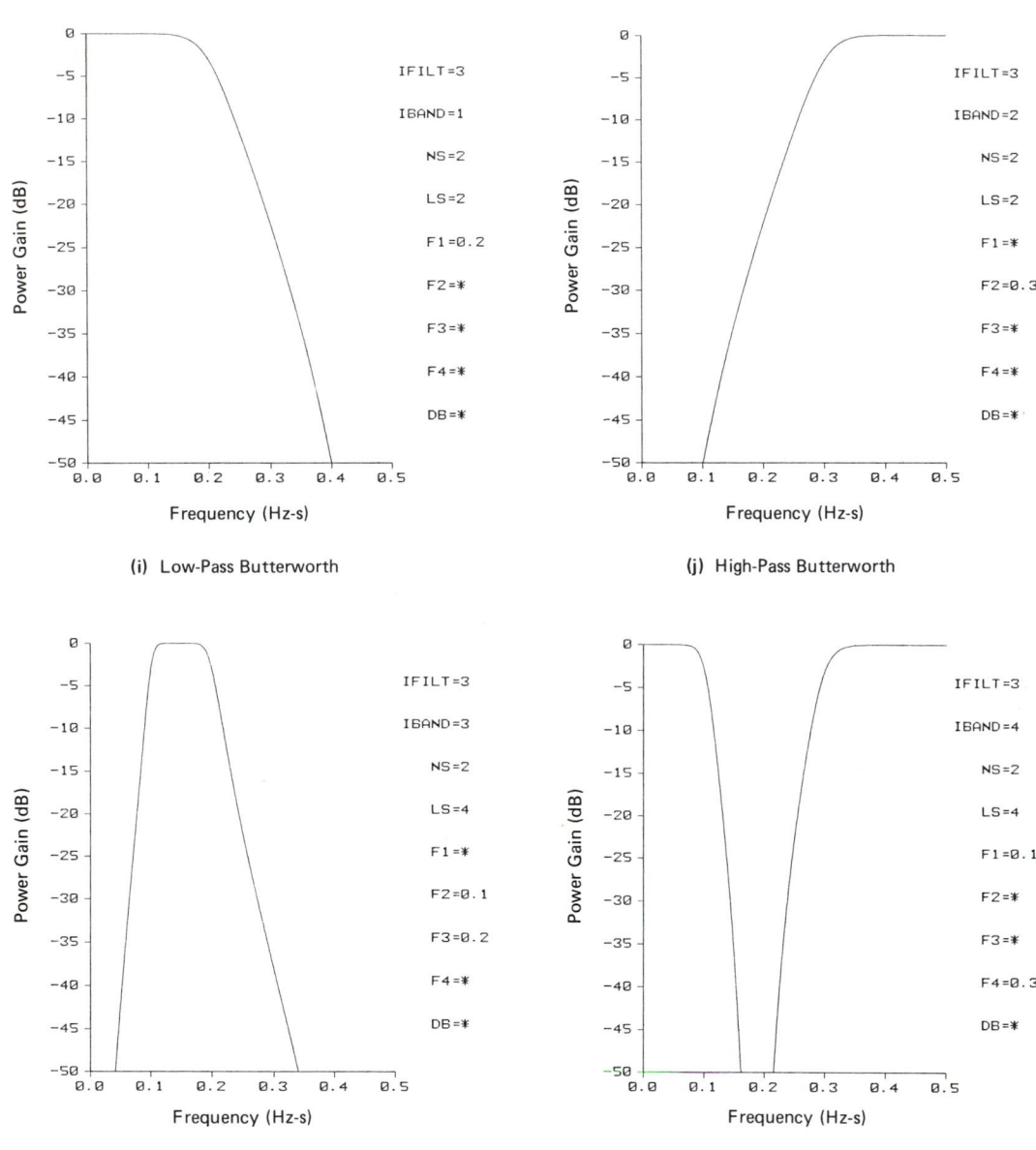

Figure 7.21(i) through (l). Butterworth examples.

The Chebyshev Type II design in Fig. 7.21(e)–(h) always has stopband ripple. In this design, increasing NS improves (increases) the gain near the end of the passband.

The Butterworth filters in Figs. 7.21(i)–(l) require only that the cutoff frequencies be specified. The power gain is always down 3 dB at the cutoff frequencies and drops in accordance with (7.36). With a given set of parameters, as illustrated in these examples, the Butterworth gain drops less steeply than the Chebyshev gain.

7.7 Summary of Calling Sequences

In this chapter we have described subroutines that enable IIR digital filter design via the conversion of normalized low-pass analog transfer functions into a specified digital $H(z)$. The calling sequences and argument lists for Chapter 7 software are summarized next.

CALL SPBILN(D,C,LN,B,A,WORK,IERROR)
CALL SPFBLT(D,C,LN,IBAND,FLN,FHN,B,A,WORK,IERROR)

Variable	Definition	Routines	
		SPBILN	SPFBLT
D(0:LN)	Analog filter $D(s)$	I	I
C(0:LN)	Analog filter $C(s)$	I	I
LN	Dimension of coefficient arrays	I	I
B(0:LN)	Digital filter $B(z)$	O	O
A(LN)	Digital filter $A(z)$	O	O
WORK(0:LN,0:LN)	Work array	X	X
IBAND	Digital filter type	—	I
	1—low-pass 2—high-pass		
	3—bandpass 4—bandstop		
FLN	Normalized low cutoff in Hz-s	—	I
FHN	Normalized high cutoff in Hz-s	—	I
IERROR	Error indicator	O	O

CALL SPBWCF(L,K,LN,D,C,IERROR)
CALL SPCHBI(L,K,LN,EP,D,C,IERROR)
CALL SPCBII(L,K,LN,SB,ATTN,D,C,IERROR)

Variable	Definition	Routines		
		SPBWCF	SPCHBI	SPCBII
L	Order of low-pass $H(s)$	I	I	I
K	Section number	I	I	I
LN	Dimension of coefficient arrays	I	I	I
D(0:LN)	Kth section $D(s)$	O	O	O
C(0:LN)	Kth section $C(s)$	O	O	O
EP	Passband ripple parameter	—	I	—
SB	Low edge of stopband	—	—	I
ATTN	Stopband attenuation	—	—	I
IERROR	Error indicator	O	O	O

CALL SPBSSL(L,PSCL,LN,D,C,IERROR)

Variable	Definition	Routine
		SPBSSL
L	Order of low-pass $H(s)$	I
PSCL	Prescaling parameter	I
LN	Dimension of coefficient arrays	I
D(0:LN)	Normalized low-pass $D(s)$ coefficients	O
C(0:LN)	Normalized low-pass $C(s)$ coefficients	O
IERROR	Error indicator	O

CALL SPIIRD(IFILT,IBAND,NS,LS,F1,F2,F3,F4,DB,B,A,IERROR)

Variable	Definition	Routine
		SPIIRD
IFILT	Digital filter type	I
	1—Chebyshev Type I	
	2—Chebyshev Type II	
	3—Butterworth	
IBAND	Digital filter characteristic	I
	1—Low-pass 2—High-pass	
	3—Bandpass 4—Bandstop	
NS	Number of sections	I
LS	First dimension of A and B arrays	I
F1–F4	Critical frequencies; see Fig. 7.21	I
DB	Stopband rejection; see Fig. 7.21	I
B(0:LS,NS)	$B(z)$ coefficients	O
A(LS,NS)	$A(z)$ coefficients	O
IERROR	Error indicator	O

In this summary, I designates a parameter that is an input to the subroutine, O denotes a value returned from the subroutine, and — indicates that the argument is not used by the routine. An X signifies work space that must be allocated but not necessarily initialized by the calling program.

7.8 Exercises

7.1. Use SPFBLT to convert the third-order transfer function

$$H(s) = \frac{1}{s^3 + 2s^2 + 2s + 1}$$

to a sixth-order bandpass filter with band edges at 1.5 kHz and 3.5 kHz and sample interval $T = 0.0001$ s. Plot the magnitude and phase-response curves.

7.2. Design a third-order lowpass Butterworth filter with cutoff frequency 1 kHz and sample interval $T = 0.0001$ s. Plot the magnitude response.

7.3. Design an eighth-order Butterworth bandstop filter with band edges at 1.5 kHz and 3.5 kHz and sample interval $T = 0.0001$ s. Plot the magnitude response.

7.4. Design a low-pass Butterworth filter with cutoff frequency 1 kHz, sample interval $T = 0.0001$ s, and a magnitude gain of less than 0.05 at 2 kHz. Plot the magnitude response.

7.5. Design a fourth-order low-pass Chebyshev Type I filter with cutoff frequency 1 kHz, sample interval $T = 0.0001$ s, and $\varepsilon = 0.20$. Plot the magnitude response.

7.6. Design an eighth-order bandpass Chebyshev Type II filter with band edges at 1.5 kHz and 3.5 kHz, sample interval $T = 0.0001$ s, and stopband characteristics determined by ATTN = 0.1 and SB = 2 rad/s. Plot the magnitude response.

7.7. Given the normalized low-pass analog transfer function,

$$H(s) = \frac{15}{s^3 + 6s^2 + 15s + 15}$$

design a digital low-pass filter with $f_L = 1.5$ kHz and $T = 0.0001$ s. Plot the magnitude and phase curves.

7.8. Design a twelfth-order bandstop Chebyshev Type II filter with band edges at 2 kHz and 3 kHz and sample interval $T = 0.0001$ s. Compare the magnitude and phase characteristics for stopband attenuation ATTN = 0.1 and ATTN = 0.05 at SB = 2 rad/s.

7.9. Compare the magnitude and phase characteristics of the Butterworth, Chebyshev Type I and Chebyshev Type II filters. Use fourth-order low-pass filters with cutoff $f_L = 2$ kHz and sample interval $T = 0.0001$ s. For the Chebyshev Type I filter, choose ε such that the magnitude response of the analog filter matches that of the analog Butterworth filter at the cutoff frequency. For the Type II filter, chose λ to match the Butterworth stopband attenuation at $f_s = 2.5$ kHz.

7.10. Given the normalized third-order transfer function in exercise 7.1, use SPBILN to convert $H(s)$ to the digital transfer function $H(z)$ (i.e., no prescaling is involved). Plot the magnitude function from 0 to 0.5 Hz. Explain the results in terms of the filter cutoff and its relationship to the mapping of the bilinear transformation. What is the implied sample interval T?

7.11. Design an eighth-order bandstop Bessel filter with a magnitude response of approximately 0.80 at band edges of 1.5 and 3.5 kHz relative to a 5-kHz Nyquist frequency. Plot the magnitude and phase (degrees) responses from 0 to 5 kHz.

7.12. Design a fourth-order low-pass Bessel filter whose magnitude response at the cutoff is equal to that of a fourth-order low-pass Butterworth filter. Use $f_L = 2$ kHz and $T = 0.0001$ s. Plot the magnitude and phase responses for both filters.

7.13. Design a second-order low-pass Chebyshev Type I filter with a passband power gain variation of 10%. Use a cutoff of 20 kHz relative to a Nyquist frequency of 50 kHz. Design a Type II filter using the same criterion, but adjust the attenuation parameters so that the response matches that of the Type I filter at twice the cutoff frequency. Plot the magnitude response of both filters. What is the approximate value of ε for the Type II filter?

7.14. Design a Chebyshev Type II filter that satisfies the following:
Stopband gain < 0.05 at 1.25 times the filter cutoff
Passband gain variation < 0.05
Low-pass filter with $f_L = 2$ kHz and $T = 0.0001$ s
Plot the magnitude and phase response of the resulting filter.

7.15. Design a Butterworth filter that satisfies the following:
Stopband gain $= 0.05$ at 1.25 times the filter cutoff
Lowpass filter with $f_L = 2$ kHz and Nyquist frequency 5 kHz
Plot the magnitude and phase response.

7.16. Design a Chebyshev Type II filter whose response matches that of the Butterworth filter in 7.15 at the cutoff and at 1.25 times the cutoff. Plot the magnitude and phase response.

7.17. Use the IIR design routine SPIIRD to design a low-pass Type II Chebyshev filter having two sections, passband from 0 to 0.2 Hz-s, stopband at 0.25 Hz-s, and 20 dB of stopband rejection. Plot the magnitude response from 0.0 to 0.5 Hz-s.

7.18. Generate the sequence $f(k) = \sin[2\pi(0.05)k] + \sin[2\pi(0.08)k]$ for $k = 0, 1 \ldots$, 299. Using SPIIRD, design a 2-section Butterworth bandstop filter with stopband from 0.06 to 0.10 Hz-s to ''notch out'' the second component of $f(k)$. Plot the power gain in decibels. Filter $f(k)$ using SPCFLT. Plot $f(k)$ before and after filtering.

7.19. Design a high-pass Butterworth filter using SPIIRD. The power gain should be down no more than 3 dB above 0.2 Hz-s and below -40 dB from 0 to 0.1 Hz-s. Plot the power gain against frequency. (*Hint:* Use equations (7.17) and (7.18) to select the number of sections (NS), with NS $\geq L/2$.)

7.9 References

[A] Ahmed, N., and T. Natarajan, *Discrete-Time Signals and Systems* (Reston, Va.: 1983).

[B] Oppenheim, Alan V., and Ronald W. Schafer, *Digital Signal Processing* (Englewood Cliffs, N.J.: Prentice-Hall, Inc., 1975).

[C] Rabiner, L. R., and B. Gold, *Theory and Application of Digital Signal Processing* (Englewood Cliffs, N.J.: Prentice-Hall, 1975).

[D] Stearns, S. D., *Digital Signal Analysis* (Rochelle Park, N.J.: Hayden Book Company, 1975).

[E] Mitra, S. K., and J. Fadavi-Ardekani, ''A New Approach to the Design of Cost-Optimal Low-Noise Digital Filters,'' *Trans. on ASSP,* ASSP-29, no. 6 (December 1981).

[F] Bolton, R., P. C. Craig, and L. C. Westphal, ''Computer-Aided Design of Recursive Digital Filters with Coefficients Having Restricted Minimal Representation,'' *IEEE Trans. on ASSP,* ASSP-29, no. 6 (December 1981).

[G] Liang, J., and R. DeFigueiredo, ''An Efficient Iterative Algorithm for Designing Optimal Recursive Digital Filters,'' *IEEE Trans. on ASSP,* ASSP-31, no. 5 (October 1983).

[H] Pusey, L. C., ''Application of Two-Channel Prediction Filtering to the Recursive Filter Design Problem,'' *IEEE Trans. on ASSP,* ASSP-31, no. 5 (October 1983).

[I] Gutknecht, M. H., J. O. Smith, and L. N. Trefethen, "The Caratheodory-Fejer Method for Recursive Digital Filter Design," *IEEE Trans. on ASSP,* ASSP-31, no. 6 (December 1983).

[J] Durani, T. S., and R. Chapman, "Optimal All-Pole Filter Design Based on Discrete Prolate Spheroidal Sequences," *IEEE Trans. on ASSP,* ASSP-32, no. 4 (August 1984).

[K] Lim, Y. C., "On the Synthesis of IIR Digital Filters Derived from Single Channel AR Lattice Network," *IEEE Trans. on ASSP,* ASSP-32, no. 4 (August 1984).

[L] Higgins, W. E., and D. C. Munson, "Noise Reduction Strategies for Digital Filters: Error Spectrum Shaping Versus the Optimal Linear State-Space Formulation," *IEEE Trans. on ASSP,* ASSP-30, no. 6 (December 1982).

[M] Mullis, C. T., and R. A. Roberts, "An Interpretation of Error Spectrum Shaping in Digital Filters," *IEEE Trans. on ASSP,* ASSP-30, no. 6 (December 1982).

[N] Kawamata, M., and T. Higuchi, "A Systematic Approach to Synthesis of Limit Cycle-Free Digital Filters," *IEEE Trans. on ASSP,* ASSP-31, no. 1 (February 1983).

[O] Gallagher, N. C., and T. A. Nodes, "Median Filters: Some Modifications and Their Properties," *IEEE Trans. on ASSP,* ASSP-30, no. 5 (October 1982).

[P] Gallagher, N. C., and R. A. Gonzalo, "State Description for the Root-Signal Set of Median Filters," *IEEE Trans. on ASSP,* ASSP-30, no. 6 (December 1982).

[Q] Gallagher, N. C., and G. L. Wise, "A Theoretical Analysis of the Properties of Median Filters," *IEEE Trans. on ASSP,* ASSP-29, no. 6 (December 1981).

[R] Schmid, C. E., "Design of IIR/FIR Filters Using a Frequency Domain Bootstrapping Technique and LPC Methods," *IEEE Trans. on ASSP,* ASSP-31, no. 4 (August 1983).

[S] Thajchayapong, P., and F. Cheevasuvit, "A Maximally Flat Group Delay Recursive Digital Filter with Improved Passband Magnitude Response," *Proc. of the IEEE,* 67, no. 12 (December 1979).

[T] Thajchayapong, P., P. Karnchanawadee, and F. Cheevasuvit, "A Recursive Digital Filter with Simultaneous Maximally Flat Magnitude and Group Delay at an Arbitrary Specified Frequency," *Proc. of the IEEE,* 67, no. 5 (May 1979).

[U] Hazra, S. N., "Linear Phase IIR Filter with Equiripple Stopband," *IEEE Trans. on ASSP,* ASSP-31, no. 3 (June 1983).

chapter 8

FIR Filter Design

8.1 Introduction

In this chapter we describe simple routines that enable the design of finite impulse response (FIR) digital filters. Our goal is to generate the $[b_n]$ coefficients, which may then be used to implement the filter whose input-output relationship is described by

$$y_k = \sum_{n=0}^{L} b_n x_{k-n} \tag{8.1}$$

A variety of techniques for FIR filter design appear in the literature [A–F]. The method described in this chapter is probably the most straightforward, since the filter coefficients are found by evaluating simple, closed-form expressions. The filters generated are, however, suboptimal in the sense that it is possible to design a lower-order FIR filter that has equally good frequency response characteristics [F]. Optimal FIR filter design typically requires the use of iterative design procedures and is, therefore, beyond the scope of this text [C, D, E].

From the discussion in Chapter 6 we know that the $[b_n]$ coefficients in (8.1) are the elements of the impulse response of the FIR filter. The related transfer function can therefore be written as

$$H(z) = B(z) = b_0 + b_1 z^{-1} + \cdots + b_L z^{-L} = \sum_{n=0}^{L} h_n z^{-n} \tag{8.2}$$

where $[h_n]$ defines the impulse response of the filter. This indicates that an FIR filter may be designed by specifying the finite sequence $[h_n]$. This approach is utilized in this chapter.

Two constraints will be imposed upon the design procedures described later. The first is that of *realizability*, or *causality*. Equivalently stated, this implies that the impulse response, which is by definition the system response to a unit sample input at time $k = 0$, is equal to zero for $k < 0$. Note that this condition is satisfied by the forms of (8.1) and (8.2).

The second restriction is that the filter must have a linear phase response. Linear phase is a desirable characteristic in any filter, digital or analog. In terms of the transfer function, this implies that $H(z) = H(z^{-1})$ [D]. In Chapter 7 we showed that in causal IIR filter design, perfectly linear phase response is not achievable, since the resulting filter would have poles outside the unit circle and would therefore be unstable. Since zeros outside the unit circle do not affect stability, the condition $H(z) = H(z^{-1})$ can be satisfied in the FIR case by requiring zeros to exist in mirror-image pairs, for example, z_1, $1/z_1$, inside and outside the unit circle [C].

In FIR filter design, an equivalent condition for linear phase response is symmetry in the coefficients such that

$$b_n = b_{L-n}, \qquad 0 \leqslant n \leqslant \text{integer}\left(\frac{L}{2}\right) \tag{8.3}$$

If L is even, the total number of coefficients, which is by definition the length of the filter, $L + 1$, is odd. In this case, there is a central sample $b_{L/2}$ about which the coefficients are symmetric. If L is odd, the upper limit on n in (8.3) is actually $(L - 1)/2$, and there is no central sample corresponding to the midpoint of the filter. The condition imposed by (8.3) results in a fixed delay of $L/2$ samples or $(L/2)T$ seconds, where T is the sample interval [C]. The corresponding phase response is given by $\theta(\omega) = -\omega(L/2)$, where ω is the normalized frequency is radians. It is apparent that this phase response is linear, that is, varies linearly with respect to frequency.

As in the design of IIR filters in Chapter 7, we first consider the design of low-pass filters and then extend the techniques developed to include high-pass, bandpass, and bandstop filter design. We begin our discussion of FIR filter design with a description of the ideal rectangular low-pass filter and its impulse response.

8.2 FIR Design via the Ideal Low-Pass Filter

The filter serving as a basis for this presentation is ideal in the sense that the magnitude response is equal to unity in the passband and zero elsewhere, and the phase response is a linear function of frequency. This can be expressed in terms of the Fourier transform as

$$H_d(e^{j\omega}) = \begin{cases} e^{-j\omega L/2} & 0 \leqslant |\omega| \leqslant \omega_c \\ 0 & \omega_c < |\omega| \leqslant \pi \end{cases} \tag{8.4}$$

where ω_c defines the normalized filter cutoff frequency in radians and $L/2$ corresponds to the delay required to satisfy the causality constraint. Note that $L/2$ is not necessarily an integer.

The ideal impulse response $h_d(n)$, which corresponds to the filter specified by (8.4), may be obtained by evaluating the inverse Fourier transform

$$h_d(n) = \frac{1}{2\pi} \int_{-\pi}^{\pi} H_d(e^{j\omega}) e^{j\omega n} \, d\omega \tag{8.5}$$

to obtain

$$h_d(n) = \frac{\sin[\omega_c \, (n - L/2)]}{\pi(n - L/2)} \tag{8.6}$$

As defined, the impulse response in (8.6) is an infinitely long sequence. To accomplish FIR filter design, we truncate the response and define

$$h(n) = \begin{cases} h_d(n) & 0 \le n \le L \\ 0 & \text{elsewhere} \end{cases} \tag{8.7}$$

Since the function in (8.6) is symmetrical about $L/2$, the condition of symmetry in the coefficients, (8.3), is satisfied and a linear phase response is achieved.

The plot in Fig. 8.1 illustrates the shape of the impulse response defined by (8.6) and (8.7) for $L = 100$ and normalized cutoff frequency $\omega_c = 0.4\pi$. Note the symmetry about the midpoint at $L/2 = 50$. The related FIR filter is defined by equating the b_n coefficients to the $h_d(n)$ samples for $0 \le n \le L$.

Assuming that a cutoff frequency in hertz, f_c, and a sample interval in seconds, T, are determined by the filtering task to be accomplished, the values in (8.6) may be computed by defining the normalized cutoff frequency $\omega_c = 2\pi f_c T$. The filter length, $L + 1$, is the only parameter remaining to be specified.

Since we are approximating the infinitely long sequence $h_d(n)$ with the truncated sequence $h(n)$ in (8.7), it is reasonable to assume that the approximation improves with increasing L. Thus, we would expect the filter response to improve in a corresponding fashion. Unfortunately, abrupt truncation of the infinite Fourier series in (8.6) results in an oscillatory effect known as the Gibbs phenomenon [A]. It has been shown that the maximum magnitude of this oscillation is relatively constant regardless of the length of the truncated sequence [D].

The Gibbs phenomenon is illustrated by the plots in Fig. 8.2. These curves

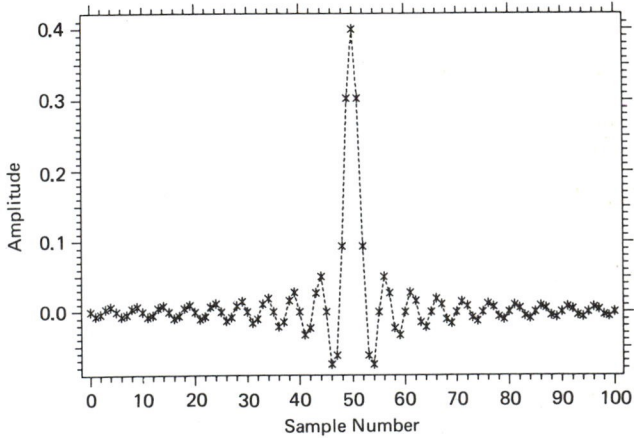

Figure 8.1. FIR impulse response for low-pass filter with $L = 100$ and normalized cutoff frequency $\omega_c = 0.4\pi$.

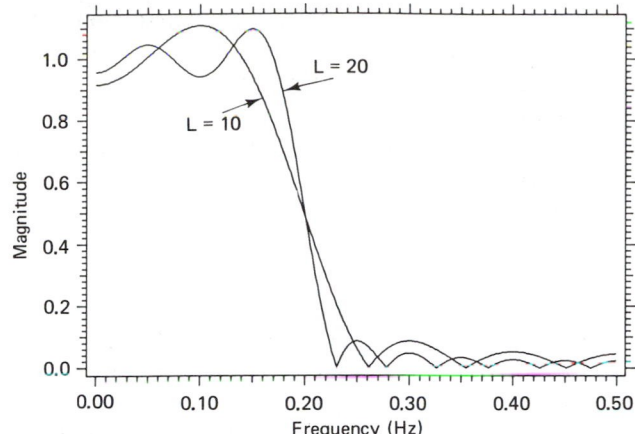

Figure 8.2. Magnitude response of low-pass FIR filter ($T = 1$ s).

correspond to the magnitude response functions of FIR filters of lengths 11 and 21, that is, $L = 10$ and $L = 20$. In both cases, the cutoff frequency was defined to be $f_c = 0.2$ Hz and the sample interval was $T = 1.0$ s. Note that the peak magnitude of the oscillation in both the passband and the stopband is approximately the same, although the order of the filters differs by a factor of 2.

Increasing the length of an FIR filter designed via (8.6) and (8.7) has four primary effects: (1) reduction in width of the transition band; (2) increase in frequency of the oscillatory response; (3) increased delay; (4) increased computational burden in the filter implementation. While a very large L will yield a filter with sharp roll-off characteristics, it may not be an acceptable design due to the oscillatory magnitude response.

Since the Gibbs oscillations result from the abrupt truncation of the infinite series, a variety of techniques is used to alleviate this problem. In the next section we describe several window functions that may be used to truncate $h_d(n)$ in (8.6) to obtain the FIR sequence. We demonstrate that whereas the magnitude response of the resulting filter is improved in the sense that it is less oscillatory, the penalty incurred is an increase in the width of the transition band.

8.3 Time-Domain Window Functions

In this section we provide a brief introduction to the topic of time-domain window functions and their use in FIR filter design. For a more complete discussion see Chapter 14, which describes routines for generating and applying windows.

Truncation of the infinite sequence as in (8.7) could also be accomplished by defining the product of two sequences

$$h(n) = h_d(n)w_R(n) \tag{8.8}$$

where $w_R(n)$ is the rectangular window function

$$w_R(n) = \begin{cases} 1 & 0 \leq n \leq L \\ 0 & \text{elsewhere} \end{cases} \tag{8.9}$$

Here we consider the use of alternate window functions that reduce the magnitude of the oscillatory response evident in Fig. 8.2.

Since the Gibbs oscillations result from the abrupt truncation described by (8.8), we might guess that the magnitude response of the filter could be improved by using a nonrectangular window with tapered ends. In order to retain the linear phase characteristic, we know that the window must also be symmetric about its midpoint. All data windows are by definition equal to zero for n outside the interval $0 \leqslant n \leqslant L$.

In addition to the rectangular window in (8.9), we consider five nonrectangular windows that may be used to truncate the ideal impulse response to obtain $h(n)$. Since the time domain characteristics of these windows are described in Chapter 14, we restrict the scope of this discussion to the frequency domain characteristics and their effects on FIR filter design.

Multiplication of two data sequences in the time domain is equivalent to convolving their transforms in the frequency domain [A]. (See Chapter 9 for details regarding convolution.) Thus the frequency response of the FIR filter that results from truncation of the ideal impulse response in (8.6) via a data window is determined by convolving the rectangular response in (8.4) with the transform of the window. The magnitude spectrum of a desirable window has two key features: (1) a narrow main lobe, since this affects the width of the transition band of the FIR filter; and (2) sidelobes that rapidly decrease in energy for increasing ω, since this results in smaller ripples in both the passband and the stopband of the filter.

In Figs. 8.3 through 8.7, we provide plots illustrating the shape of the spectral magnitude of each of the five alternative windows compared to that of the rectangular window defined in (8.9). The data was normalized such that the peak of the rectangular window magnitude spectrum corresponded to 0 dB and the minimum was clipped at -100 dB to allow direct comparison between the plots. Use of the decibel scale enhances the illustration of the varying sidelobe characteristics.

It is apparent that whereas the tapered window functions have better frequency characteristics in terms of low sidelobe levels, the price paid is an increase in the main-lobe width. This translates directly to an increase in the transition

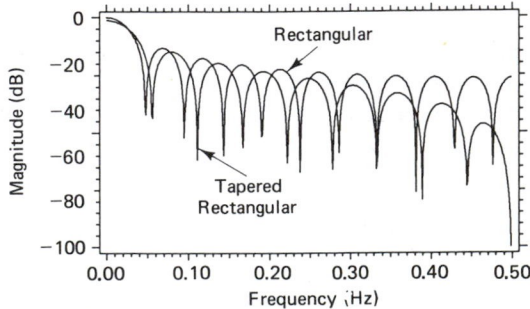

Figure 8.3. Frequency-domain characteristics of rectangular and tapered rectangular data windows.

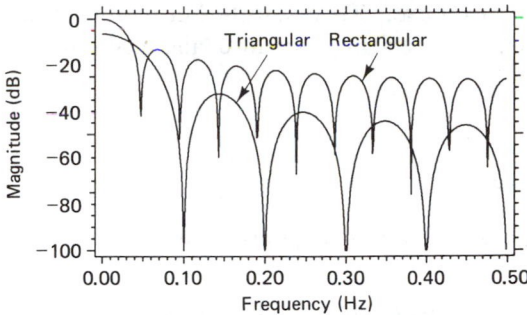

Figure 8.4. Frequency-domain characteristics of rectangular and triangular data windows.

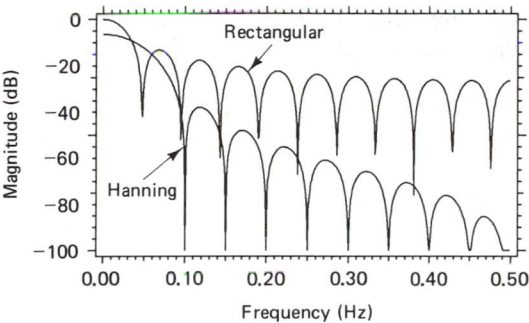

Figure 8.5. Frequency-domain characteristics of rectangular and Hanning data windows.

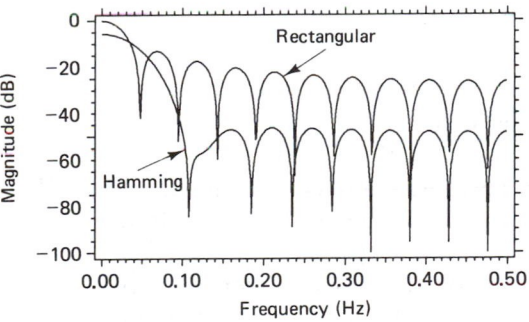

Figure 8.6. Frequency-domain characteristics of rectangular and Hamming data windows.

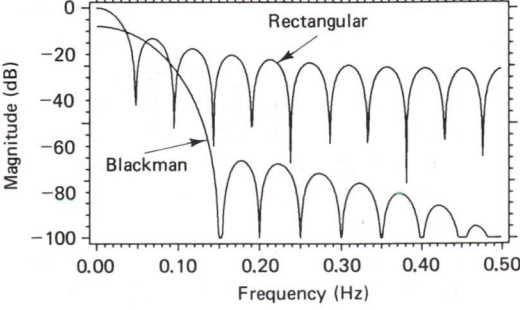

Figure 8.7. Frequency-domain characteristics of rectangular and Blackman data windows.

bandwidth of the FIR filter response. To illustrate better the effects of these windows, we now formalize the design procedure and provide a set of test programs.

8.4 Low-Pass FIR Filter-Design Routine

We now generalize the lowpass FIR filter design procedure to include the window function $w(n)$:

$$b_n = h(n) = h_d(n)w(n), \qquad 0 \leq n \leq L \qquad (8.10)$$

The $[b_n]$ coefficients are used to implement the difference equation in (8.1), $h_d(n)$ is the ideal impulse-response sequence in (8.6), and $w(n)$ is any window function. The magnitude response of the resulting filter will depend upon the shape of the window $w(n)$.

As in previous chapters, we begin by describing a general FIR low-pass filter-design subroutine SPFIRL and its parameter list. The calling sequence is

CALL SPFIRL(L,FCN,IWNDO,B,IERROR)

L	= last index of B(0:L). (Length of filter = L + 1.)
FCN	= normalized cutoff frequency in hertz-seconds
IWNDO	= data window in (8.10):

        1—rectangular   2—tapered rectangular
        3—triangular   4—Hanning
        5—Hamming   6—Blackman

B(0:L)	= digital filter coefficients—returned
IERROR	= 0 no errors detected
	1 invalid filter length (L $\leq$ 0)
	2 invalid IWNDO parameter (IWNDO < 1 or IWNDO > 6)
	3 invalid cutoff frequency (FCN $\leq$ 0.0 or FCN $\leq$ 0.5)

This routine designs a normalized FIR low-pass filter; thus, the cutoff frequency, FCN, must be between 0.0 and 0.5 Hz-s. This value is computed as FCN = f_cT, where f_c is the filter cutoff in hertz and T is the sample interval in seconds.

In the six examples that follow, only the IWNDO parameter is altered, thereby enabling direct comparison of the effects of the window selection. In each case, the target filter has a cutoff at f_c = 200 Hz and a sample interval of T = 0.001 s. Thus, the normalized cutoff frequency is FCN = 0.2 Hz-s. The filter length is specified by L = 20. The SPGAIN function from Chapter 5 is used to compute the magnitude response of the resulting digital filters. Note that this was accomplished by defining the denominator coefficient A(1) = 0.0 with LA = 1.

In test program SPA0801, the rectangular window was used to truncate the infinite impulse response of the ideal low-pass filter. The program listing is provided, together with the resulting filter coefficients and a plot of the magnitude response. Note the symmetry in the coefficients, which achieves linear phase response. The oscillatory effect due to the Gibbs phenomenon is evident in the magnitude response plot in Fig. 8.8.

```
              PROGRAM SPA0801
       C-DESIGN FIR LOWPASS FILTER USING RECTANGULAR WINDOW
              DIMENSION B(0:20),A(1),AMP(0:500)
              COMPLEX SPGAIN
              DATA FC/200./,T/.001/,L/20/,IWNDO/1/,A/0./
              FCN=FC*T
              CALL SPFIRL(L,FCN,IWNDO,B,IERROR)
              IF(IERROR.NE.0)THEN
                PRINT *,' ERROR RETURNED = ',IERROR
                STOP
              ENDIF
              DO 1 I=0,500
                AMP(I)=ABS(SPGAIN(B,A,L,1,I*.5/500.))
            1 CONTINUE
              CALL MCPLOT(0.,1.,AMP,501,0,0,1,1,3,2,5.,3.6)
              PRINT 100,(N,B(N),N+7,B(N+7),N+14,B(N+14),N=0,6)
        100   FORMAT(' N       B(N)         N       B(N)         N       B(N)',
            +         /,(I2,F10.5,I8,F10.5,I8,F10.5))
              END


       $ RUN SPA0801
       N       B(N)         N       B(N)         N       B(N)
       0     0.00000        7    -0.06237       14    -0.07568
       1    -0.03364        8     0.09355       15     0.00000
       2    -0.02339        9     0.30273       16     0.05046
       3     0.02673       10     0.40000       17     0.02673
       4     0.05046       11     0.30273       18    -0.02339
       5     0.00000       12     0.09355       19    -0.03364
       6    -0.07568       13    -0.06237       20     0.00000
       $
```

In the next example, SPA0802, the tapered rectangular window is used to truncate the ideal impulse-response sequence. The only program change required is to define IWNDO = 2 in the call to SPFIRL. A program listing and the corresponding results are provided. Note in Fig. 8.9 that use of the tapered

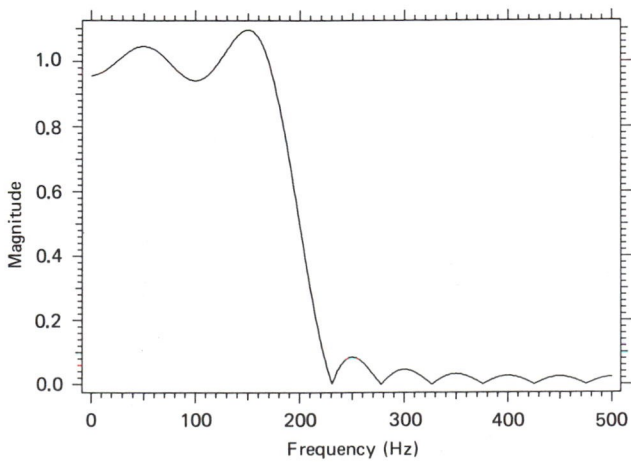

Figure 8.8. SPA0801: Low-pass FIR filter designed via the rectangular window; $f_c = 200$ Hz; $T = 0.001$ s.

window results in a slight amplitude reduction in the oscillatory response and that the frequency roll-off characteristics are essentially maintained.

```
                   PROGRAM SPA0802
           C-DESIGN FIR LOWPASS FILTER USING TAPERED RECTANGULAR WINDOW
                   DIMENSION B(0:20),A(1),AMP(0:500)
                   COMPLEX SPGAIN
                   DATA FC/200./,T/.001/,L/20/,IWNDO/2/,A/0./
                   FCN=FC*T
                   CALL SPFIRL(L,FCN,IWNDO,B,IERROR)
                   IF(IERROR.NE.0)THEN
                     PRINT *,' ERROR RETURNED = ',IERROR
                     STOP
                   ENDIF
                   DO 1 I=0,500
                     AMP(I)=ABS(SPGAIN(B,A,L,1,I*.5/500.))
                 1 CONTINUE
                   CALL MCPLOT(0.,1.,AMP,501,0,0,1,1,3,2,5.,3.6)
                   PRINT 100,(N,B(N),N+7,B(N+7),N+14,B(N+14),N=0,6)
               100 FORMAT(' N       B(N)         N       B(N)        N       B(N)',
                 +      /,(I2,F10.5,I8,F10.5,I7,F10.5))
                   END


               $ RUN SPA0802
               N      B(N)       N      B(N)       N      B(N)
               0    0.00000      7  -0.06237     14  -0.07568
               1   -0.01682      8   0.09355     15   0.00000
               2   -0.02339      9   0.30273     16   0.05046
               3    0.02673     10   0.40000     17   0.02673
               4    0.05046     11   0.30273     18  -0.02339
               5    0.00000     12   0.09355     19  -0.01682
               6   -0.07568     13  -0.06237     20   0.00000
               $
```

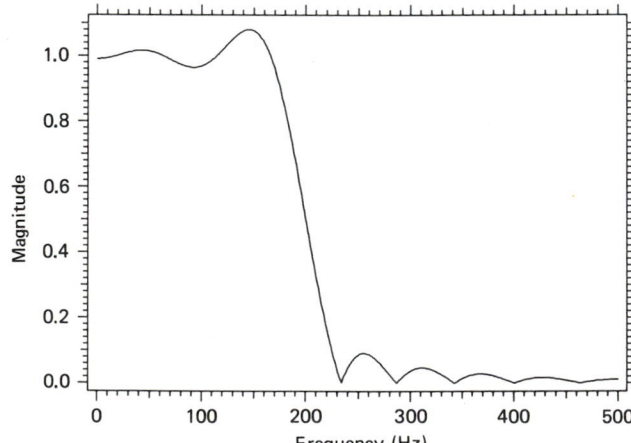

Figure 8.9. SPA0802: Low-pass FIR filter designed via the tapered rectangular window; $f_C = 200$ Hz; $T = 0.001$ s.

Program SPA0803 specifies that the triangular window is to be used to truncate $h_d(n)$. It is apparent from the curve in Fig. 8.10 that this window degrades the filter magnitude response somewhat. While the oscillatory response is nearly eliminated, the passband gain is reduced and the stopband gain is increased.

```
            PROGRAM SPA0803
C-DESIGN FIR LOWPASS FILTER USING TRIANGULAR WINDOW
            DIMENSION B(0:20),A(1),AMP(0:500)
            COMPLEX SPGAIN
            DATA FC/200./,T/.001/,L/20/,IWNDO/3/,A/0./
            FCN=FC*T
            CALL SPFIRL(L,FCN,IWNDO,B,IERROR)
            IF(IERROR.NE.0)THEN
              PRINT *,' ERROR RETURNED = ',IERROR
              STOP
            ENDIF
            DO 1 I=0,500
              AMP(I)=ABS(SPGAIN(B,A,L,1,I*.5/500.))
        1   CONTINUE
            CALL MCPLOT(0.,1.,AMP,501,0,0,1,1,3,2,5.,3.6)
            PRINT 100,(N,B(N),N+7,B(N+7),N+14,B(N+14),N=0,6)
      100   FORMAT(' N        B(N)        N       B(N)       N      B(N)',
        +            /,(I2,F10.5,I8,F10.5,I7,F10.5))
            END


        $ RUN SPA0803
        N        B(N)        N       B(N)       N      B(N)
        0     0.00000     7   -0.04366    14   -0.04541
        1    -0.00336     8    0.07484    15    0.00000
        2    -0.00468     9    0.27246    16    0.02018
        3     0.00802    10    0.40000    17    0.00802
        4     0.02018    11    0.27246    18   -0.00468
        5     0.00000    12    0.07484    19   -0.00336
        6    -0.04541    13   -0.04366    20    0.00000
        $
```

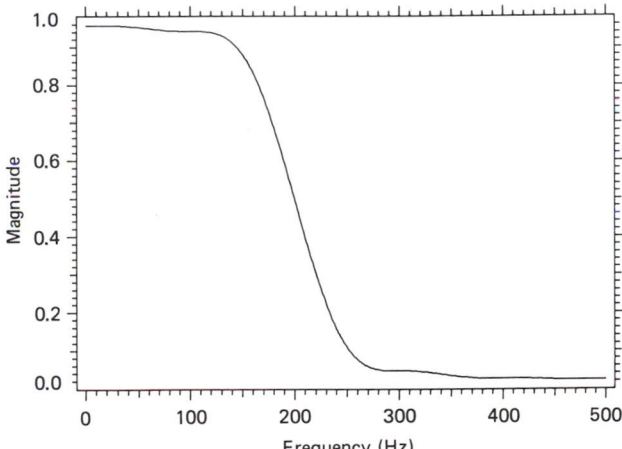

Figure 8.10. SPA0803: Low-pass FIR filter designed via the triangular window; f_c = 200 Hz; T = 0.001 s.

The Hanning window was used in SPA0804. Once again the only program change required is in the initialization of the IWNDO parameter. In the magnitude response plot in Fig. 8.11, we see that this window clearly attenuates the Gibbs oscillations, but the transition bandwidth is approximately double that of the filter designed using the rectangular window. This could be predicted from the frequency response plots in Fig. 8.5, since the width of the main lobe of the Hanning window response is twice that of the main lobe of the rectangular window response.

```
                    PROGRAM SPA0804
        C-DESIGN FIR LOWPASS FILTER USING HANNING WINDOW
                    DIMENSION B(0:20),A(1),AMP(0:500)
                    COMPLEX SPGAIN
                    DATA FC/200./,T/.001/,L/20/,IWNDO/4/,A/0./
                    FCN=FC*T
                    CALL SPFIRL(L,FCN,IWNDO,B,IERROR)
                    IF(IERROR.NE.0)THEN
                      PRINT *,' ERROR RETURNED = ',IERROR
                      STOP
                    ENDIF
                    DO 1 I=0,500
                      AMP(I)=ABS(SPGAIN(B,A,L,1,I*.5/500.))
               1    CONTINUE
                    CALL MCPLOT(0.,1.,AMP,501,0,0,1,1,3,2,5.,3.6)
                    PRINT 100,(N,B(N),N+7,B(N+7),N+14,B(N+14),N=0,6)
             100    FORMAT(' N      B(N)          N      B(N)         N      B(N)',
               +             /,(I2,F10.5,I8,F10.5,I7,F10.5))
                    END

               $ RUN SPA0804
               N      B(N)          N      B(N)         N      B(N)
               0    0.00000        7   -0.04951       14  -0.04953
               1   -0.00082        8    0.08462       15   0.00000
               2   -0.00223        9    0.29532       16   0.01743
               3    0.00551       10    0.40000       17   0.00551
               4    0.01743       11    0.29532       18  -0.00223
               5    0.00000       12    0.08462       19  -0.00082
               6   -0.04953       13   -0.04951       20   0.00000
               $
```

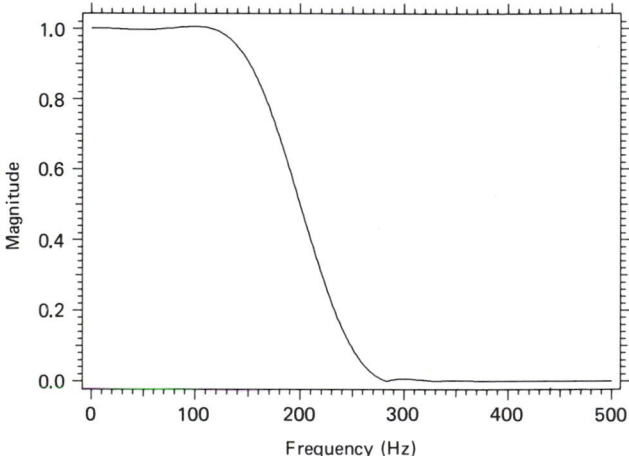

Figure 8.11. SPA0804: Low-pass FIR filter designed via the Hanning window; $f_C = 200$ Hz; $T = 0.001$ s.

In program SPA0805 the Hamming window was used to truncate the ideal impulse response. The response of this filter, shown in Fig. 8.12, is very similar to the one obtained via the Hanning window (Fig. 8.11). Both the passband and the stopband are slightly flatter, but the transition bandwidth is nearly identical in this case.

```
          PROGRAM SPA0805
C-DESIGN FIR LOWPASS FILTER USING HAMMING WINDOW
          DIMENSION B(0:20),A(1),AMP(0:500)
          COMPLEX SPGAIN
          DATA FC/200./,T/.001/,L/20/,IWNDO/5/,A/0./
          FCN=FC*T
          CALL SPFIRL(L,FCN,IWNDO,B,IERROR)
          IF(IERROR.NE.0)THEN
            PRINT *,' ERROR RETURNED = ',IERROR
            STOP
          ENDIF
          DO 1 I=0,500
            AMP(I)=ABS(SPGAIN(B,A,L,1,I*.5/500.))
      1   CONTINUE
          CALL MCPLOT(0.,1.,AMP,501,0,0,1,1,3,2,5.,3.6)
          PRINT 100,(N,B(N),N+7,B(N+7),N+14,B(N+14),N=0,6)
    100   FORMAT(' N      B(N)          N      B(N)          N      B(N)',
         +        /,(I2,F10.5,I8,F10.5,I7,F10.5))
          END
```

```
    $ RUN SPA0805
    N      B(N)         N      B(N)         N      B(N)
    0    0.00000        7   -0.05054       14   -0.05163
    1   -0.00345        8    0.08533       15    0.00000
    2   -0.00393        9    0.29592       16    0.02007
    3    0.00721       10    0.40000       17    0.00721
    4    0.02007       11    0.29592       18   -0.00393
    5    0.00000       12    0.08533       19   -0.00345
    6   -0.05163       13   -0.05054       20    0.00000
    $
```

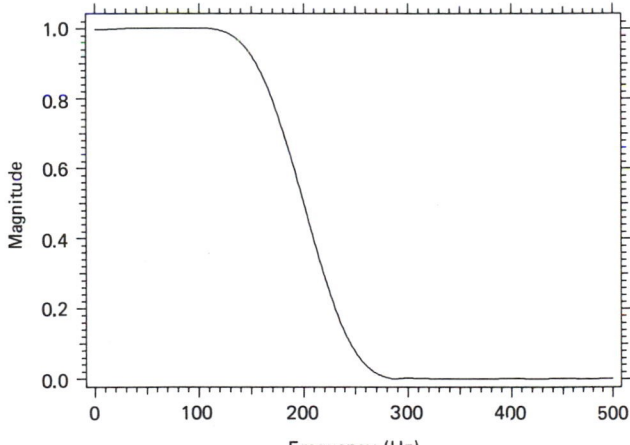

Figure 8.12. SPA0805: Low-pass FIR filter designed via the Hamming window; $f_c = 200$ Hz; $T = 0.001$ s.

In the final example, SPA0806, the Blackman window is used to truncate the impulse response. The resulting magnitude response, illustrated in Fig. 8.13, is similar in shape to that obtained via the Hamming window in Fig. 8.12, but it is clear that the transition band is wider in this case.

```
                    PROGRAM SPA0806
      C-DESIGN FIR LOWPASS FILTER USING BLACKMAN WINDOW
                    DIMENSION B(0:20),A(1),AMP(0:500)
                    COMPLEX SPGAIN
                    DATA FC/200./,T/.001/,L/20/,IWNDO/6/,A/0./
                    FCN=FC*T
                    CALL SPFIRL(L,FCN,IWNDO,B,IERROR)
                    IF(IERROR.NE.0)THEN
                      PRINT *,' ERROR RETURNED = ',IERROR
                      STOP
                    ENDIF
                    DO 1 I=0,500
                      AMP(I)=ABS(SPGAIN(B,A,L,1,I*.5/500.))
            1       CONTINUE
                    CALL MCPLOT(0.,1.,AMP,501,0,0,1,1,3,2,5.,3.6)
                    PRINT 100,(N,B(N),N+7,B(N+7),N+14,B(N+14),N=0,6)
         100        FORMAT(' N        B(N)        N       B(N)        N          B(N)',
              +          /,(I2,F10.5,I8,F10.5,I7,F10.5))
                    END


            $ RUN SPA0806
            N      B(N)        N      B(N)        N      B(N)
            0    0.00000       7   -0.04298      14   -0.03858
            1   -0.00031       8    0.07944      15    0.00000
            2   -0.00094       9    0.29070      16    0.01013
            3    0.00271      10    0.40000      17    0.00271
            4    0.01013      11    0.29070      18   -0.00094
            5    0.00000      12    0.07944      19   -0.00031
            6   -0.03858      13   -0.04298      20    0.00000
            $
```

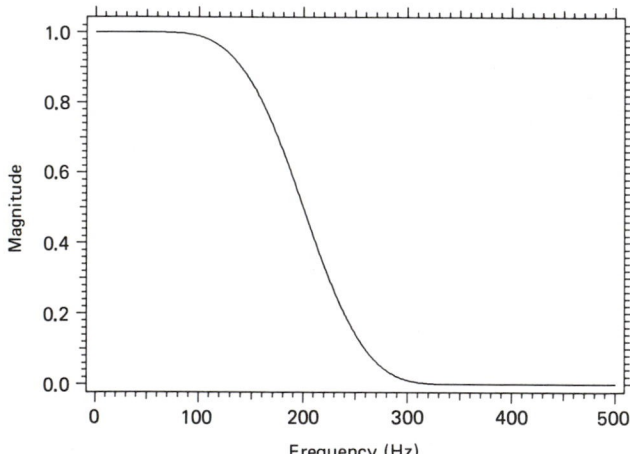

Figure 8.13. SPA0806: Low-pass FIR filter designed via the Blackman window; $f_C = 200$ Hz; $T = 0.001$ s.

In this section we have provided a general subroutine SPFIRL, which enables low-pass FIR filter design. We now extend the procedure to encompass high-pass, bandpass, and bandstop FIR filter design.

8.5 Generalized FIR Digital Filter Design

The design procedure presented in Section 8.2 is easily extended to enable the design of other types of FIR filters. In each case, we begin by defining an ideal magnitude response function, $H_d(e^{j\omega})$. The FIR filter is then designed by evaluating the inverse Fourier transform to obtain the ideal (infinite) impulse-response sequence, $h_d(n)$, and windowing that response to obtain $h(n)$ or, equivalently, the $[b_n]$ filter coefficients.

As an example, we first consider FIR high-pass filters. In terms of the normalized ideal filter, we define

$$H_d(e^{j\omega}) = \begin{cases} e^{-j\omega L/2}, & \omega_c \le |\omega| \le \pi \\ 0, & 0 \le |\omega| < \omega_c \end{cases} \tag{8.11}$$

By evaluating the inverse Fourier transform in (8.5), we obtain the ideal impulse-response sequence

$$h_d(n) = \frac{\sin[\pi(n - L/2)]}{\pi(n - L/2)} - \frac{\sin[\omega_c(n - L/2)]}{\pi(n - L/2)} \tag{8.12}$$

where ω_c is again the normalized cutoff frequency (sampling frequency $= 2\pi$) and $L/2$ is the delay required for causality.

In a similar manner, we can define an ideal bandpass filter response as

$$H_d(e^{j\omega}) = \begin{cases} e^{-j\omega L/2}, & \omega_L \le |\omega| \le \omega_H \\ 0, & \text{elsewhere} \end{cases} \tag{8.13}$$

where ω_L and ω_H define the low and high edges of the passband and $L/2$ is once again the delay. When the response in (8.13) is inverse-transformed, we find that the ideal bandpass impulse response is given by

$$h_d(n) = \frac{\sin[\omega_H(n - L/2)]}{\pi(n - L/2)} - \frac{\sin[\omega_L(n - L/2)]}{\pi(n - L/2)} \tag{8.14}$$

This response could be interpreted as the difference between two low-pass filter functions, one with cutoff at ω_L and the other with cutoff at ω_H. The result is a filter whose passband extends from ω_L to ω_H.

The final category to be considered is the bandstop filter. In this case we define the ideal frequency response as

$$H_d(e^{j\omega}) = \begin{cases} e^{-j\omega L/2}, & |\omega| \le \omega_L, |\omega| \ge \omega_H \\ 0, & \text{elsewhere} \end{cases} \tag{8.15}$$

where ω_L and ω_H now specify the low and high edges of the stopband. Taking the inverse Fourier transform of (8.15) yields

$$h_d(n) = \frac{\sin[\omega_L(n - L/2)]}{\pi(n - L/2)} + \frac{\sin[\pi(n - L/2)]}{\pi(n - L/2)} - \frac{\sin[\omega_H(n - L/2)]}{\pi(n - L/2)} \tag{8.16}$$

By grouping the last two terms in (8.16) and comparing the form to (8.12), we see that we have a high-pass filter whose passband begins at ω_H. The total response is therefore due to the combination of this high-pass characteristic with the low-pass response defined by the first term. Thus, the stopband extends from ω_L to ω_H, as desired.

We can now design a high-pass, bandpass, or bandstop FIR filter by simply windowing the appropriate ideal impulse response from (8.12), (8.14), or (8.16).

Thus, the design procedure is identical to that described previously for low-pass FIR filter design.

To facilitate the design process, we provide a subroutine SPFIRD, which is basically a generalization of SPFIRL. The calling sequence is

CALL SPFIRD(L,IBAND,FLN,FHN,IWNDO,B,IERROR)

L   = last index of B(0:L). (Length of filter = L + 1.)
IBAND   = type of filter
         1—low-pass   2—high-pass
         3—bandpass   4—bandstop
FLN   = normalized cutoff frequency in Hz-s (low)
FHN   = normalized cutoff frequency in Hz-s (high: bandpass, bandstop)
IWNDO   = data window
         1—rectangular   2—tapered rectangular
         3—triangular   4—Hanning
         5—Hamming   6—Blackman
B(0:L)   = digital filter coefficients—returned
IERROR   = 0 no errors detected
         1 invalid filter length ($L \leq 0$)
         2 invalid window specification (IWNDO < 1 or > 6)
         3 invalid filter type (IBAND < 1 or > 4)
         4 invalid normalized cutoff frequency FLN or FHN

The IERROR = 4 return indicates that either one of the normalized cutoff frequencies is outside the range 0.0 < FLN, FHN < 0.5, or that FLN $\geq$ FHN. Parameter FHN is used only for bandpass and bandstop filter design.

Note that there is some redundancy between SPFIRL and SPFIRD, since either routine may be used to design low-pass filters. Three simple examples are now provided to demonstrate the use of SPFIRD.

The task accomplished by test program SPA0807 is the design of a high-pass filter with a cutoff frequency of 2.5 kHz and a sample interval of 0.0001 s; thus the normalized filter cutoff is FLN = 0.25 Hz-s. The length of the FIR filter is specified by L = 20 and the rectangular window is used to truncate the ideal impulse-response function. A plot of the resulting magnitude response is provided in Fig. 8.14. Note that this curve exhibits the same oscillatory characteristics as the low-pass filter designed via the rectangular window (Fig. 8.8).

A bandpass filter is designed in program SPA0808. In this case the Hamming window is used to modify and truncate the ideal impulse response. The band edges are specified to be at 1.5 and 3.5 kHz and the Nyquist frequency is 5 kHz; thus FLN = 0.15 and FHN = 0.35 Hz-s. The filter length is again specified by L = 20. Since only the initialization portion of the program has been changed from SPA0807, only that part is listed. From the magnitude response illustrated in Fig. 8.15, we see that the design objectives have been achieved.

```
            PROGRAM SPA0807
C-DESIGN FIR HIGHPASS FILTER USING RECTANGULAR WINDOW
            DIMENSION B(0:20),A(1),AMP(0:500)
            COMPLEX SPGAIN
            DATA FL/2500./,T/.0001/,L/20/,IWNDO/1/,A/0./,IBAND/2/
            FLN=FL*T
            CALL SPFIRD(L,IBAND,FLN,0.,IWNDO,B,IERROR)
            IF(IERROR.NE.0)THEN
              PRINT *,' ERROR RETURNED = ',IERROR
              STOP
            ENDIF
            DO 1 I=0,500
              AMP(I)=ABS(SPGAIN(B,A,L,1,I*.5/500.))
       1    CONTINUE
            CALL MCPLOT(0.,.01,AMP,501,0,0,1,1,4,2,5.,3.6)
            PRINT 100,(N,B(N),N+7,B(N+7),N+14,B(N+14),N=0,6)
     100    FORMAT(' N     B(N)        N      B(N)        N      B(N)',/,
         +         (I2,F10.5,I7,F10.5,I7,F10.5))
            END

         $ RUN SPA0807
         N      B(N)        N      B(N)        N      B(N)
         0    0.00000       7    0.10610      14    0.00000
         1   -0.03537       8    0.00000      15   -0.06366
         2    0.00000       9   -0.31831      16    0.00000
         3    0.04547      10    0.50000      17    0.04547
         4    0.00000      11   -0.31831      18    0.00000
         5   -0.06366      12    0.00000      19   -0.03537
         6    0.00000      13    0.10610      20    0.00000
         $
```

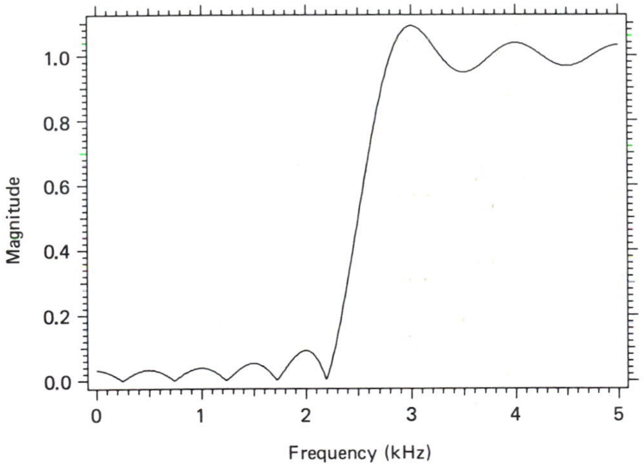

Figure 8.14. SPA0807: High-pass FIR filter designed via the rectangular window; $f_L = 2.5$ kHz; $T = 0.0001$ s.

```
            PROGRAM SPA0808
C-DESIGN FIR BANDPASS FILTER USING HAMMING WINDOW
            DIMENSION B(0:20),A(1),AMP(0:500)
            COMPLEX SPGAIN
            DATA FL/1500./,T/.0001/,L/20/,IWNDO/5/,A/0./,IBAND/3/
            DATA FH/3500./
            FLN=FL*T
            FHN=FH*T
            CALL SPFIRD(L,IBAND,FLN,FHN,IWNDO,B,IERROR)
            IF(IERROR.NE.0)THEN
              PRINT *,' ERROR RETURNED = ',IERROR
              STOP
            ENDIF
            DO 1 I=0,500
              AMP(I)=ABS(SPGAIN(B,A,L,1,I*.5/500.))
       1    CONTINUE
            CALL MCPLOT(0.,.01,AMP,501,0,0,1,1,4,2,5.,3.6)
            PRINT 100,(N,B(N),N+7,B(N+7),N+14,B(N+14),N=0,6)
     100    FORMAT(' N     B(N)        N      B(N)        N      B(N)',/,
         +         (I2,F10.5,I7,F10.5,I7,F10.5))
            END
```

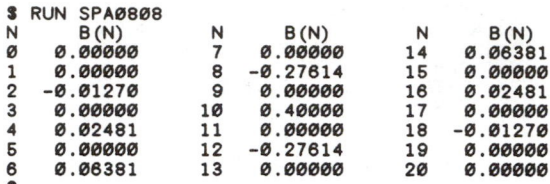

```
$ RUN SPA0808
N      B(N)       N      B(N)       N      B(N)
0    0.00000      7    0.00000     14    0.06381
1    0.00000      8   -0.27614     15    0.00000
2   -0.01270      9    0.00000     16    0.02481
3    0.00000     10    0.40000     17    0.00000
4    0.02481     11    0.00000     18   -0.01270
5    0.00000     12   -0.27614     19    0.00000
6    0.06381     13    0.00000     20    0.00000
$
```

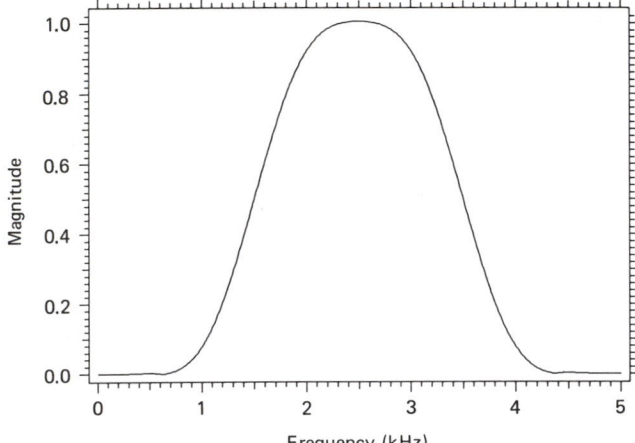

Figure 8.15. SPA0808: Bandpass FIR filter designed via the Hamming window; f_L = 1.5 kHz; f_H = 3.5 kHz; T = 0.0001 s.

The Hanning window was used to design a bandstop filter in SPA0809. In this example, the band edges are again at 1.5 and 3.5 kHz relative to a Nyquist frequency of 5 kHz. The magnitude response shown in Fig. 8.16 again satisfies the filter specifications.

```
        PROGRAM SPA0809
C-DESIGN FIR BANDSTOP FILTER USING HANNING WINDOW
        DIMENSION B(0:20),A(1),AMP(0:500)
        COMPLEX SPGAIN
        DATA FL/1500./,T/.0001/,L/20/,IWNDO/4/,A/0./,IBAND/4/
        DATA FH/3500./
        FLN=FL*T
        FHN=FH*T
        CALL SPFIRD(L,IBAND,FLN,FHN,IWNDO,B,IERROR)
        IF(IERROR.NE.0)THEN
          PRINT *,' ERROR RETURNED = ',IERROR
          STOP
        ENDIF
        DO 1 I=0,500
          AMP(I)=ABS(SPGAIN(B,A,L,1,I*.5/500.))
   1    CONTINUE
        CALL MCPLOT(0.,.01,AMP,501,0,0,1,1,4,2,5.,3.6)
        PRINT 100,(N,B(N),N+7,B(N+7),N+14,B(N+14),N=0,6)
 100    FORMAT(' N       B(N)       N      B(N)       N      B(N)',/,
     +          (I2,F10.5,I7,F10.5,I7,F10.5))
        END

$ RUN SPA0809
N      B(N)       N      B(N)       N      B(N)
0    0.00000      7    0.00000     14   -0.06123
1    0.00000      8    0.27382     15    0.00000
2    0.00723      9    0.00000     16   -0.02155
3    0.00000     10    0.60000     17    0.00000
4   -0.02155     11    0.00000     18    0.00723
5    0.00000     12    0.27382     19    0.00000
6   -0.06123     13    0.00000     20    0.00000
$
```

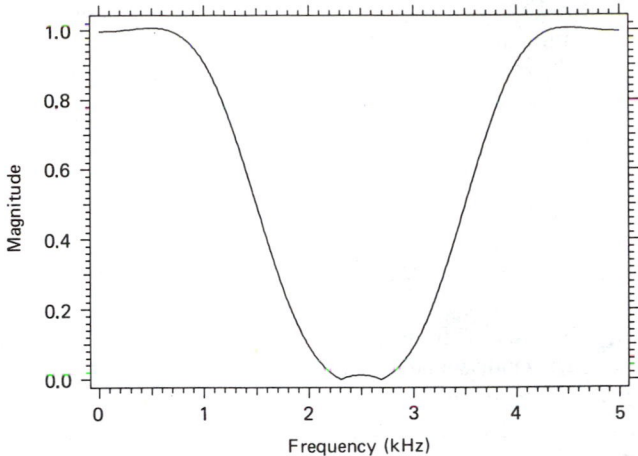

Figure 8.16. SPA0809: Bandstop FIR filter designed via the Hanning window; $f_L = 1.5$ kHz; $f_H = 3.5$ kHz; $T = 0.0001$ s.

The examples in this section demonstrate the ease with which any type of FIR filter can be designed. It is again apparent that the response characteristics vary depending upon the shape of the window used to shape and truncate the ideal infinite impulse response. Selection of this window as well as specification of the filter order, sample interval, and cutoff characteristics depend upon the specific application for which the filter is being designed.

8.6 Summary of Calling Sequences

Two subroutines that enable the design of FIR digital filters have been introduced in this chapter. Their parameter lists and calling sequences are summarized as follows.

CALL SPFIRL(L,FCN,IWNDO,B,IERROR)
CALL SPFIRD(L,IBAND,FLN,FHN,IWNDO,B,IERROR)

Variable	Definition	Routines	
		SPFIRL	SPFIRD
L	Last index of B(0:L). (Length of filter = L + 1.)	I	I
FCN	Normalized cutoff in Hz-s	I	—
IWNDO	Data window	I	I
	1—rectangular 2—tapered rectangular		
	3—triangular 4—Hanning		
	5—Hamming 6—Blackman		
B(0:L)	Digital filter coefficients	O	O
IBAND	Filter type	—	I
	1—low-pass 2—high-pass		
	3—bandpass 4—bandstop		
FLN	Normalized cutoff (low)	—	I
FHN	Normalized cutoff (high: bandpass, bandstop)	—	I
IERROR	Return error indicator	O	O

In this summary, I indicates that the variable is an input to the routine, O denotes an output from the routine, and — means that the variable is not used by the subroutine.

8.7 Exercises

8.1. Design a low-pass filter with cutoff frequency at 10 kHz relative to a Nyquist frequency of 40 kHz. Use the rectangular window to truncate the impulse response and a filter of length specified by $L = 40$. Plot the magnitude and phase responses from 0 to the Nyquist frequency.

8.2. A high-pass filter is specified by a cutoff frequency of 0.25 Hz and sample interval equal to 1 s. Use the Hamming window and compare the magnitude responses of filters designed with $L = 10$, $L = 20$ and $L = 40$.

8.3. Design a bandpass filter with band edges at 200 and 300 Hz and a Nyquist frequency of 500 Hz. Use the Hanning window and set $L = 49$. Plot the magnitude response as a function of frequency. Examine the filter coefficients to determine that there is no sample that corresponds to the midpoint of the impulse response.

8.4. Given an input data sequence $x_k = \sin[2\pi(100)kT] + 0.2r_k$, where the sample interval T is 0.001 s and r_k is a zero mean white noise sequence with unit power density, design an FIR filter to reduce the noise in $[x_k]$. Use the Hamming window and set $L = 100$. The filter bandwidth is to be 50 Hz. Plot the filter magnitude response from 0 to the Nyquist frequency. Filter a 1000-point segment of $[x_k]$ using the routines in Chapter 6. Plot the first 200 samples of the input and output data sequences.

8.5. Design a digital filter that could be used to eliminate 60-Hz interference from a signal that has been sampled such that the Nyquist frequency is 100 Hz. Assume that the input signal has no information content beyond 50 Hz. Choose the filter type and defend your selection. Choose a window and filter order such that the passband magnitude gain is 1 ± 0.1 and the gain at 60 Hz is less than 0.05. Plot the resulting magnitude characteristics.

8.6. Bandstop filter specifications are given by: lower band edge at half the Nyquist frequency and upper band edge at three-quarters of the Nyquist frequency with sample interval equal to 0.1 s. The filter order is $L = 30$ and the triangular window is to be used. Plot the resulting magnitude response characteristics from 0 to the Nyquist frequency.

8.7. A recorded signal contains low-frequency sinusoidal interference at 0.1 times the Nyquist frequency. Design a normalized high-pass filter to eliminate this interference (i.e., assume the sample interval is 1 s). Assume the filter order is $L = 200$ and the Hamming window is to be used. Also assume that the desired signal contains frequencies no lower than 0.2 times the Nyquist frequency and choose the filter cutoff to preserve the passband response while attenuating the interference. Plot the resulting magnitude-response characteristics from 0 to the Nyquist frequency.

8.8. Work exercise 8.4 using the tapered rectangular window.

8.9. Work exercise 8.2 using the tapered rectangular window.

8.10. Work exercise 8.6 using the Hanning window.

8.11. Work exercise 8.1 using the triangular window.

8.12. Work exercise 8.2 using the triangular window.

8.13. Work exercise 8.6 using the Hamming window.

8.14. Work exercise 8.7 with $L = 100$ and use the tapered rectangular window.

8.15. Work exercise 8.4 with $L = 100$.

8.16. Design a bandpass filter with the passband extending from 2 to 4 kHz relative to a Nyquist frequency of 6 kHz. Use $L = 100$ and the Hanning window. Plot the magnitude response.

8.17. Work exercise 8.16 using the tapered rectangular window.

8.18. Design a bandstop filter with the stopband extending from 2.5 to 3.5 kHz and a Nyquist frequency of 6 kHz. Use $L = 200$ and the Blackman window. Plot the magnitude response.

8.8 References

[A] Stearns, S. D., *Digital Signal Analysis* (Rochelle Park, N.J.: Hayden, 1975).

[B] Ahmed, N., and T. Natarajan, *Discrete-Time Signals and Systems* (Englewood Cliffs, N.J.: Prentice-Hall, Inc., 1983).

[C] Rabiner, L. R., and B. Gold, *Theory and Application of Digital Signal Processing* (Englewood Cliffs, N.J.: Prentice-Hall, Inc., 1975).

[D] Oppenheim, Alan V., and Ronald W. Schafer, *Digital Signal Processing* (Englewood Cliffs, N.J.: Prentice-Hall, Inc., 1975).

[E] Antoniou, A., *Digital Filters: Analysis and Design* (New York: McGraw-Hill, 1979).

[F] Crochiere, R. E., and L. R. Rabiner, *Multirate Digital Signal Processing* (Englewood Cliffs, N.J.: Prentice-Hall, Inc., 1983).

chapter 9

Fast Convolution
and Correlation

9.1 Introduction

Convolution and correlation are closely related operations that are basic to many areas of digital signal processing. As discussed previously in this text as well as other texts on digital filters [A, B], the convolution of two time series is equivalent to the product of the transforms of the two time series. Thus, as indicated in Fig. 9.1, the output of a linear system may be found either as a product of transforms, $[Y_m] = [H_m X_m]$, or as the convolution of the input time series $[x_k]$ with the impulse response $[h_k]$, denoted in the figure by $[y_k] = \text{conv } \{[x_k], [h_k]\}$. As seen in equation (6.11), convolution may be described roughly as the integral of the product of two data sequences, with one of the sequences reversed and shifted relative to the other.

The correlation function of two data sequences is the same as the convolution without the reversal of one of the sequences. It may therefore be described as the integral of the product of two sequences with one shifted relative to the other or as the average product of the two sequences with one sequence shifted. The correlation function is used as a measure of how well two sequences are correlated, or agree with one another, or are like one another.

We next describe convolution and correlation in more concrete terms, along with some standard notation that simplifies the discussion in this chapter.

9.2 Notation and Basic Theory

In this chapter we assume that the two time series to be either convolved or correlated each have M samples and can be designated as follows:

$$[x_k] = [x_0 \, x_1 \, x_2 \cdots x_{M-1}]$$
$$[y_k] = [y_0 \, y_1 \, y_2 \cdots y_{M-1}]$$

(9.1)

This notation excludes the use of negative indexes such as those found in convolutions associated with noncausal finite impulse response (FIR) filters; never-

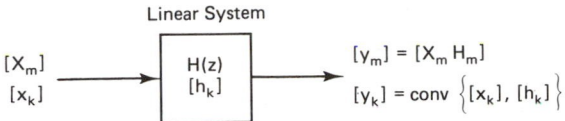

Linear System

$[X_m]$ → H(z), $[h_k]$ → $[Y_m] = [X_m H_m]$

$[y_k] = \text{conv}\left\{[x_k], [h_k]\right\}$

Figure 9.1. The output of a linear system may be found as a product of transforms or as a convolution of time series.

theless it is more convenient in this chapter, as it was also in Chapter 8. The noncausal FIR impulse response can always be described using the form of (9.1) by imposing a suitable delay as described in Section 5.6. The notation also implies that both sequences have the same length, which may not be the case and indeed is rarely the case with FIR filtering, where the data sequence is usually much longer than the impulse response. When the sequences are of different lengths, we use the convention of adding zeros to the right end of the shorter sequence. For example, if an FIR filter with impulse response $[h_k]$ and length 4 were processing an input signal $[x_k]$ of length $M = 9$, we would have

$$[x_k] = [x_0 \ x_1 \ x_2 \ x_3 \ x_4 \ x_5 \ x_6 \ x_7 \ x_8]$$
$$[y_k] = [h_0 \ h_1 \ h_2 \ h_3 \ 0 \ 0 \ 0 \ 0 \ 0] \tag{9.2}$$

The identification of x with one sequence and y with the other in this discussion is arbitrary.

Having defined the data sequences for use in this chapter, we can now define the convolution and correlation functions in more precise terms and show how these two functions are found by taking products of DFTs. For these purposes we will assume that, unless otherwise specified, all data sequences are extended with zeros both to the left and to the right. For later use, we also define the *periodic extension* $[\overline{x}_k]$ of any sequence $[x_k]$ to be the sequence in (9.1) repeated infinitely in both directions, left and right. Thus, with $M = 4$ for example, we might have the following:

$$\text{Zero extension: } [x_k] = [. \ . \ . \ 0 \ 0 \ 0 \ a \ b \ c \ d \ 0 \ 0 \ 0 \ . \ . \ .] \tag{9.3}$$

$$\text{Periodic extension: } [\overline{x}_k] = [. \ . \ . \ b \ c \ d \ a \ b \ c \ d \ a \ b \ c \ . \ . \ .] \tag{9.4}$$

The convolution function of two sequences $[x_k]$ and $[y_k]$ is obtained as just described, that is, just as in the FIR filtering applications in Chapters 6 and 8. We think of the convolution function as another data sequence with M samples and define it by

$$c_{xy}(n) = \text{conv}\{[x_k], [y_k]\}$$
$$\triangleq \sum_{k=0}^{M-1} x_k y_{n-k}, \qquad n = 0, 1, ..., M - 1 \tag{9.5}$$

The correlation function of $[x_k]$ and $[y_k]$ is similar to (9.5). It is designated by $r_{xy}(n)$ and is defined by

$$r_{xy}(n) = \text{corr}\{[x_k], [y_k]\}$$
$$\triangleq \sum_{k=0}^{M-1} x_k y_{k+n}; \qquad n = 0, 1, ..., M - 1 \tag{9.6}$$

When $[x_k]$ and $[y_k]$ are different sequences, $r_{xy}(n)$ is the cross-correlation function. When $[x_k]$ and $[y_k]$ are the same sequence, $r_{xx}(n)$ is the autocorrelation function of $[x_k]$.

These functions, $c_{xy}(n)$ and $r_{xy}(n)$, are sometimes defined differently in the literature. We note particularly that the correlation function $r_{xy}(n)$ is not defined here to be an estimate of the overall mean lagged product $E[x_k y_{k+n}]$. Instead, since each sum in (9.6) has $M - n$ nonzero products because of the zero extension in (9.3), the best unbiased estimate of $E[x_k y_{k+n}]$ is

$$\phi_{xy}(n) = \text{estimated mean lagged product}$$

$$= \frac{r_{xy}(n)}{M - n}, \qquad n = 0, 1, ..., M - 1 \tag{9.7}$$

Thus, having obtained $[\phi_{xy}(n)]$, we can obtain the estimated mean lagged products in a simple manner.

In the remainder of this discussion we use *zero-padded* versions of the time series in (9.1). Zero-padding consists of extending the sequence length from M in (9.1) to N, where N is usually a power of 2, by appending zeros to the sequence, as follows:

$$[x_k] = \underbrace{[x_0 x_1 \cdots x_{M-1} \quad \overbrace{0\ 0 \cdots 0}^{N - M \text{ zeros}}]}_{N \text{ samples}} \tag{9.8}$$

The rules for zero and periodic extension in (9.3) and (9.4) still apply to the N-sample zero-padded sequence. Zero-padding is used to lengthen the original sequence for purposes described later, and also to increase the sequence length to a power of 2 for Fast Fourier Transform (FFT) purposes. Note that when zero extension is used, the limits in (9.5) and (9.6) may be changed from $M - 1$ to $N - 1$ without altering the results, but that (9.7) is still correct as it stands. In what follows, we use N to designate the length, generally a power of 2, of a sequence that has been padded with $N - M$ zeros. (We also allow the case where $N = M$ and there are no extra zeros.)

With the functions defined in (9.5) and (9.6) with M changed to N, the basis for fast convolution and correlation is in the following relationships:

$$[X_m Y_m] = \text{DFT}\,[c_{x\bar{y}}(n)] \tag{9.9}$$

$$[X_m^* Y_m] = \text{DFT}\,[r_{x\bar{y}}(n)] \tag{9.10}$$

(As before, the asterisk denotes the complex conjugate.) Thus the products of the Discrete Fourier Transform (DFT) components of $[x_k]$ and $[y_k]$ form the DFT of $[c_{x\bar{y}}(n)]$, and the products of the components of the conjugate of the DFT of $[x_k]$ times the components of the DFT of $[y_k]$ form the DFT of $[r_{x\bar{y}}(n)]$. We recall from (9.4) that $[\bar{y}_k]$ is the periodic extension of $[y_k]$ so that $c_{x\bar{y}}$ and $r_{x\bar{y}}$ differ from c_{xy} and r_{xy}, but before addressing this problem, we first show that (9.9) and (9.10) are correct. Taking the inverse DFT of (9.9) in accordance with equation (3.10), we have

$$\text{DFT}^{-1}[X_m Y_m] = \frac{1}{N} \sum_{m=0}^{N-1} X_m Y_m e^{j(2\pi mn/N)}, \qquad n = 0, 1, \ldots, N-1$$

We now substitute the forward DFT in (3.5) for X_m and Y_m and manipulate the result:

$$\text{DFT}^{-1}[X_m Y_m] = \frac{1}{N} \sum_{m=0}^{N-1} \sum_{k=0}^{N-1} \sum_{i=0}^{N-1} x_k y_i \, e^{j[2\pi m(n-k-i)/N]}$$

$$= \frac{1}{N} \sum_{k=0}^{N-1} x_k \sum_{i=0}^{N-1} y_i \sum_{m=0}^{N-1} e^{j[2\pi m(n-k-i)/N]} \qquad (9.11)$$

$$= \sum_{k=0}^{N-1} x_k \bar{y}_{n-k} = c_{x\bar{y}}(n)$$

The second line in (9.11) follows from the first by interchanging the order of summation. The final sum in this line is N wherever $i = n - k \pm$ (multiple of N) and zero otherwise. Therefore, the only nonzero term in the second sum is the term where either $i = n - k$ or $i = n - k + N$. Thus the result follows in the third line, with the periodic extension of $[y_k]$ being necessary to include terms where $n - k$ is negative. Thus we have shown the validity of (9.9). The validity of (9.10) can be shown similarly by using the conjugate X_m^* in place of X_m and also changing $-k$ to $+k$ in (9.11). See exercise 9.19.

If we are to make use of (9.9) and (9.10) to convolve and correlate time series, we must find a way, using zero-padding, to make $c_{x\bar{y}}(n)$ in (9.11) equal to $c_{xy}(n)$ in (9.5); the same is true for $r_{x\bar{y}}(n)$ and $r_{xy}(n)$. Let us examine this problem first for the correlation functions $r_{x\bar{y}}(n)$ and $r_{xy}(n)$, since these functions are easier to illustrate.

Suppose we have the sequences $[x_k]$ and $[y_k]$ shown in Fig. 9.2(a). As just discussed, we extend these sequences with zeros until they are (1) of equal length and (2) a power of 2 in length. In this illustration, the resulting length is 8 as in Fig. 9.2(a). Then, Fig. 9.2(b) shows how these sequences are aligned for computing $r_{xy}(5)$ on the left and $r_{x\bar{y}}(5)$ on the right. These two correlation functions differ because, on the right, the term $x_3 y_8 = x_3 y_0$ is included due to the periodicity of the $\bar{y}_k$ with period $= 8$. That is, in Fig. 9.2(b),

$$\text{For } N = 8: \quad r_{xy}(5) = x_0 y_5 \qquad (9.12)$$

$$r_{x\bar{y}}(5) = x_0 y_5 + x_3 y_0 \neq r_{xy}(5)$$

Therefore, the inverse DFT of $[X_m^* Y_m]$ in (9.9) would not give the correct correlation function $r_{xy}(5)$ with $N = 8$.

To make use of (9.10) in the computation of $[r_{xy}(n)]$, we raise N to the next power of 2, that is, make $N = 16$ by appending zeros to $[x_k]$ and $[y_k]$. Then, as shown in Fig. 9.2(c), the correlation of $[x_k]$ with $[\bar{y}_k]$ gives the correct result, that is,

$$\text{For } N = 16: \quad r_{xy}(5) = x_0 y_5 = r_{x\bar{y}}(5) \qquad (9.13)$$

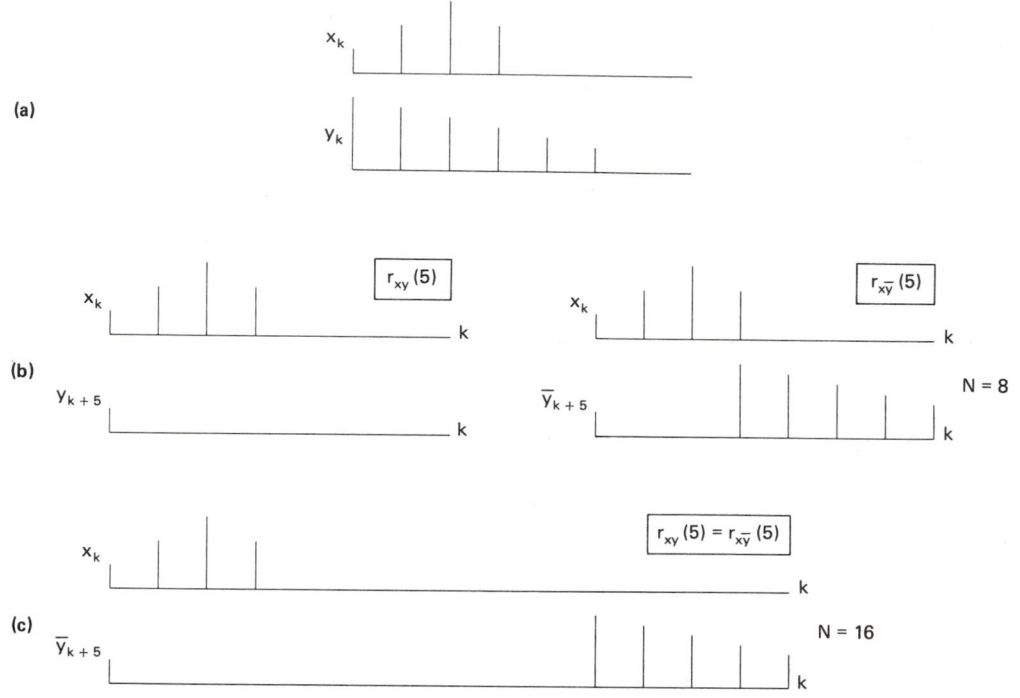

Figure 9.2. Alignment of $[x_k]$ and $[y_k]$ for correlation with $n = 5$. (a) Original sequences; $N = 8$. (b) Alignment for $r_{xy}(5)$ and $[r_{x\bar{y}}(5)]$ with $N = 8$. (c) Alignment for $r_{xy}(5) = r_{x\bar{y}}(5)$ with $N = 16$.

Thus, to make $[r_{xy}(n)] = [r_{x\bar{y}}(n)]$, we must pad the original data sequences with enough zeros so that the circular correlation caused by the periodicity of $[\bar{y}_k]$ does not involve samples beyond the first period of $[\bar{y}_k]$. Generalizing from Fig. 9.2, we see that (assuming $y_0 \neq 0$) when $[y_k]$ is shifted n places to the left, we must have n zeros on the right end of $[x_k]$ in order to prevent an overlap. More generally, if y_K is the first nonzero sample in $[y_k]$, we must have $n - K$ zeros on the right end of $[x_k]$. Thus, if $n$max is the largest shift of interest, we have the following rule:

> To compute $[r_{xy}(n)]$ for $0 \leq n \leq n\text{max} \leq N - 1$ using the inverse DFT of $[X_m^* Y_m]$, the sequence lengths (N) should be equal and $[x_k]$ should end with at least $n\text{max} - K$ zeros, where y_K is the first nonzero sample in $[y_k]$. To use the FFT routine in this book, N must also be a power of 2.
>
> (9.14)

So, in (9.14) we have a rule for preparing data sequences for fast correlation.

Before turning to convolution, there is one other point to consider. In (9.6) and (9.14) we allow only positive values of the shift parameter n. Suppose we want a function computed around $n = 0$ instead, say with $-n\text{max}/2 \leq n \leq n\text{max}/2$, assuming $n\text{max}$ is even. To obtain this type of correlation function via fast transforms, we can first shift $[y_k]$ to the right to produce a new sequence, say $[v_k]$, and then compute $r_{xv}(n)$. Thus, using $[y_k]$ as in (9.3) with leading zeros, we have

$$v_k \triangleq y_{k-n\text{max}/2}, \qquad k = 0, 1, \ldots, N - 1 \tag{9.15}$$

$$r_{xv}(n) = \sum_{k=0}^{N-1} x_k v_{k+n}$$

$$= \sum_{k=0}^{N-1} x_k y_{k+n-n\text{max}/2} = r_{xy}(n - n\text{max}/2), \qquad n = 0, 1, \ldots, n\text{max} \tag{9.16}$$

So $[r_{xv}(n)]$ gives the desired values of $r_{xy}(n)$. The only extra requirement here is that N be large enough so that nonzero samples are not shifted beyond $k = N - 1$ in creating $[v_k]$ from $[y_k]$. The general rule for this method is as follows:

> To compute $[r_{xy}(n)]$ for $-n\text{max}/2 \leq n \leq n\text{max}/2$, first make sure that $[y_k]$ ends with at least $n\text{max}/2$ zeros. Then form $[v_k]$ as in (9.15), assuming $[y_k]$ is extended on the left with zeros. Then compute $[r_{xv}(n)]$ as in (9.14). The result is $r_{xy}(n - n\text{max}/2)$ for $0 \leq n \leq n\text{max}$.
$$\tag{9.17}$$

There is also an alternative form of (9.17) for computing the correlation function around $n = 0$ that derives from the definition of the DFT in equation (3.5). Suppose we take the DFT of $[x_{-k}]$ with k going from 0 to $N - 1$. Then in (3.5), we see that the sign of the exponent is reversed. Thus, in general,

$$\text{DFT}[x_{-k}] = [X_m^*] \tag{9.18}$$

Now suppose that N is large enough and $[x_k]$ ends with enough zeros that (9.14) is satisfied, so that $r_{xy}(n)$ may be used in place of $r_{x\bar{y}}(n)$ in (9.9). Then we have

$$\text{DFT}[r_{xy}(-n)] = [(X_m^* Y_m)^*]$$

$$= [Y_m^* X_m] \tag{9.19}$$

$$= \text{DFT}[r_{yx}(n)]$$

From the relationship in (9.19) we conclude that

$$r_{xy}(-n) = r_{yx}(n) \tag{9.20}$$

This result tells us that we can compute $r_{xy}(-n)$ by computing $r_{yx}(n)$, that is, simply by exchanging the two data sequences. The rule is as follows:

An alternative method for computing $[r_{xy}(n)]$ for $-n\mathrm{max}$ $\leq n \leq n\mathrm{max}$ is to compute first $r_{xy}(n)$ for $0 \leq n \leq$ $n\mathrm{max}$, then $r_{yx}(n)$ for $0 \leq n \leq n\mathrm{max}$, being sure that (9.14) is satisfied in both cases. Then combine the two results, setting $r_{xy}(-n) = r_{yx}(n)$. (9.21)

An interesting corollary to (9.20) gives the symmetry of the autocorrelation function about $n = 0$. If we set $x = y$, we obtain

$$r_{xx}(-n) = r_{xx}(n) \tag{9.22}$$

Turning now from correlation to convolution, we have in Fig. 9.3 an illustration similar to Fig. 9.2 but this time for $c_{xy}(0)$ instead of $r_{xy}(5)$. Note that the y sequences are reversed in accordance with (9.5), that is, they begin at $k = 0$ because $n = 0$ in this example and then proceed to the left. In Fig. 9.3(b), we

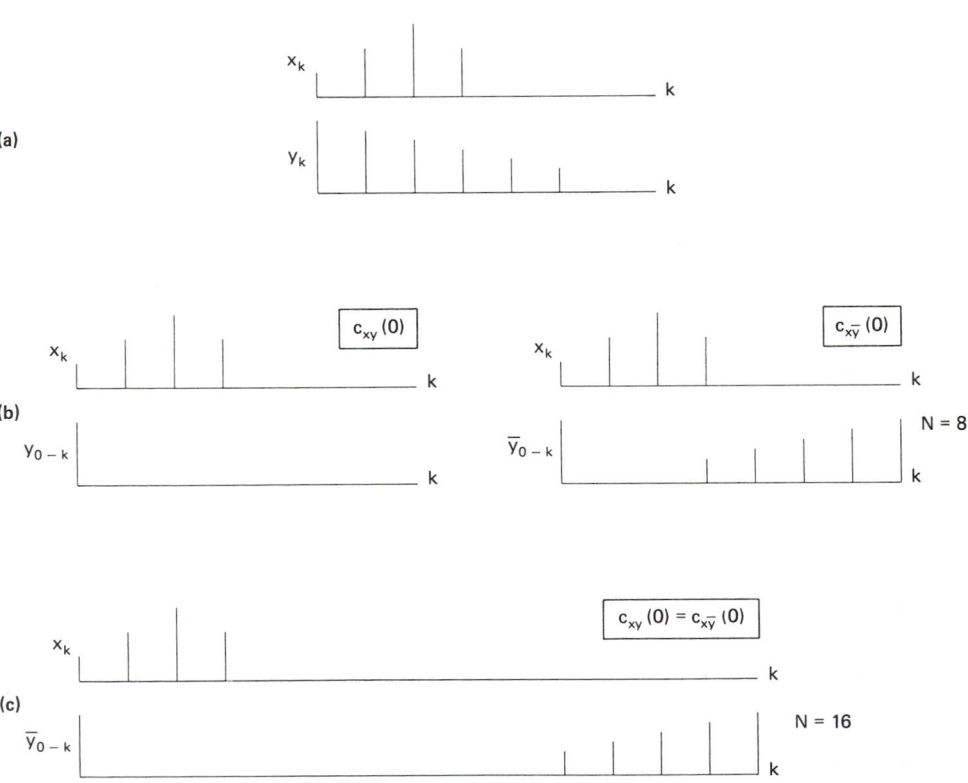

Figure 9.3. Alignment of $[x_k]$ and $[y_k]$ for convolution with $n = 0$. (a) Original sequences; $N = 8$. (b) Alignment for $c_{xy}(0)$ and $[c_{\bar{y}}(0)]$ with $N = 8$. (c) Alignment for $c_{xy}(0) = c_{x\bar{y}}(0)$ with $N = 16$.

see that $c_{xy}(0)$ and $c_{x\bar{y}}(0)$ differ for $N = 8$ because $[x_k]$ does not end in enough zeros to allow the next period of $[y_k]$ to begin beyond the nonzero part of $[x_k]$. In Fig. 9.3(c) with $N = 16$, $[x_k]$ now ends in enough zeros, and $c_{xy}(0) = c_{x\bar{y}}(0)$.

For convolution in general, suppose y_K is the last nonzero sample in $[y_k]$. Generalizing from Fig. 9.3(b) and (c), we observe that overlap will be avoided in the computation of $[c_{xy}(n)]$ if $[x_k]$ ends with at least $K - n$ zeros and further that this requirement becomes less severe as n increases. Therefore, the rule analogous to (9.14) is

To compute $[c_{xy}(n)]$ for $0 \leq n\mathrm{min} \leq n \leq N - 1$ using the inverse DFT of $[X_m Y_m]$, the sequence lengths (N) should be equal and $[x_k]$ should end with at least $K - n\mathrm{min}$ zeros, where y_K is the last nonzero sample in $[y_k]$. To use the FFT routine in this book, N must also be a power of 2.

(9.23)

In this rule, $n\mathrm{min}$ is the minimum shift of interest in (9.5) and is usually zero.

9.3 Routines for Fast Convolution and Correlation

We saw in the preceding section how the correlation and convolution functions are related to products of DFTs. Now we can discuss two routines, SPCORR and SPCONV, that use FFT products for fast, efficient computation of these functions. As in previous chapters, the two routines are short and simple to use and are designed to be highly portable.

Each of the routines uses an FFT product internally, as described in the preceding section, but the user does not need to be specifically aware of the form of the product to use a routine. The main purpose of the discussion in the preceding section was to develop the rules for padding the sequences with zeros. As we saw in (9.14) and (9.23), the sequence labeled $[x_k]$ must end in at least the required number of zeros in order for each routine to work correctly. The user must assure that this requirement is met, although each routine indicates an error if the zero padding is inadequate. Reviewing (9.14) and (9.23), we can see that switching the identity of $[x_k]$ and $[y_k]$ can sometimes shorten the required sequence lengths.

We consider first the fast convolution routine SPCONV, which is listed in Appendix A. The calling sequence for SPCONV is

$$\text{CALL SPCONV(X,Y,L,NMIN,IERROR)} \tag{9.24}$$

| X,Y | = two distinct data vectors to be convolved |
| L | = last index of both X(0:L) and Y(0:L); must be (power of 2) + 1; see (9.26) |

NMIN   = minimum shift of interest, as used in (9.23)

IERROR = error indicator: 0 if no error detected; 1 if $L - 1$ is not a power of 2, 2 if NMIN is out of range, or 3 if zero padding is inadequate

Upon execution, the data vectors X and Y are altered. The convolution function $c_{xy}(n)$ in (9.5) replaces X in accordance with

$$c_{xy}(n) \rightarrow x_n, \qquad n\text{min} \leqslant n \leqslant N - 1 \qquad (9.25)$$

where N is $L - 1$ and is computed internally by the routine. The routine calls SPFFTR, the FFT routine for real data described in Chapter 3, to perform an N-point FFT on the X and Y data. Then, after taking the FFT product, the SPIFTR routine is used to produce the correlation function in X(NMIN) through X($N-1$) in accordance with (9.25). The elements of X outside of the index range from XMIN through $N - 1$ are not useful outputs.

The arrays X and Y in SPCONV, which initially contain the data sequences $[x_k]$ and $[y_k]$, must be padded with zeros in accordance with (9.23). However, note that N, not L, is the actual sequence length used by the routine and that X($N-1$) is the last element in the sequence. The array elements (N) and $(N+1) = (L)$ are used internally. Therefore, the zero-padding requirement in (9.23) becomes the following:

Zero-padding requirement for SPCONV:
The original sequences must be padded with zeros such that both have length N, a power of 2, and furthermore, letting

$Y(K) = $ last nonzero element in Y(0) ...Y($N-1$)

N must be large enough that

$X(N-K+\text{NMIN}) = \cdots = X(N-1) = 0$

The last array index, L, which is specified in the calling sequence, must be $N + 1$.

(9.26)

We note that (9.26) is always satisfied if the original sequences of length M are doubled in length, that is, padded with at least M zeros, because then, in the worst case where $N = 2M$, $K = M - 1$, and NMIN = 0, X($M + 1$) through X($N-1$) will all be zeros, as required in (9.26). In the next section we give some examples to illustrate (9.26) as well as the computation of the convolution function as in (9.25).

The fast correlation routine is called SPCORR and is very similar to the convolution routine just described. A complete listing appears in Appendix A.

The calling sequence is

$$\text{CALL SPCORR(X,Y,L,ITYPE,NMAX,IERROR)} \qquad (9.27)$$

X,Y   = data vectors to be correlated; for autocorrelation, X and Y are the same vector

L   = last index of both X(0:L) and Y(0:L); must be (power of 2) + 1; see (9.29)

ITYPE   = type of correlation: 0 for autocorrelation, 1 for cross correlation

NMAX   = maximum shift, as used in (9.14)

IERROR = error indicator: 0 if no error detected; 1 if $L - 1$ is not a power of 2, 2 if NMAX is out of range, or 3 if zero-padding is inadequate

As before, the data vector(s) are altered during execution. The correlation function $r_{xy}(n)$ (in (9.6)) replaces the original X data in accordance with

$$r_{xy}(n) \rightarrow x_n, \qquad 0 \leqslant n \leqslant n\text{max} \qquad (9.28)$$

Also as before, SPCORR computes N as $L - 1$ and then uses N-point FFTs internally. The resulting rule for zero-padding, which is a restatement of (9.14), is as follows.

Zero-padding requirement for SPCORR:
The original sequences must be padded with zeros such that both have length N, a power of 2, and furthermore, letting
$$Y(K) = \text{first nonzero element in } Y(0)\dots Y(N-1) \qquad (9.29)$$
N must be large enough that
$$X(N+K-\text{NMAX}) = \cdots = X(N-1) = 0$$
The last array element, L, which is specified in the calling sequence, must be $N + 1$.

As above, we note again that the zero-padding requirement is always satisfied if the original sequences of length M are doubled in length, that is, padded with at least M zeros.

The next section contains examples of the use of SPCORR as well as SPCONV. The details on how the routines operate using FFT products may be found in section 9.5 on theory of operation.

9.4 Examples of Correlation and Convolution

We first consider some simple examples of correlation that can be used to check for the proper operation of SPCORR and also to illustrate correlation in situations that are easy to check by hand. In the first example we compute the autocorrelation function of the following sequence with $M = 7$:

$$[x_k] = [0 \quad 1 \quad 2 \ -1 \quad 0 \quad 0 \quad 0] \tag{9.30}$$

When the autocorrelation function $r_{xx}(n)$ is computed for $n = 0$ through 2, the sequence (extended with zeros) is in effect shifted left n places, multiplied by the unshifted sequence, and integrated as follows:

$$
\begin{array}{lllllll}
n = 0: & 0 & 1 & 2 \ -1 & 0 & 0 & 0 \\
       & 0 & 1 & 2 \ -1 & 0 & 0 & 0
\end{array}
\qquad r_{xx}(0) = 6 \tag{9.31}
$$

$$
\begin{array}{lllllll}
n = 1: & 0 & 1 & 2 \ -1 & 0 & 0 & 0 \\
       & 1 & 2 \ -1 & 0 & 0 & 0 & 0
\end{array}
\qquad r_{xx}(1) = 0 \tag{9.32}
$$

$$
\begin{array}{lllllll}
n = 2: & 0 & 1 & 2 \ -1 & 0 & 0 & 0 \\
       & 2 \ -1 & 0 & 0 & 0 & 0 & 0
\end{array}
\qquad r_{xx}(2) = -1 \tag{9.33}
$$

When n is greater than 2, we can see that the nonzero elements in the sequence no longer line up, and $r_{xx}(n) = 0$. This example is repeated in SPA0901. The data vector in (9.30) is stored in X(0:9), and at execution $r_{xx}(n)$ is computed and printed for $n = 0$ through 5. The important thing to note in this example is the application of the rule in (9.29). In the calling sequence of SPCORR, L = 9 and NMAX = 5. Therefore, $N = 8$ in (9.29) and, with $K = 1$, $X(4) = \cdots = X(7) = 0$ satisfies the condition in (9.29). We see that this zero-padding condition is met in SPA0901 and also that if NMAX were increased to 6, an error would result (see exercise 9.20).

```
            PROGRAM SPA0901
     C-SIMPLE EXAMPLE OF AUTOCORRELATION.
            DIMENSION X(0:9)
            DATA X/0.,1.,2.,-1.,0.,0.,0.,0.,0.,0./
            CALL SPCORR(X,X,9,0,5,IERR)
            PRINT 1,IERR,(K,X(K),K=0,5)
          1 FORMAT(I2/(I3,F6.1))
            STOP
            END
     $ RUN SPA0901
       0
       0    6.0
       1    0.0
       2   -1.0
       3    0.0
       4    0.0
       5    0.0
     FORTRAN STOP
```

Finally, in SPA0901 note that the data-sequence length used by SPCORR is a power of 2 (8 in this case). The extra two array locations X(8) and X(9) are needed internally by the FFT routine, as described in Chapter 3. Thus, as just stated, the rule in (9.29) is satisfied with $N = 8$, $K = 1$, and NMAX = 5.

In the second simple correlation example in SPA0902, we use SPCORR to find the cross correlation function of two data sequences, as follows:

$$[x_k] = [0 \quad 1 \quad 2 \ -1 \quad 0 \quad 0 \quad 0]$$

$$[y_k] = [1 \ -1 \quad 1 \ -1 \quad 0 \quad 0 \quad 0] \tag{9.34}$$

$$[r_{xy}(n)] = [2 \ -1 \ -1 \quad 0 \quad 0]$$

The data sequences are in X(0:9) and Y(0:9), and SPCORR is called with L = 9 and ITYPE = 1 for cross correlation in this case. The result is $r_{xy}(n)$ in (9.34) for $n = 0$ through NMAX = 4, which is verified when SPA0902 is executed. The result in (9.34) can easily be seen by imagining $[y_k]$ shifted n places to the left underneath $[x_k]$ and then summing the products to obtain $r_{xy}(n)$. Again, an important feature to note in SPA0902 is the application of the rule in (9.29). We see that NMAX = 4 is the largest possible value of NMAX in this example. See Exercise 9.21. The error indicator, IERROR, is printed by SPA0901 and SPA0902 and is zero in both cases. If NMAX were increased in either program, IERROR would be 1.

```
       PROGRAM SPA0902
C-SIMPLE EXAMPLE OF CROSS CORRELATION.
       DIMENSION X(0:9),Y(0:9)
       DATA X/0.,1.,2.,-1.,0.,0.,0.,0.,0.,0./
       DATA Y/1.,-1.,1.,-1.,0.,0.,0.,0.,0.,0./
       CALL SPCORR(X,Y,9,1,4,IERR)
       PRINT 1,IERR,(K,X(K),K=0,4)
     1 FORMAT(I2/(I3,F6.1))
       STOP
       END
$ RUN SPA0902
 0
    0    2.0
    1   -1.0
    2   -1.0
    3    0.0
    4    0.0
FORTRAN STOP
```

For a more realistic example of correlation, we take a sequence of 1500 samples of a unit sine wave in uniform white noise given by

$$x_k = \sin\left(\frac{2\pi k}{25}\right) + \sqrt{60}\,(r_k - 0.5), \qquad 0 \leqslant k \leqslant 1499 \qquad (9.35)$$

where r_k is obtained by invoking the function SPRAND described in Chapter 4. From equation (4.13) we can see that the white noise in (9.35) has a mean-squared value (power) of 5.0, compared with 0.5 for the sine wave, and that the signal-to-noise ratio (SNR) is therefore

$$\begin{aligned} \text{SNR} &= \frac{0.5}{5.0} = 0.1 \\ &= -10 \text{ dB} \end{aligned} \qquad (9.36)$$

The sequence in (9.35) is produced in the DO 1 loop of SPA0903 after initializing ISEED to 100, and is stored in the array X. The first call to PY (one of the plotting routines described in Appendix B) then plots the first 301 samples of the sequence. The plot is shown in Fig. 9.4, and we note that the sine wave is just barely discernible in the white noise, which is typical when the SNR is -10 dB.

```
      PROGRAM SPA0903
C-AUTOCORRELATION OF 1500 SAMPLES OF A SINE WAVE IN NOISE.
      DIMENSION X(0:2049),WORK(301)
      PI=4.*ATAN(1.)
      ISEED=100
      DO 1 K=0,1499
        X(K)=SIN(2.*PI*K/25.)+SQRT(12.*5.)*(SPRAND(ISEED)-.5)
    1 CONTINUE
      CALL PY(0.,1.,X,301,1,0,WORK,.1,.1,.9,.9)
      DO 2 K=1500,2047
        X(K)=0.
    2 CONTINUE
      CALL SPCORR(X,X,2049,0,300,IERR)
      IF(IERR.NE.0) PAUSE
      CALL PY(0.,1.,X,301,1,0,WORK,.1,.1,.9,.9)
      STOP
      END
```

To consider the zero-padding requirement in this example, suppose we wish to compute $r_{xx}(n)$ for $n = 0$ through 300, that is, $n$max $= 300$. We know that our sequence length (L + 1) for SPCORR must be two plus a power of 2. The next power of 2 above the data-sequence length (1500) is $N = 2048$, and in (9.27) we see that $N = 2048$ and $L = 2049$ easily satisfies the zero-padding requirement with NMAX $= 300$. Therefore, the array X in SPA0903 is dimensioned with 2050 elements, that is, X(0:2049). In the DO 2 loop, zeros are appended to the end of the sequence through $k = 2047$. The final two elements need not be initialized because they are used internally for temporary storage, as explained previously with SPA0901.

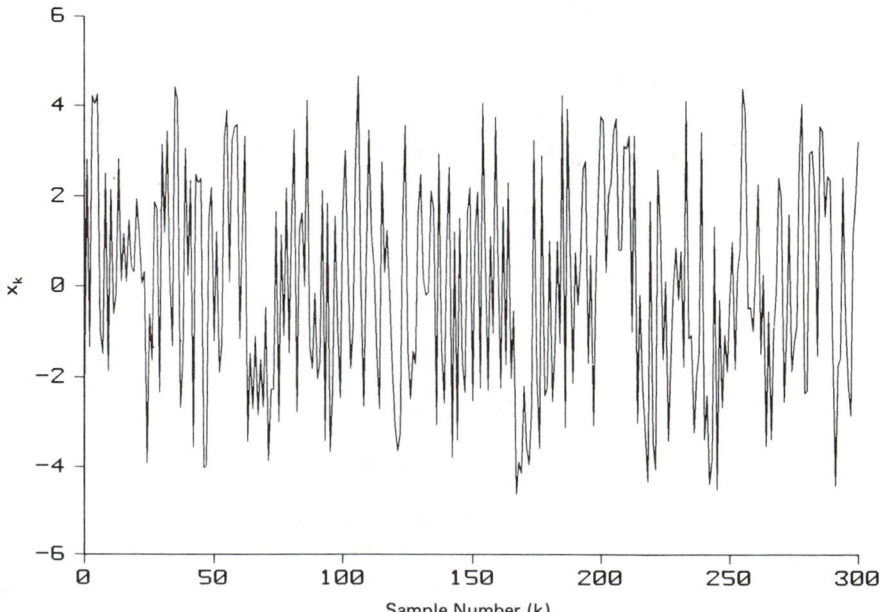

Figure 9.4. First 301 samples of $[x_k]$ in SPA0903. The plot consists of 12 cycles of a unit sine wave added to uniform white noise with SNR $= 0.1$.

Finally, in SPA0903, SPCORR is called in the autocorrelation mode with ITYPE = 0, and with NMAX = 300. After SPCORR computes the required autocorrelation function $[r_{xx}(n)]$ in X(0) through X(300), the second call to PY causes $[r_{xx}(n)]$ to be plotted. The plot is shown in Fig. 9.5. One of the properties of the autocorrelation function, easily seen in the definition (9.6) as well as in (9.22), is the symmetry about $n = 0$, that is,

$$r_{xx}(n) = r_{xx}(-n) \tag{9.37}$$

and so $r_{xx}(n)$ is plotted only for $0 \leq n \leq n\text{max}$ in Fig. 9.5.

The autocorrelation plot in Fig. 9.5 consists essentially of an impulse at $n = 0$ due to the white noise in $[x_k]$, which is uncorrelated for $n \neq 0$, plus a cosine component due to the sine wave in $[x_k]$. We observe that the sine wave is much more evident in Fig. 9.5 than in Fig. 9.4, due to the averaging effect in equation (9.6). Thus, as we might expect, the correlation function is sometimes used in conjunction with the spectrum to help identify periodic components.

As indicated by the vertical scale in Fig. 9.5, $r_{xx}(n)$ is the correlation function in (9.6) and not the mean lagged product in (9.7). After calling SPCORR, we could easily have added a loop in SPA0903 to implement (9.7) and convert $[r_{xx}(n)]$ into $[\phi_{xx}(n)]$.

Having illustrated correlation, we turn now to two simple examples of the use of SPCONV to obtain the convolution function. For the first example, which can be used to verify the operation of a new copy of SPCONV, we use the two

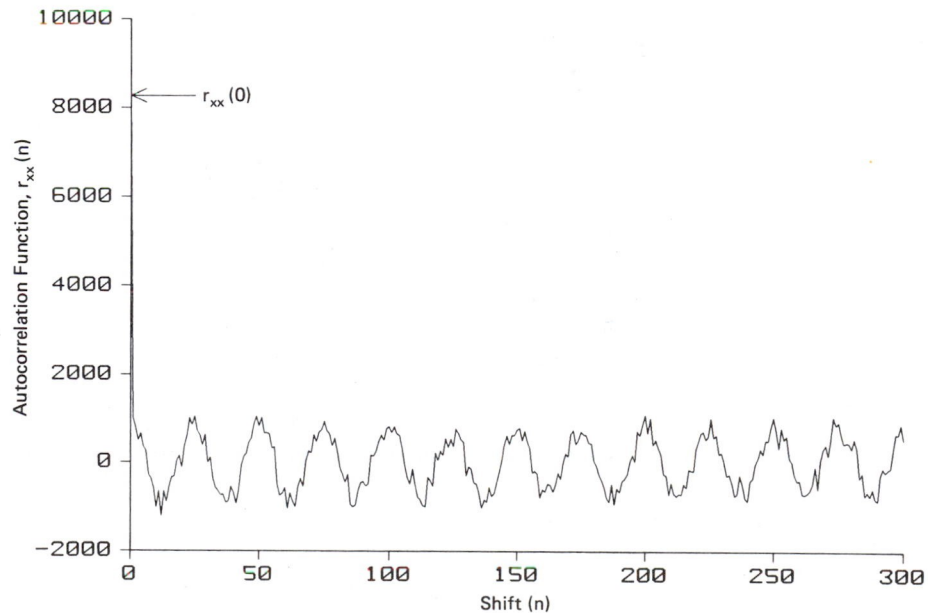

Figure 9.5. Autocorrelation function of the data sequence in (9.35), part of which is plotted in Fig. 9.4.

sequences in (9.34):

$$[x_k] = [0 \quad 1 \quad 2 \ -1 \quad 0 \quad 0 \quad 0]$$

$$[y_k] = [1 \ -1 \quad 1 \ -1 \quad 0 \quad 0 \quad 0] \qquad (9.38)$$

$$[c_{xy}(n)] = [0 \quad 1 \quad 1 \ -2 \quad 2 \ -3 \quad 1 \quad 0]$$

The result here, $[c_{xy}(n)]$, can be visualized in a manner similar to (9.34) if $[y_k]$ is reversed and then aligned so that y_0 is under x_0 for $n = 0$, under x_1 for $n = 1$, and so on. Or, we can obtain $[c_{xy}(n)]$ from (9.5) with $N = 8$ and $[x_k]$ and $[y_k]$ extended, as usual, with zeros. The same computation is made by calling SPCONV in the example given in SPA0904.

The data sequences in SPA0904 are defined in the DATA statements with $L = 9$ being the last array index. If we choose $n\text{min} = 0$ to obtain the complete convolution sequence, then the rule in (9.26) tells us in this case that the sequence $[x_0, \ldots, x_{N-1}]$ must end with at least three zeros, because $K = 3$ in this example. Thus, x_5, x_6, and x_7 must be zeros, and we see in (9.38) that this is so. The result in (9.38), $c_{xy}(n)$ for $n = 0$ through 7, is printed by SPA0904 upon execution.

```
        PROGRAM SPA0904
C-SIMPLE EXAMPLE OF CONVOLUTION.
        DIMENSION X(0:9),Y(0:9)
        DATA X/0.,1.,2.,-1.,0.,0.,0.,0.,0.,0./
        DATA Y/1.,-1.,1.,-1.,0.,0.,0.,0.,0.,0./
        CALL SPCONV(X,Y,9,0,IERR)
        IF(IERR.NE.0) PAUSE
        PRINT 1,(K,X(K),K=0,7)
      1 FORMAT(I3,F6.1)
        STOP
        END
$ RUN SPA0904
    0    0.0
    1    1.0
    2    1.0
    3   -2.0
    4    2.0
    5   -3.0
    6    1.0
    7    0.0
FORTRAN STOP
```

Our final example in this section provides a somewhat more realistic example of convolution. In this example we compute the output of a linear system, given the input sequence $[x_k]$ and the system's impulse response, as shown in Fig. 9.6. All the signals are generated in SPA0905 and plotted in Figs. 9.7 through 9.9. The input signal is a damped sine wave, and to make the example more realistic we assume that it describes an electromagnetic source in volts.

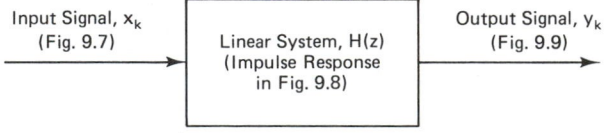

Input Signal, x_k (Fig. 9.7) → Linear System, H(z) (Impulse Response in Fig. 9.8) → Output Signal, y_k (Fig. 9.9)

Figure 9.6. Relation of signals generated in SPA0905 and plotted in Figs. 9.7 through 9.9.

The impulse response of the linear system, $H(z)$, is generated in SPA0905 and plotted in Fig. 9.8. It represents the response of an electromagnetic communication channel between two points to a unit impulse and illustrates an idealization of the multipath effect. The impulses in Fig. 9.8 represent various reflectors, that is, different paths from transmitter to receiver, that cause the input signal to be added with different delays and at different strengths at the output. (The equally spaced delays are somewhat artificial but make the illustration simpler.)

```
      PROGRAM SPA0905
C-MORE OR LESS REALISTIC EXAMPLE OF CONVOLUTION.
      DIMENSION X(0:2049),Y(0:2049),COEFF(0:7),WORK(1000)
      DATA COEFF/8.,4.,7.,2.,6.,3.,1.,1./
      PI=4.*ATAN(1.)
      DO 1 K=0,2047
        X(K)=EXP(-K/50.)*SIN(2.*PI*K/50.)
        IF(K.GT.400) X(K)=0.
        Y(K)=0.
        IF(K.LE.700.AND.MOD(K,100).EQ.0) Y(K)=COEFF(K/100)
    1 CONTINUE
      CALL PY(0.,1.,X,1000,1,0,WORK,.1,.1,.9,.9)
      CALL PY(0.,1.,Y,1000,1,0,WORK,.1,.1,.9,.9)
      CALL SPCONV(X,Y,2049,0,IERR)
      IF(IERR.NE.0) PAUSE
      CALL PY(0.,1.,X,1000,1,0,WORK,.1,.1,.9,.9)
      STOP
      END
```

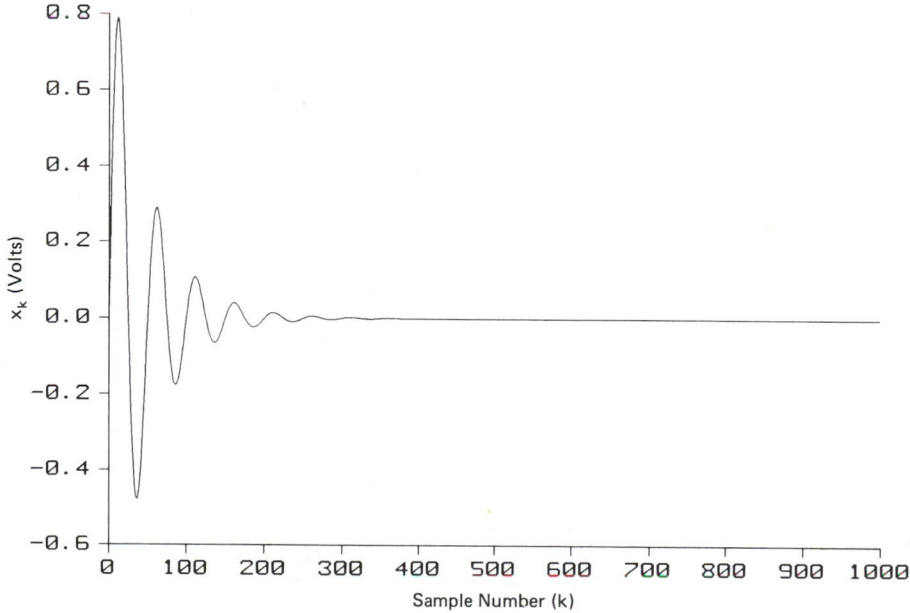

Figure 9.7.   Input signal in Fig. 9.6, generated in SPA0905.

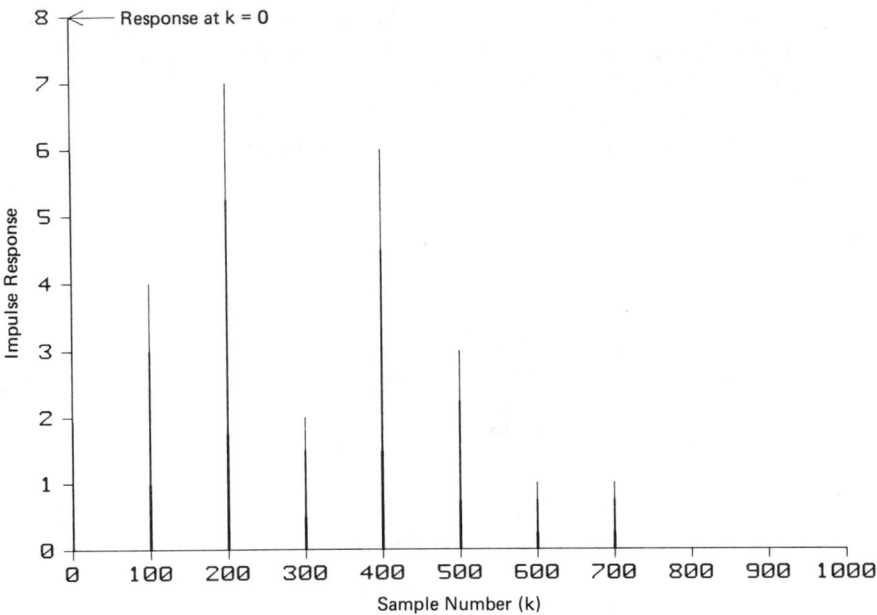

Figure 9.8. Impulse response of the linear system in Fig. 9.6, generated in SPA0905.

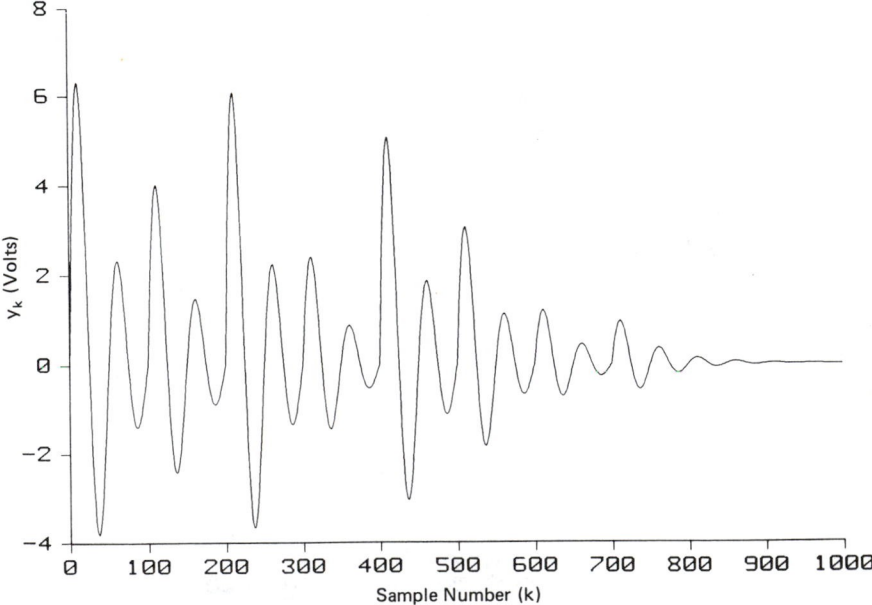

Figure 9.9. Output of the linear system in Fig. 9.6, generated in SPA0905, which is the convolution of the impulse response in Fig. 9.8 with the input in Fig. 9.7.

The input and impulse response signals are generated in the X and Y arrays in the DO 1 loop in SPA0905. Note that x_k is set to zero for $k > 400$ to provide more than enough zero-padding with $N = 2048$ in this case. Before calling SPCONV, PY is called twice to produce Figs. 9.7 and 9.8. Then, upon execution, SPCONV replaces X(0) through X(2047) with the convolution $[c_{xy}(n)]$. The final call to PY plots $c_{xy}(n)$ for $n = 0$ through 999 in Fig. 9.9. This final waveform represents the output, y_k, at the receiving end of the multipath channel.

9.5 Further Theory

In Section 4.8 we discussed the correlation function and its relationship to the power spectrum. The discussion there was similar to the discussion in Section 9.2 of this chapter. The two discussions were different, however, in their assumptions about the length of the time series. The discussion in Chapter 4 assumed infinite stationary time series, whereas in this chapter we are using finite sequences taken from infinite stationary time series.

Thus, we think of the correlation formulas in this chapter as finite versions (or estimates) of the formulas in Chapter 4, and of course the analogous formulas exist for convolution functions. For example, we compare (4.27) with (9.6) substituted into (9.7):

$$\text{True version:} \quad \phi_{xy}(n) = \text{E}\,[x_k y_{k+n}] \tag{9.39}$$

$$\text{Finite version:} \quad \phi_{xy}(n) = \frac{1}{M-n} \sum_{k=0}^{M-1} x_k y_{k+n} \tag{9.40}$$

Given the extended stationary sequences of data samples defined in (9.3), the finite version in (9.40) is our best unbiased estimate of the exact version in (9.39).

Comparing (4.31) with (9.10), we again see similar relationships involving the power spectrum. Finite versions of (4.31) are also seen in the definitions of the periodogram in (4.1) and of the cross periodogram in (4.21). Each formula has its use in the context in which it is given, but we emphasize again here the relationship found in (4.31), that the power spectrum is the transform of the correlation function. Thus, as in (9.10), the DFT of the cross correlation estimate $\phi_{xy}(n)$ in this chapter is an estimate of the cross power spectrum $\Phi_{xy}(z)$ and the DFT of $\phi_{xx}(n)$ is an estimate of $\Phi_{xx}(z)$. Similarly, as in Chapter 4, the integral of the power spectrum, or total power in $[x_k]$, is $\phi_{xx}(0)$, which is the average squared value of $[x_k]$, and so on.

Thus, the theory in this book relating the correlation functions with other functions is found primarily in Chapter 4. In this chapter we have emphasized mainly the theory concerning the fast computation of correlation functions.

Considering the code in Appendix A for the subroutines SPCORR and SPCONV, we see that the listings for the two routines are very similar. Looking first at the code for SPCORR, we see that the sequence length is set as

$$N = L - 1 \tag{9.41}$$

where L is the last index in the arrays X(0:L) and Y(0:L). Then, in statements 1 through 6, the routine checks to see that N is a power of 2 and also that the rule (9.29) is satisfied. Then the data sequence or sequences are transformed using SPFFTR and, in the DO 7 loop, the product of transforms in (9.10) is implemented. In anticipation of the call to SPIFTR, which produces the correlation function, the transform product is also scaled by $1/N$ in this same loop.

The operation of SPCONV is similar to that just described, and the differences between the two routines are straightforward. In SPCONV the rule in (9.26) instead of (9.29) is checked, and the product of transforms in (9.9) instead of (9.10) is implemented. We also note again that ITYPE is not used in SPCONV because autoconvolution is not a useful operation.

9.6 Summary of Calling Sequences

In this chapter we have described two subroutines, SPCONV and SPCORR, for fast convolution and correlation. The calling sequences are similar. We summarize them here with variable definitions:

$$\text{CALL SPCONV (X,Y,L,NMIN,IERROR)}$$
$$\text{CALL SPCORR (X,Y,L,ITYPE,NMAX,IERROR)}$$

Variable	Definition	Input/Output	Remarks
X	Data vector, X(0:L)	I,O	Output replaces input
Y	Data vector, Y(0:L)	I	Not used for autocorrelation
L	Last index of X and Y	I	L = (power of 2) + 1
ITYPE	Type of correlation	I	0 = auto, 1 = cross
NMIN	Minimum shift (n) of interest in (9.5)	I	$0 \leq \text{NMIN} < \text{L} - 1$ Usually, NMIN = 0.
NMAX	Maximum shift (n) of interest in (9.6)	I	$0 \leq \text{NMAX} < \text{L} - 1$ Usually, NMAX > 0.
IERROR	Error indicator	O	0 — no error 1 — L − 1 is not a power of 2 2 — NMIN or NMAX out of range 3 — Inadequate zero-padding; see (9.26) or (9.29)

As in previous chapters, I indicates an input to the subroutine and O indicates an output computed during execution.

9.7 Exercises

9.1. Two data sequences are given as follows:
$$[x_k] = [0 \quad 1 \quad 2 \quad 3 \quad 2 \quad 1 \quad 0 \quad 0]$$
$$[y_k] = [0 \quad 1 \quad -1 \quad 1 \quad -1 \quad 0 \quad 0 \quad 0]$$

Compute by hand the convolution function $c_{xy}(n)$ for $n = 0$ through 7, assuming zero extension on both sequences as in equation (9.3).

9.2. Do exercise 9.1 for the cross-correlation function $r_{xy}(n)$, for $n = 0$ through 4.

9.3. Compute by hand the complete autocorrelation function of

$$[x_k] = [1 \;\; -1 \;\; 2 \;\; 0 \;\; -1 \;\; 0 \;\; 0 \;\; 0]$$

assuming zero extension as in (9.3).

9.4. Compute by hand the correlation of $[x_k]$ in exercise 9.3 with the unit impulse function,

$$[y_k] = [1 \;\; 0 \;\; 0 \;\; 0 \;\; 0 \;\; 0 \;\; 0 \;\; 0]$$

9.5. Compute by hand the correlation of $[x_k]$ in exercise 9.3 with the unit step function,

$$[y_k] = [1 \;\; 1 \;\; 1 \;\; 1 \;\; 1 \;\; 1 \;\; 1 \;\; 1]$$

Assume zero extension on the left and 1s extended to the right.

9.6. Compute by hand the convolution of $[x_k]$ in exercise 9.3 with $[y_k]$ in exercise 9.4. Assume zero extension as in equation (9.3).

9.7. Compute by hand the convolution of $[x_k]$ in exercise 9.3 with $[y_k]$ in exercise 9.5. Assume zero extension as in equation (9.3).

9.8. Two data sequences with $N = 8$ are given as follows:

$$[x_k] = [1 \;\; 2 \;\; 3 \;\; 2 \;\; 1 \;\; 0 \;\; 0 \;\; 0]$$
$$[y_k] = [0 \;\; 1 \;\; 2 \;\; 1 \;\; 0 \;\; -1 \;\; 0 \;\; 0]$$

Write these sequences underneath each other in a manner similar to Fig. 9.2(b) to show that $r_{xy}(5)$ and $r_{x\bar{y}}(5)$ are not equal. Then do the same with $N = 16$ and zero extension to show that $r_{xy}(5)$ and $r_{x\bar{y}}(5)$ are equal in this case.

9.9. Using the two sequences in exercise 9.1, demonstrate the relationship in equation (9.20), that is, $r_{xy}(-n) = r_{yx}(n)$, using $n = 2$.

9.10. Write the two sequences in exercise 9.8 underneath each other in a manner similar to Fig. 9.3(b) to show that $c_{xy}(0)$ and $c_{x\bar{y}}(0)$ are not equal. Then do the same with $N = 16$ and zero extension to show that $c_{xy}(0) = c_{x\bar{y}}(0)$ in this case.

9.11. If we apply the rule (9.26) to the sequences in exercise 1 with NMIN = 0, what value of L would be used with SPCONV?

9.12. Use SPCONV in a program with $L = 33$ to produce $[c_{xy}(n)]$ in exercise 1.

9.13. Use SPCONV in a program to produce $[c_{xy}(n)]$ for $n = 0$ through 10 for the sequences in exercise 9.8. Use the minimum possible value of L.

9.14. Use SPCONV to compute 100 samples of the output of the filter shown here:

$$\sin\left(\frac{2\pi k}{16}\right) \longrightarrow \boxed{0.25 + 0.5z^{-1} + 0.25z^{-2}} \longrightarrow$$

Plot the input and output together. Using equation (5.9) in Chapter 5, check to see that the amplitude gain is correct.

9.15. Modify SPA0905 to compute the response of $H(z) = 0.4 + 0.2z^{-25} + 0.4z^{-50}$ to the input signal in Fig. 9.7, and show the result.

9.16. Develop and express a shifting rule that allows SPCONV to be used to compute the output of a noncausal FIR filter. (*Hint:* See Section 5.6.)

9.17. Using the rule in exercise 9.16, do exercise 9.14 with the transfer function changed to $0.25z^{-1} + 0.5 + 0.25z$. Again, plot the input and output waveforms together. Explain the difference in results.

9.18. Use SPCONV to compute 600 samples of the output of the following system:

$$e^{-k/50} \sin\left(\frac{2\pi k}{50}\right) \longrightarrow \boxed{\frac{z}{z^2 - 1.8z + 0.85}} \longrightarrow$$

To accomplish the computation, note that the infinite impulse response of the IIR filter decreases with sample number and may be considered to be 0 after, say, 50 samples in this system.

9.19. Use a development similar to (9.11) to show that the DFT product $[X_m^* Y_m]$ is the DFT of $[r_{x\bar{y}}(n)]$.

9.20. Suppose we have the following data vector with eight samples:

$$[x_k] = [0 \quad 1 \quad 2 \quad -1 \quad 0 \quad 0 \quad 0 \quad 0]$$

By comparing $r_{xx}(n)$ with $r_{x\bar{x}}(n)$, show how the rule (9.14) applies, first with $n\text{max} = 5$ and then with $n\text{max} = 6$. Also, show how the rule (9.29) applies with a suitable value of L. Finally, run SPCORR both ways to check the operation of IERROR.

9.21. Run SPA0902 with NMAX = 5 to check the proper operation of IERROR. Using a diagram similar to Fig. 9.2, show why NMAX = 5 is not valid.

9.22. Write a program similar to SPA0903 but with the sequence length equal to 500 instead of 1500. Make L as small as possible. Execute the program and compare the result with Fig. 9.5.

9.23. Write a single program that uses SPCORR to check the answers to exercises 9.3 and 9.4. In the data array X(0:L) in the program, make L as small as possible.

9.24. Using the two data sequences $[x_k]$ and $[y_k]$ in exercise 9.8, write a program that calls SPCORR to compute and print the correlation function $r_{xy}(n)$ for $-7 \leq n \leq 7$. Use rule (9.17) with L no larger than necessary.

9.25. Do exercise 9.24 using rule (9.21).

9.26. Use SPCORR in a program to estimate the mean lagged product $\phi_{xy}(n)$ in equation (9.7), when $[x_k]$ is a unit cosine wave and $[y_k]$ is a unit square wave symmetric about $k = 0$, both with 18 samples per cycle. Use 5 cycles of $[x_k]$ and $[y_k]$ in the computation and plot just one cycle of $\phi_{xy}(n)$, that is, for $-9 \leq n \leq 9$. (*Note:* For the unit square wave, use $y_k = 1$ for $0 \leq k < 5$, -1 for $5 \leq k < 14$, 1 for $14 \leq k < 23$, and so on.)

9.27. Use SPCORR in a program to plot the mean lagged product $\phi_{xx}(n)$ for $-20 \leq n \leq 20$, when $[x_k]$ is a uniform white random sequence with mean-squared value $E[x_k^2] = 2.0$. First use a sequence of 256 samples and then a sequence of 1024 samples. Plot $\phi_{xx}(n)$ for both sequences and comment on the comparison between the two plots. Use ISEED = 100 for both sequences.

9.8 References

[A] Oppenheim, A. V., A. S. Willsky, and I. T. Young, *Signals and Systems* (Englewood Cliffs, N.J.: Prentice-Hall, Inc., 1983), Chapter 4.

[B] Ahmed, N., and T. Natarajan, *Discrete-Time Signals and Systems* (Reston, Va.: Reston, 1983), Chapter 5.

[C] Gold, B., and C. M. Rader, *Digital Processing of Signals* (New York: McGraw-Hill, 1969), Chapter 7 (by T. G. Stockham, Jr.).

[D] Bendat, J. S., and A. G. Piersol, *Random Data: Analysis and Measurement Procedures* (New York: Wiley-Interscience, 1971), Section 9.7.

[E] Brigham, E. O., *The Fast Fourier Transform* (Englewood Cliffs, N.J.: Prentice-Hall, 1974), Chapter 13.

[F] Hamming, R. W., *Digital Filters* (Englewood Cliffs, N.J.: Prentice-Hall, Inc., 1977), Section 5.6.

[G] Maurice, R. D. A., *Convolution and Fourier Transforms for Communications Engineers* (New York: John Wiley Halsted Press, 1976).

[H] Papoulis, A., *Circuits and Systems* (New York: Holt, Rinehart and Winston, 1980), Chapter 4.

[I] Tretter, S. A., *Introduction to Discrete-Time Signal Processing* (New York: Wiley, 1976), Section 10.5.

[J] Stearns, S. D., *Digital Signal Analysis* (Rochelle Park, N.J.: Hayden, 1975), Chapters 7, 13.

chapter 10

Decimation and Interpolation Routines

10.1 Introduction

In this chapter we introduce simple routines for decreasing or increasing the sampling rate in a given data sequence. When the sampling rate is decreased, the process is called *decimation* because the original sample set is decimated, or reduced in number. When the sampling rate is increased, *interpolation* is required to place new samples between the original samples and thereby increase the number of samples.

The usual applications of these operations in engineering and signal processing are in altering the sampling rate of recorded data, where the original sampling rate is determined by the sampling theorem as described in Chapter 2 as well as by other economical and physical considerations. We envision the situation shown in Fig. 10.1, where a physical quantity is sampled with sampling rate $1/T$ samples per second and converted to a digital sequence. In practical situations the sampling rate is determined by hardware capabilities, quantization and noise considerations, the total recording capacity, and economic factors as well as by frequency content.

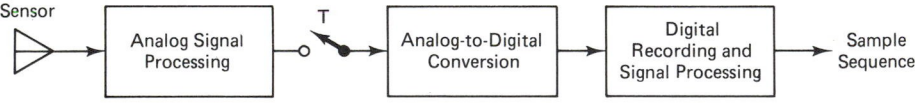

Figure 10.1. Conversion of physical data to a sample sequence.

Most important, however, is the assumption that the sampling theorem is satisfied in Fig. 10.1. We assume that the analog signal processing section contains low-pass filtering, either as a natural part of the sensor or in the form of a specific device, that limits the spectrum of its output to frequencies below $1/2T$. Otherwise, no amount of digital processing could remove the aliasing effects from the output sample sequence.

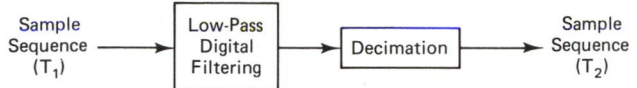

Figure 10.2. diagram: Sample Sequence (T_1) → Low-Pass Digital Filtering → Decimation → Sample Sequence (T_2)

Figure 10.2. Low-pass filtering and decimation to change the time step from T_1 to T_2, with $T_2 > T_1$.

With the sampling theorem satisfied at the sampler in Fig. 10.1, we can decimate the data sequence, as shown in Fig. 10.2. Low-pass digital filtering is used, if necessary, prior to the decimation process to limit frequencies in the data to $1/(2T_2)$, thus maintaining the conditions of the sampling theorem. In the decimation process the sampling rate is decreased by increasing the time step from T_1 to T_2 in accordance with

$$T_2 = RT_1 \qquad (10.1)$$

where ratio R is the *time-step ratio* and is greater than 1. If R is an integer, decimation consists of discarding $R - 1$ out of every R samples, as illustrated with $R = 3$ in Fig. 10.3.

If R is not an integer, interpolation is necessary in order to determine some of the new samples. Suppose $R = 2.2$, as illustrated in Fig. 10.4. Then four out of every five samples in the decimated sequence are between samples in the original sequence and must therefore be found via interpolation.

Thus, interpolation is needed for decimation when the decimation ratio R

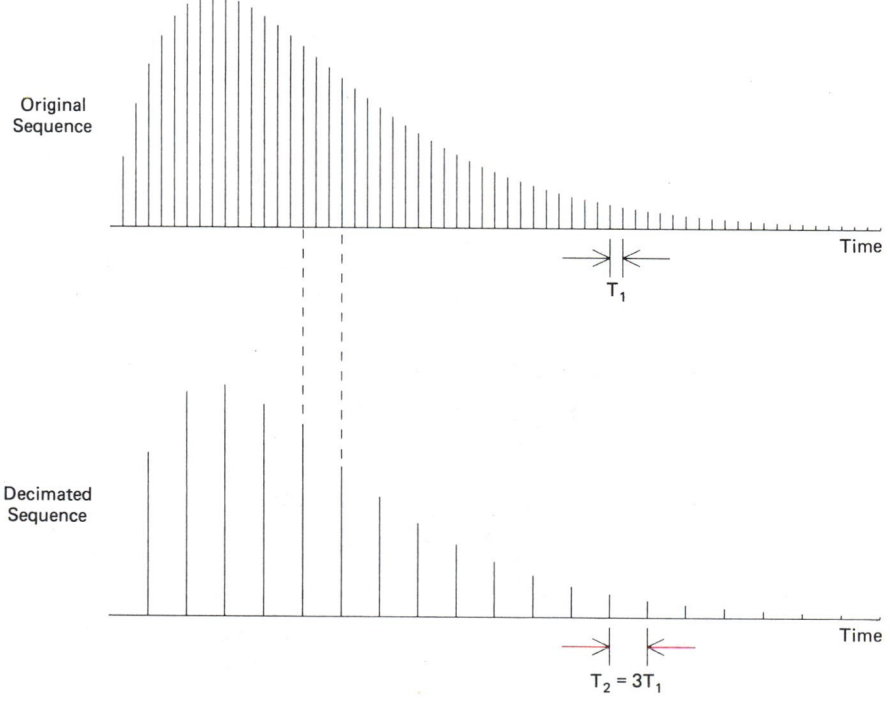

Figure 10.3. Decimation with time-step ratio $R = 3$. All samples in the decimated sequence coincide with samples in the original sequence.

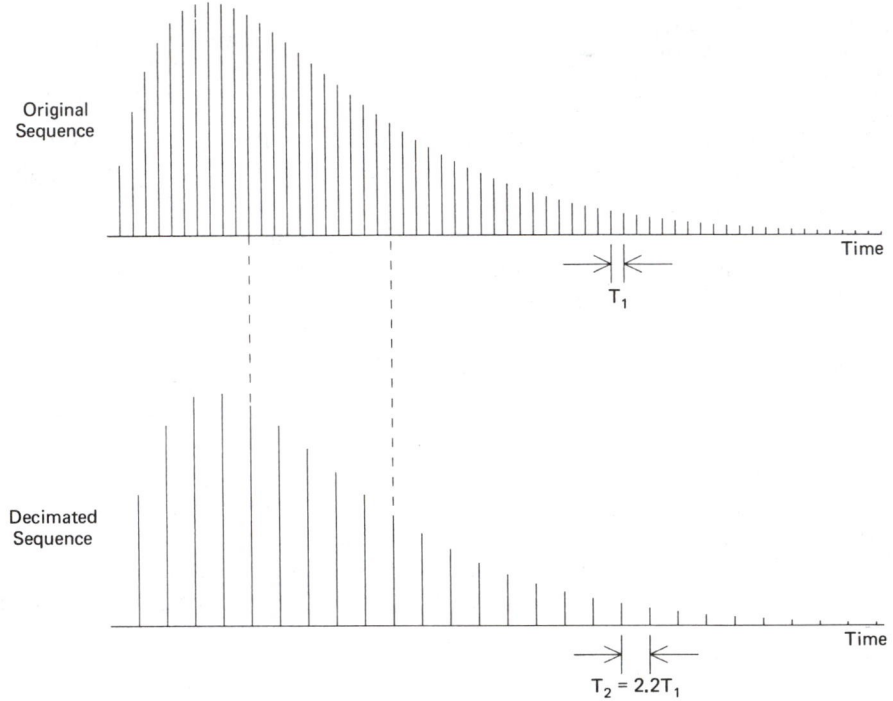

Figure 10.4. Decimation with time-step ratio $R = 2.2$. Every fifth sample in the decimated sequence coincides with a sample in the original sequence.

is not an integer. Interpolation is also often used to prepare data sequences for subsequent digital-to-analog (D/A) conversion, as in the two examples illustrated in Fig. 10.5. Speech signals are often most conveniently converted from digital sequences to analog waveforms (D/A conversion) using zero-order hold circuits [A–C], which are circuits that hold the circuit sample value until the arrival of the next sample as illustrated in Fig. 10.6. Digital plotters, on the other hand, usually construct straight lines between sample points and thus use the first-order type of reconstruction illustrated in Fig. 10.7.

Figures 10.6 and 10.7 suggest that to produce a reasonable reconstruction of the original waveform using zero- or first-order hold, the sampling rate must be increased to many times the required rate of twice the highest frequency.

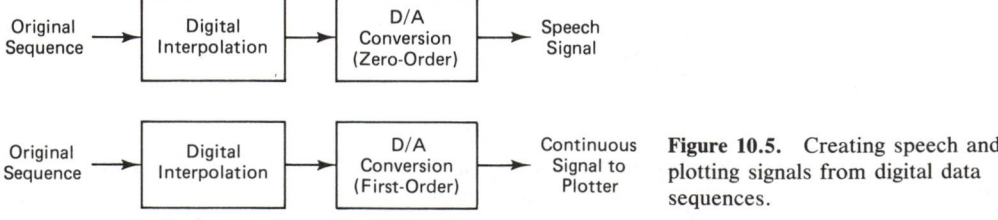

Figure 10.5. Creating speech and plotting signals from digital data sequences.

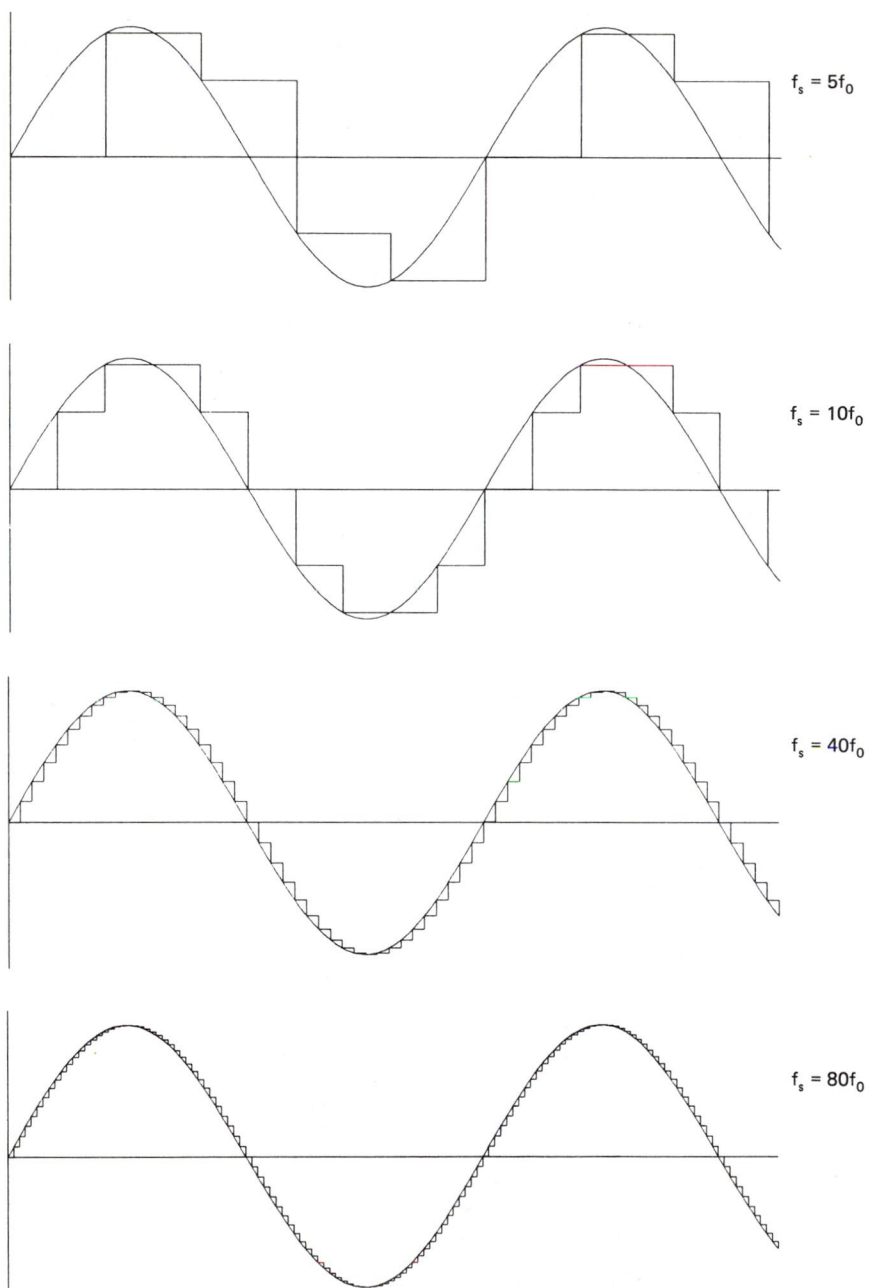

Figure 10.6. Zero-order-hold reconstruction of a sine wave at frequency f_0, showing improvement as the sampling rate, $f_s = 1/T$, increases.

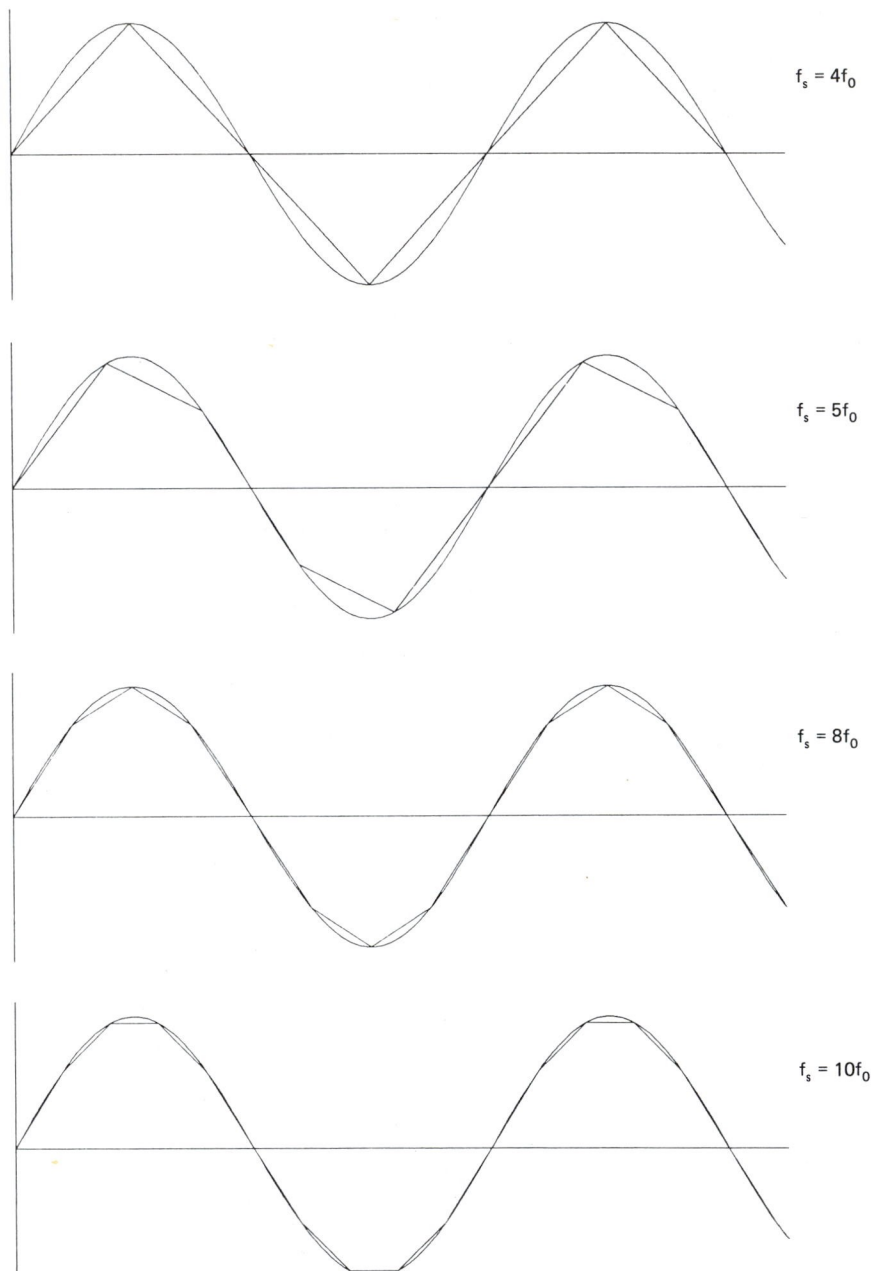

$f_s = 4f_0$

$f_s = 5f_0$

$f_s = 8f_0$

$f_s = 10f_0$

Figure 10.7. First-order-hold reconstruction of a sine wave at frequency f_0, showing improvement as the sampling rate, $f_s = 1/T$, increases.

Factors of 100 or 10 times the highest frequency would not be unreasonable for zero-order or first-order hold reproduction, respectively. Thus, as in Fig. 10.5, digital interpolation of the data sequence before D/A conversion is often necessary.

Digital interpolation is also useful in other applications. For example, we may wish to adjust the sampling rate of a digital data sequence or to delay the sequence for less than one time step in order to synchronize it to another sequence. After a short discussion of decimation, we concentrate in this chapter mainly on interpolation methods, that is, methods for producing "correct" data points between the given samples.

10.2 Linear Decimation and Interpolation

The first-order hold reconstruction in Fig. 10.7 is an example of linear interpolation. By drawing a straight line segment between each pair of sample points, the plotter constructs a continuous linear interpolation. A digital linear interpolation routine is therefore a routine that finds points at specified intervals along these line segments.

Suppose our original data sequence consists of $LX1 + 1$ samples, $X(0)$, $X(1), \ldots, X(LX1)$, and is stored in the array designated $X(0:LX)$, with $LX1 \leq LX$. Decimation would decrease the length of X below $LX1 + 1$, whereas interpolation would add new samples between the original samples and thus increase the length above $LX1 + 1$. For this reason it is convenient to accomplish linear decimation and interpolation in separate routines, so that the original data vector can be either reduced or increased in length.

In linear decimation we change the time step from T_1 to $T_2 = RT$, as in (10.1) and, when R is not an integer, we compute the new samples using linear interpolation. The routine for linear decimation is named SPDECI and its calling sequence is

$$\text{CALL SPDECI (X, LX, RATIO, LX2, IERROR)} \tag{10.2}$$

$X(0:LX)$ = data vector
LX = last index in original sequence, $X(0), \cdots, X(LX)$; input
$RATIO$ = decimation ratio (R); $1.0 \leq RATIO \leq LX$; input
$LX2$ = last index of decimated sequence, computed as $LX/RATIO$
$IERROR$ = error indicator, 0 if no error detected, or 1 if RATIO is out of range

The decimation routine is called with the original data sequence in the array X, with LX the index of the last sample, so that the sequence has $LX + 1$ samples from $X(0)$ through $X(LX)$. Decimation is done in place, and the decimated sequence, which is no longer than the original sequence, occupies the first part of the data vector from $X(0)$ through $X(LX2)$, where LX2 is the index of the last output sample and is computed internally. The remaining sequence elements,

X(LX2 + 1) through X(LX), are set to zeros in order to facilitate the taking of transforms or other subsequent operations on the data.

Two simple examples of linear decimation are presented in the execution of SPA1001. The two DO loops in the program cause the 10-element data sequence, given as [0, 2, 4, 6, 8, 8, 6, 4, 2, 0], to be stored in the vector X before calling SPDECI. The first call to SPDECI with RATIO = 2.0 then effectively doubles the time step by taking every other sample to form the sequence [0, 4, 8, 6, 2], which is aligned (in real time) in the following manner with the original sequence:

$$\begin{array}{lccccccccccc}
\text{Original:} & 0 & 2 & 4 & 6 & 8 & 8 & 6 & 4 & 2 & 0 \\
& \downarrow & & \downarrow & & \downarrow & & \downarrow & & \downarrow & \\
\text{Decimated:} & 0 & & 4 & & 8 & & 6 & & 2 &
\end{array} \qquad (10.3)$$

The output sequence, the lower sequence in (10.3), is the first sequence printed after RUN SPA1001, the execution command. In this case we note that LX2, the output sequence length, is 4 and that each of the remaining elements, X(5) through X(9), is set to 0 by the routine.

```
      PROGRAM SPA1001
C-TWO SIMPLE EXAMPLES OF DECIMATION.
      DIMENSION DATA(0:9),X(0:9)
      DATA DATA/0.,2.,4.,6.,8.,8.,6.,4.,2.,0./
      DO 1 K=0,9
        X(K)=DATA(K)
    1 CONTINUE
      CALL SPDECI(X,9,2.0,LX2,IERROR)
      PRINT 2,IERROR,LX2,X
    2 FORMAT(2I2,4X,10F6.2)
      DO 3 K=0,9
        X(K)=DATA(K)
    3 CONTINUE
      CALL SPDECI(X,9,1.5,LX2,IERROR)
      PRINT 2,IERROR,LX2,X
      STOP
      END
$ RUN SPA1001
 0  4      0.00  4.00  8.00  6.00  2.00  0.00  0.00  0.00  0.00  0.00
 0  6      0.00  3.00  6.00  8.00  6.00  3.00  0.00  0.00  0.00  0.00
FORTRAN STOP
```

The second call to SPDECI in SPA1001 is the same as the first but this time with RATIO = 1.5. With the fractional time step ratio, the original and decimated sequences are aligned as follows:

$$\begin{array}{lccccccccccc}
\text{Original:} & 0 & 2 & 4 & 6 & 8 & 8 & 6 & 4 & 2 & 0 \\
& \downarrow & & & \downarrow & & & \downarrow & & & \downarrow \\
\text{Decimated:} & 0 & & 3 & & 6 & & 8 & & 6 & & 3 & & 0
\end{array} \qquad (10.4)$$

In this case the output sample x_1 is interpolated halfway between the original samples x_1 and x_2, output x_2 is equal to the original x_3, and so on. Note that LX2, the last index of the decimated sequence, is 4 for the first example in (10.3) and 6 for the second example in (10.4). In the output below SPA1001, we note that the remaining elements of X, X(LX2 + 1) through X(LX), are again set to 0s.

The linear interpolation routine, which is similar to the decimation routine just described, is called SPLINT. Its calling sequence is

$$\text{CALL SPLINT(X, LX, LX1, RATIO, LX2, IERROR)} \qquad (10.5)$$

$X(0:LX)$ = data vector, with $LX \geq LX1/RATIO$
LX = last index of X array
$LX1$ = last index of original sequence, $X(0) \cdots X(LX1)$; input
(*Note:* $X(LX1+1)$ through $X(LX)$ can be anything initially)
$RATIO$ = interpolation (time-step) ratio (R), between 0.0 and 1.0; input
$LX2$ = last index of interpolated sequence, computed as $LX1/RATIO$
$IERROR$ = error indicator: 0 if no error detected; 1 if RATIO is out of limits, or 2 if LX is too small

The interpolation and decimation calling sequences, (10.5) and (10.2), are similar; however, the important difference in the operation of the two routines is that interpolation adds samples to the original sequence, whereas decimation deletes samples. Therefore, to allow space for the extra samples generated during interpolation, the initial data vector $X(0:LX)$ usually contains zero padding when used with the interpolation routine. For example, we could have the following sequences with $LX = 10$, $LX1 = 5$, and $RATIO = 0.5$:

$$
\begin{array}{lccccccccccc}
\text{Original:} & 2 & 4 & 6 & 6 & 8 & 6 & 0 & 0 & 0 & 0 & 0 \\
& \downarrow & \downarrow & \downarrow & \downarrow & \downarrow & \downarrow & & & & & \\
\text{Interpolated:} & 2 & 3 & 4 & 5 & 6 & 6 & 6 & 7 & 8 & 7 & 6 \\
\end{array} \qquad (10.6)
$$

In this example there are 11 elements in the data vector X, and initially the original data consists of 6 samples padded with 5 zeros. Then, after interpolation with $RATIO = 0.5$, the interpolated sequence, which is twice as dense as the original sequence, replaces all 11 elements of X. We can see that interpolation stops at element 5 of the original sequence in this case, and the routine would therefore be called with $LX1 = 5$ to designate the end of the interpolation range in the original data. In general, LX1 is set in this manner, to designate the end of the data sequence in the array $X(0:LX)$. Then LX2, the last index in the interpolated array, is computed in accordance with

$$LX2 = LX1/RATIO \qquad (10.7)$$

Thus, LX2 is the largest integer less than $LX1/RATIO$, and interpolation beyond the last output sample, $X(LX2)$, is not allowed.

Two examples of the use of SPLINT are presented in SPA1002. As in SPA1001, the original data sequence is in the DATA vector and the two output sequences are computed in $X(0:10)$ and printed below the program, along with IERROR and LX2 for both cases. In the first example with $RATIO = 0.5$, we have a repeat of the result in (10.6). We note again that with $LX1 = 5$, interpolation stops at the element $X(5)$ in the original sequence. The second example in SPA1002 is similar, but this time with $RATIO = 0.75$. Comparing the original and interpolated sequences, we can see that the results are again correct, and that interpolation in this case stops just prior to the end of the original sequence, halfway between the samples 8.0 and 6.0, with LX2 = 6.

```
            PROGRAM SPA1002
C-TWO SIMPLE EXAMPLES OF LINEAR INTERPOLATION.
            DIMENSION DATA(0:10),X(0:10)
            DATA DATA/2.,4.,6.,6.,8.,6.,0.,0.,0.,0.,0./
            DO 1 K=0,10
              X(K)=DATA(K)
          1 CONTINUE
            CALL SPLINT(X,10,5,0.5,LX2,IERROR)
            PRINT 2,IERROR,LX2,X
          2 FORMAT(I2,I3,3X,11F5.1)
            DO 3 K=0,10
              X(K)=DATA(K)
          3 CONTINUE
            CALL SPLINT(X,10,5,0.75,LX2,IERROR)
            PRINT 2,IERROR,LX2,X
            STOP
            END
$ RUN SPA1002
  0 10      2.0  3.0  4.0  5.0  6.0  6.0  6.0  7.0  8.0  7.0  6.0
  0  6      2.0  3.5  5.0  6.0  6.0  7.5  7.0  0.0  0.0  0.0  0.0
FORTRAN STOP
```

10.3 Interpolation Using Zero Insertion

We have seen some simple examples of linear interpolation, which is easily accomplished in the short subroutine SPLINT. Linear interpolation, however, is not generally desirable with bandlimited data. It is useless, for example, in the plotter application shown in Fig. 10.5, because the plotter itself draws straight lines between the sample points and is thus a continuous linear interpolator. Linear interpolation with RATIO equal to the reciprocal of an integer prior to plotting would not alter the plot at all.

Suppose we have a digitized waveform representing real physical data as in Fig. 10.1 and we know the data is limited to frequencies below one-half the sampling rate. In other words, the original waveform has no spectral content at or above $1/2T$ hertz, where T is the digitizing time step. Suppose further that the digitized data is plotted (or reconstructed in some other way) in a manner such that the spectrum of the interpolated sequence, consisting of the digitized data plus a large number of intermediate points, is also limited to frequencies below $1/2T$. Then the interpolated sequence should exactly match the original continuous waveform.

An illustration of this argument is presented in Fig. 10.8. First, in Fig. 10.8(a), we have the original continuous waveform, recorded for 1 s and limited to frequencies below 12.5 Hz. The waveform in Fig. 10.8(a) is digitized at 25 samples per second, and the entire digital sequence consists of 26 samples. Next, in Fig. 10.8(b), we have a reconstruction using linear interpolation, and we note the jagged appearance resulting from the connected line segments. Finally, in Fig. 10.8(c), we have a smooth reconstruction that is itself limited to frequencies below 12.5 Hz. We note that this latter reconstruction appears to be smooth and very similar to the original waveform in Fig. 10.8(a).

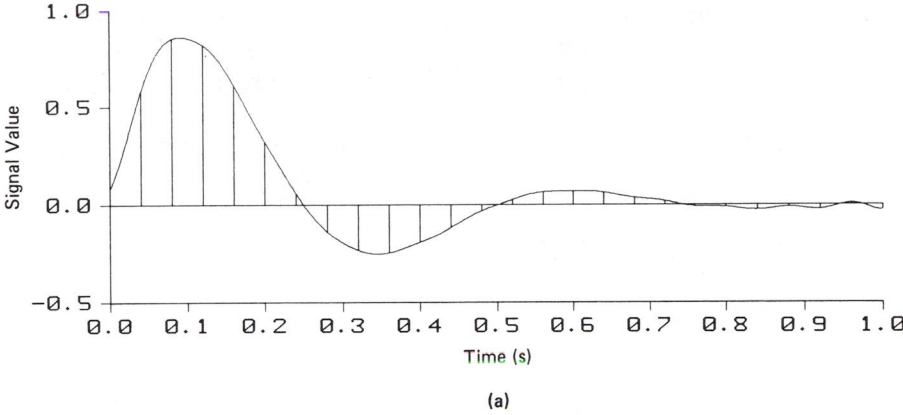

(a)

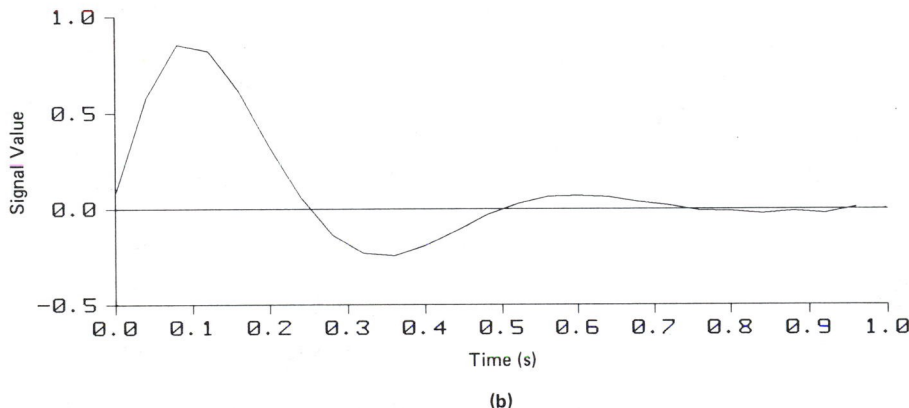

(b)

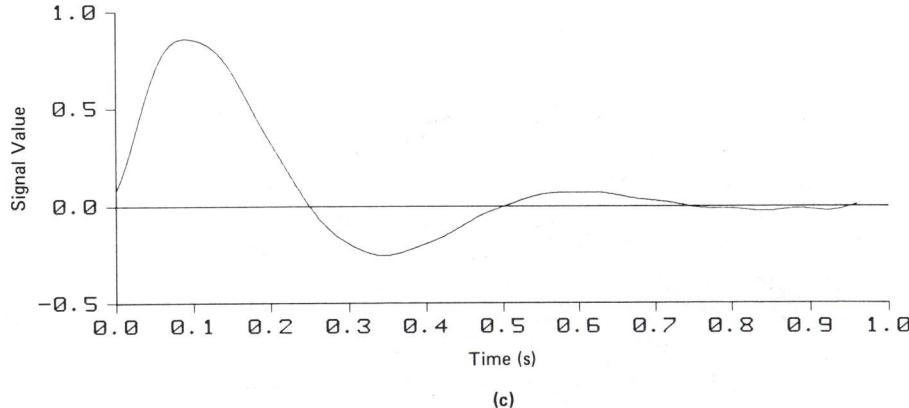

(c)

Figure 10.8.   (a) One-second record of a waveform digitized at 25 samples per second ($T = 0.04$). Continuous waveform has no spectral content above 12.5 Hz. (b) Reconstruction using linear interpolation. (c) Frequency-limited reconstruction using zero insertion, interpolated at 400 samples per second ($T = 0.0025$).

The smooth reconstruction in Fig. 10.8(c) was made using the zero-insertion method, which is now described. After the method and the use of the associated routine SPZINT are understood, the reader can reproduce Fig. 10.8 by working Exercise 10.9.

Interpolation via zero insertion involves the insertion of a fixed number of zeros between each pair of elements in the data sequence or, equivalently, the insertion of a sequence of zeros at the end of the discrete Fourier transform (DFT) of the data sequence. Let us begin with a sequence $[x_k]$ having L elements and insert Z zeros between each pair of elements and after the final element, to produce a sequence $[y_k]$ having $L(Z + 1)$ elements, as in the following example:

$$[x_k] = [x_0 \, x_1 \, x_2 \, \cdots \, x_{L-1}] \tag{10.7}$$

$$[y_k] = [x_0 \, \underbrace{0 \cdots 0}_{Z \text{ zeros}} \, x_1 \, \underbrace{0 \cdots 0}_{Z \text{ zeros}} \, \cdots \, x_{L-1} \underbrace{0 \cdots 0}_{Z \text{ zeros}}] \tag{10.8}$$

To see how zero insertion can be used to interpolate between the samples of $[x_k]$, we must now examine the spectrum of $[y_k]$ in (10.8). From equation (3.5), the DFT of $[y_k]$ is

$$Y_m = \sum_{k=0}^{L(Z+1)-1} y_k \, e^{-j[2\pi mk/L(Z+1)]}, \qquad m = 0, 1, \cdots \frac{L(Z+1)}{2} \tag{10.9}$$

From (10.8) we note that most of the terms in this sum are zeros. If we define a new index i such that

$$k = (Z + 1)i \tag{10.10}$$

then, with $y_k = x_i$ as in (10.8), we can rewrite (10.9) as follows:

$$Y_m = \sum_{i=0}^{L-1} x_i \, e^{-j[2\pi m(Z+1)i/L(Z+1)]}$$

$$= \sum_{i=0}^{L-1} x_i \, e^{-j(2\pi mi/L)} \tag{10.11}$$

$$= X_m, \qquad m = 0, 1, \ldots, \frac{L(Z+1)}{2}$$

In this result the DFT of $[y_k]$ is seen to be exactly the DFT of $[x_k]$ but repeated Z times beyond $m = L/2$ due to the extent of the frequency index, m. Thus, noting that the DFT of $[x_k]$ has an index range from $m = 0$ to $L/2$, we have the important result that the DFT of a sequence with zeros inserted contains the DFT of the original sequence, plus a periodic extension of the latter.*

An example of the foregoing result is shown in Fig. 10.9. To plot this figure we expanded a 128-sample sequence from an exponentially decaying sinusoid

---

* See equation (3.5).

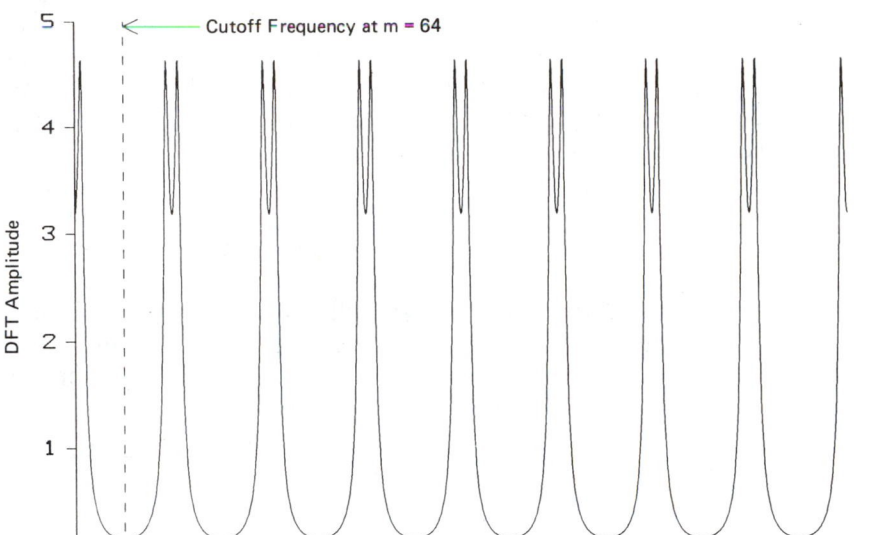

Figure 10.9. Amplitude spectrum of a sequence with zeros inserted between samples. Original sequence had $L = 128$ samples of a waveform similar to Fig. 10.8(a) and was expanded with $Z = 15$ zeros between samples as in (10.8), giving $L(Z + 1) = 2048$ samples in the resultant sequence.

like the waveform in Fig. 10.8, using $Z = 15$ zeros between samples as in (10.8). The amplitude of the DFT of the sequence with zeros inserted having 2048 samples is plotted in Fig. 10.9. We note that the DFT of the original 128-sample sequence extends from $m = 0$ to $m = 64$ and is then repeated 15 times from $m = 64$ to $m = 128$, $m = 128$ to $m = 192$, and so on. Eight of the repetitions are reflections of the original in accordance with (10.11); however, the important point in Fig. 10.9 is that if we eliminate all of the spectrum above the cutoff frequency at $m = 64$ shown in the figure, we obtain the spectrum of the original sequence. Thus, the inverse DFT of the low-passed (truncated) version of Fig. 10.9 produces a 2048-point interpolated sequence having the spectrum of the original 128-point sequence.

To summarize thus far, as illustrated in Fig. 10.10, the insertion of Z zeros between samples followed by low-pass filtering produces a new time series $[y_k]$

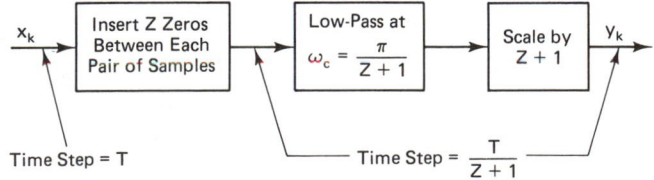

Figure 10.10. Real-time, spectrum-preserving interpolation using zero insertion between samples.

having the same spectrum as $[x_k]$ and a shorter time step, $T/(Z + 1)$. The scaling by $Z + 1$ shown in Fig. 10.10 results from equation (3.10), if we think of doing the low-pass filtering in the frequency domain. The inverse DFT needed to produce $[y_k]$ involves scaling by $1/(L(Z + 1))$, whereas the inverse DFT of the original sequence requires scaling by only $1/L$.

The scheme in Fig. 10.10 could be used in real time with a time-domain low-pass filter. However, as before, we wish to assume here that the entire sequence $[x_k]$ is available to our interpolation routine. In this case, Fig. 10.11 applies. As seen in the development of (10.11), we can append zeros to $[X_m]$ without altering the spectrum of $[x_k]$. Thus the inverse DFT of this modified spectrum, with proper scaling, is a time series $[y_k]$ with the original spectrum of $[x_k]$ but with more samples. As shown in Fig. 10.11, the time-step ratio decreases with the number (K) of zeros appended to $[X_m]$. The scaling factor $(L + K)/L$ is correct assuming that the inverse DFT includes scaling by $1/(L + K)$, as in (3.10).

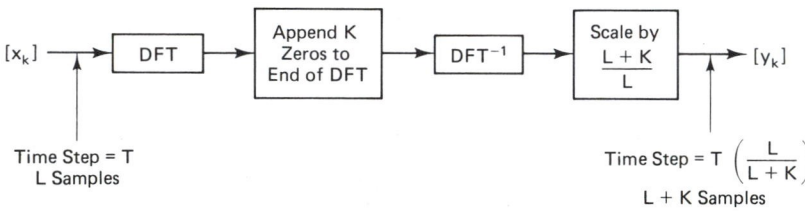

Figure 10.11.   Spectrum-preserving interpolation equivalent to Fig. 10.10 using zero-padding at the end of the DFT.

A routine called SPZINT incorporates the idea in Fig. 10.11. SPZINT, however, uses the FFT routines in Chapter 3 for fast execution, and so Fig. 10.11 must be modified slightly into Fig. 10.12, where initial zero extension is provided (if necessary) to make the data sequence length a power of 2. Also, the time-step ratio is the reciprocal of a power of 2, so that the sequence length for SPIFTR is also a power of 2. Since SPIFTR does not include scaling by $1/N_2$, the scaling factor is $1/N$ in Fig. 10.12, equivalent to $1/L$ in Fig. 10.11.

The calling sequence for SPZINT has all the arguments described previously,

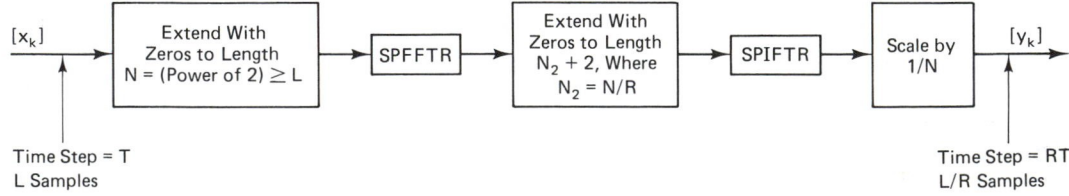

Figure 10.12.   Spectrum-preserving interpolation, similar to Fig. 10.11, produced by SPZINT. To accommodate the FFT routines, N and N_2 are powers of 2 and R is the reciprocal of a power of 2.

but now the time-step ratio is restricted to reciprocal powers of 2. The sequence is

$$\text{CALL SPZINT (X, LX, LX1, RATIO, LX2, IERROR)} \qquad (10.12)$$

$X(0{:}LX) \quad =$ data vector, with $LX > [(\text{smallest power of 2}) > LX1]/RATIO$

$LX \qquad = $ last index of X array

$LX1 \qquad =$ last index of original data sequence, $X(0)$ through $X(LX1)$

$RATIO \quad =$ ratio of time steps (R), T_2/T_1; must be $1/(\text{power of 2})$

$LX2 \qquad =$ last index of output data sequence $X(0)$ through $X(LX2)$, computed internally

$IERROR =$ error indicator: 0 if no error detected; 1 if RATIO is invalid, or 2 if LX is too small

The subroutine is called with the original sequence stored in $X(0)$ through $X(LX1)$. Generally, LX is much larger than LX1, and in any case the elements $X(LX1+1)$ through $X(LX)$ can be any values when SPZINT is called. To make the calling sequence in (10.12) easier to understand, we use an example based on the previous discussion of Fig. 10.8. Suppose we have a 25-point sequence $X(0)$ through $X(24)$, and we specify a total vector length $(LX+1)$ equal to 521. Then the calling sequence could be

$$\text{CALL SPZINT (X, 520, 24, 1./16., LX2, IE)} \qquad (10.13)$$

We note that $LX = 520$, $LX1 = 24$, and $RATIO = 1/16$ are inputs, whereas LX2 and IE are outputs of the routine. Given LX, LX1, and a validated value for RATIO, the routine computes two internal sequence lengths, N and N_2, as

$$N = (\text{smallest power of 2}) > LX1 = 32 \qquad (10.14)$$

$$N_2 = N/RATIO = 512 \qquad (10.15)$$

The routine then checks for the condition

$$LX > N_2 \qquad (10.16)$$

which must be met in order to proceed. If (10.16) is satisfied, the routine proceeds to determine the final index, LX2, of the output sequence, corresponding with LX1 in the input sequence. As seen in (10.8) with $LX1 = L - 1$, this index is

$$LX2 = LX1/RATIO \qquad (10.17)$$
$$= 384$$

Note that the routine does not interpolate (extrapolate) beyond the last data point, that is, beyond x_{L-1} in (10.8).

Having determined these parameters, the routine proceeds as in Fig. 10.12, using the routines SPFFTR and SPIFTR to take the required forward and inverse DFTs. The result is a longer interpolated sequence, $X(0)$ through $X(LX2)$, like the sequence plotted in Fig. 10.8(c). With 385 instead of 25 points, the plot appears as a smooth curve.

We have specified that RATIO must be the reciprocal of a power of 2, that is,

$$\text{Allowed values of RATIO} = \frac{1}{2}, \frac{1}{4}, \frac{1}{8}, \dots \qquad (10.18)$$

and thus the choice of time-step ratios with SPZINT appears to be very sparse. In the next section we show that this restriction is not as bad as it first appears to be.

A very simple and artificial example of the use of SPZINT is given in SPA1003, where the routine is used to interpolate the sequence used in (10.6). This example produces a result different from the linear interpolation in (10.6) and can be used as a simple check for the proper operation of SPZINT. See also exercises 10.14 and 10.15.

```
        PROGRAM SPA1003
C-SIMPLE EXAMPLE OF INTERPOLATION USING SPZINT.
        DIMENSION X(0:17)
        DATA X/2.,4.,6.,6.,8.,6.,12*0./
        CALL SPZINT(X,17,6,0.5,LX2,IERROR)
        PRINT 1,IERROR,LX2,(X(K),K=0,LX2)
      1 FORMAT(2I4/(10F7.3))
        STOP
        END
$ RUN SPA1003
      0   12
   2.000   2.801   4.000   5.307   6.000   5.964   6.000   6.837   8.000   8.027
   6.000   2.693   0.000
FORTRAN STOP
```

Another artificial but more interesting example is given in SPA1004. This example illustrates the effectiveness of the reconstruction of bandlimited data by starting with 16 samples of a sine wave over exactly four periods. The samples are plotted in Fig. 10.13(a). When SPZINT takes the DFT, there are no components (except rounding error) above one-half the sampling rate; in fact there is only one nonzero component in the entire DFT, as described in exercises 3.30–3.32. Therefore, the interpolation gives points on a sine wave, and we note that this is due to the lack of leakage caused by the fortuitous choice of LX1 in the program. The call to SPZINT as well as the printed output below SPA1004 show that the time-step ratio was $T_2/T_1 = \frac{1}{16} = 0.0625$ and that the last sample index increased from LX1 = 15 to LX2 = 240. The resulting 241-point sequence is plotted in Fig. 10.13(b). We can see that this sequence corresponds with the sine function computed in SPA1004, which was offset in phase by 0.5 rad, and that the smooth interpolated sequence begins at the first sample and ends with the last sample in the original sequence.

```
        PROGRAM SPA1004
C-MORE INTERESTING EXAMPLE OF INTERPOLATION USING SPZINT.
C-RESULTS ARE PLOTTED IN FIGURE 10.13.
        DIMENSION X(0:257)
        PI=4.*ATAN(1.)
        DO 1 K=0,15
          X(K)=SIN(2.*PI*K/4.+0.5)
```

```
      1 CONTINUE
        CALL SPZINT(X,257,15,1./16.,LX2,IERROR)
        PRINT 2,IERROR,LX2
      2 FORMAT(' IERROR,LX2=',2I4)
        STOP
        END
   $ RUN SPA1004
    IERROR,LX2=    0 240
   FORTRAN STOP
```

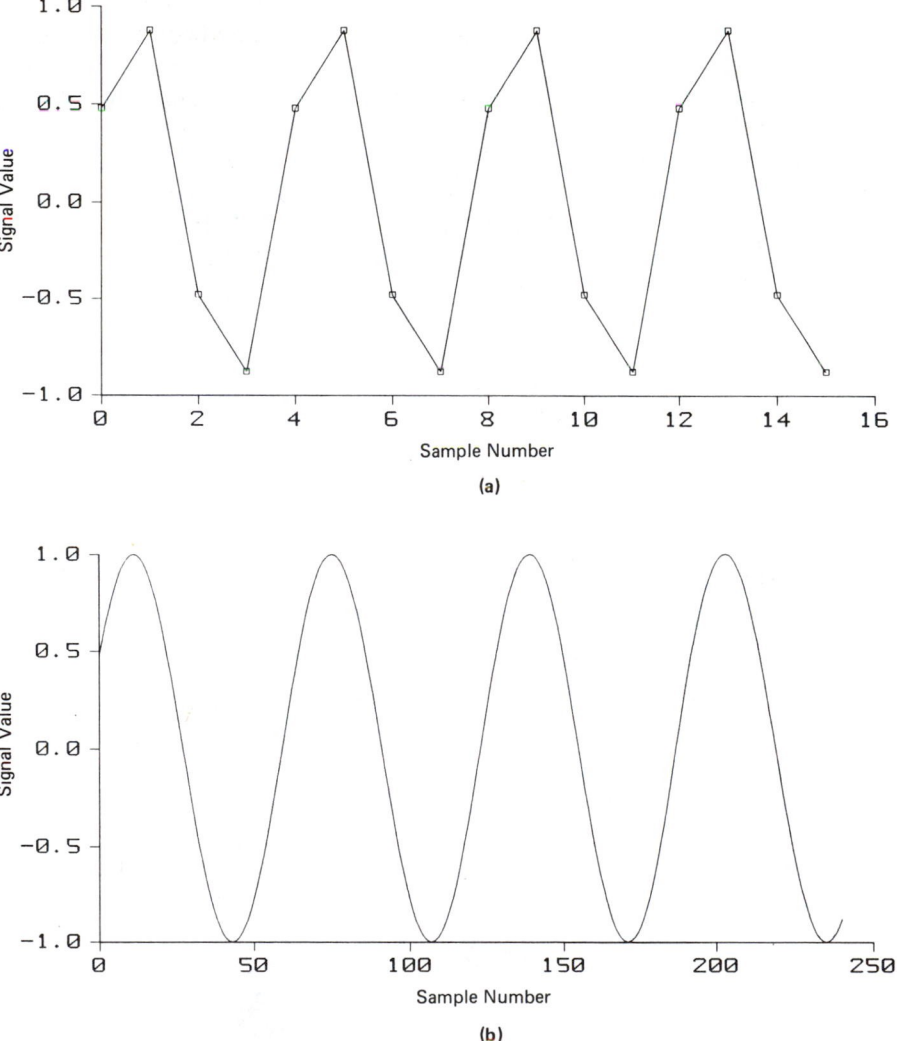

(a)

(b)

Figure 10.13.   Plots of data sequences in SPA1004. (a) Original sequence of 16 points. (b) Interpolated sequence of 241 points.

10.4 Further Theory and Applications

In this section we present some additional remarks on the operation of the routines described in Sections 10.2 and 10.3. The linear decimation and interpolation routines are so simple that they need very little further discussion. As illustrated in Fig. 10.14, a simple rule of proportions is used to find an intermediate value, $x(k + \delta)$, between two successive elements, $x(k)$ and $x(k + 1)$, in a sequence. This simple rule is incorporated in the statements with "X(K) =" in the DO loops of SPDECI and SPLINT. In SPDECI the DO loop proceeds from K = 1 to LX and the sequence is compressed by pulling back the samples from right to left in the sequence. Samples beyond the end of the sequence, that is, beyond K = LX2, are set to zero. In SPLINT, on the other hand, the DO loop proceeds from K = LX to 1 because the sequence is expanded with the added in-between samples, and again samples beyond K = LX2 are set to zero.

The operation of SPZINT follows the steps outlined in the preceding section. The routine computes N and N_2 as in (10.14) and (10.15). If the spectral array size ($N_2 + 2$) is larger than the data array size (LX + 1), an error is indicated.

We noted that the restriction (10.18) on RATIO is rather severe. Even if zero-filling between samples were used explicitly rather than implicitly by extending the DFT with zeros, RATIO would still be restricted in this way in order to produce a spectrum with length equal to a power of 2. Suppose the sequence in (10.8) with Z zeros between each pair of samples is extended with K additional

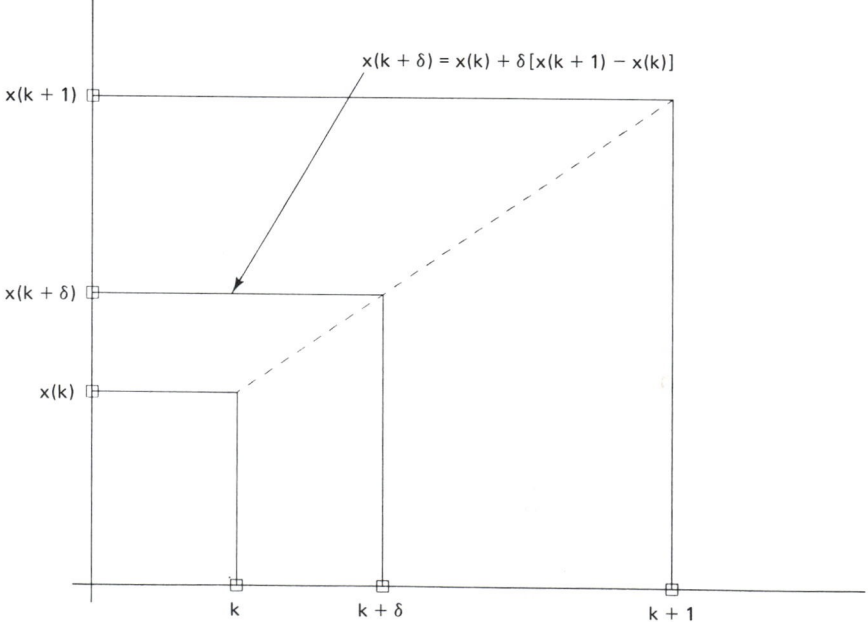

Figure 10.14.   Linear interpolation between two sequence values.

zeros as follows to make $N = (Z + 1)L + K$ a power of 2:

$$N = \text{power of 2 samples}$$

$$[y_k] = [x_0 \;\; \underbrace{0 \ldots 0}_{Z \text{ zeros}} \;\; x_1 \;\; \underbrace{0 \ldots 0}_{Z \text{ zeros}} \;\; x_2 \ldots x_{L-1} \;\; \underbrace{0 \ldots 0}_{Z \text{ zeros}} \;\; \underbrace{0 \ldots 0}_{K \text{ zeros}}] \qquad (10.19)$$

Then, similar to (10.9) through (10.11), we see that the DFT is

$$Y_m = \sum_{k=0}^{N-1} y_k \, e^{-j(2\pi mk/N)}$$

$$= \sum_{i=0}^{L-1} x_i \, e^{-j[2\pi mi(Z+1)/N]}, \qquad m = 0, 1, \ldots, \frac{N}{2} \qquad (10.20)$$

$$= X_m, \qquad m = 0, 1, \ldots, \frac{N}{2(Z+1)} \qquad (10.21)$$

Thus, in (10.21), we have a result similar to (10.11) that is, however, exact only when $(Z + 1)$ divides evenly into $N/2$—that is, only when $(Z + 1)$ is a power of 2 or when RATIO $= 1/(Z + 1)$ is the reciprocal of a power of 2.

So in any case, RATIO is restricted to be the reciprocal of a power of 2 in SPZINT. Having seen how SPZINT operates, we can now see how to use the spectrum-preserving properties of SPZINT along with SPDECI to achieve other rational values for RATIO, for either interpolation or decimation.* The essential theory for this approach is presented by Crochiere and Rabiner [D].

Suppose we wish to change the time step (T) or the inverse of the sampling rate (S) by a ratio given by

$$\text{RATIO} = \frac{T_2}{T_1} = \frac{S_1}{S_2} \qquad (10.22)$$

$$= \frac{I}{J}$$

where I and J are integers and J is a power of 2. Then, with RATIO a fraction in this form, we have an increase in time step (decimation) with $I > J$ and a decrease in time step (interpolation) with $I < J$.

In either case, Fig. 10.15 applies. We first use SPZINT to interpolate the original sequence X(0), . . . , X(LX1) with time step T_1 using the integer ratio $1/J$. To do this we set the array size LX as shown in Fig. 10.15 and we set RATIO $= 1.0/J$. This fixes N and RATIO $= 1/J$ correctly in accordance with (10.14) through (10.16). We execute SPZINT, and then we use SPDECI to decimate the output of SPZINT with RATIO $= I$. The result is a new sequence X(0), . . ., X(LX3), with the final overall ratio equal to I/J. Note that LX2, not LX, is the correct original sequence length to use in the call to SPDECI in Fig. 10.15 and that the final sequence length is LX3.

---

* In the case of decimation, the spectrum is not preserved if the sampling theorem is violated.

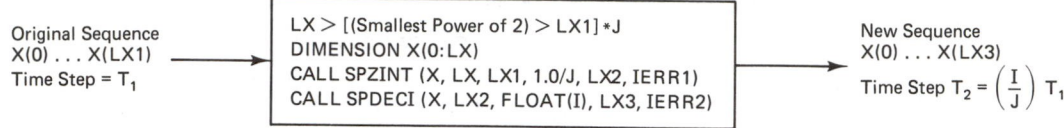

Figure 10.15. Spectrum-preserving interpolation or decimation with the time-step ratio $T_2/T_1 = I/J$, where I and J are integers and J is a power of 2.

An example of the operation in Fig. 10.15 is presented in the program SPA1005. In this example the original sequence, with LX1 = 6, consists of the first 7 samples of the sequence in (10.6). The time step is changed from T_1 to $T_2 = 0.75T_1$. Thus, in this example, $I/J = \frac{3}{4}$ is achievable because $J = 4$ is a power of 2. The intermediate sequence with LX2 = 24, obtained after the call to SPZINT, is printed first below the program, and then the final sequence with LX3 = 8 is printed. We note in the final sequence that the elements $[x_k]$ with k a multiple of 4 are unchanged from the original sequence.

```
        PROGRAM SPA1005
C-SIMPLE EXAMPLE OF INTERPOLATION WITH TIME-STEP RATIO=I/J=3/4.
        DIMENSION X(0:33)
        DATA X/2.,4.,6.,6.,8.,6.,28*0./
        NFILL=3
        CALL SPZINT(X,33,6,1./4.,LX2,IERROR)
        PRINT 1,IERROR,LX2,(X(K),K=0,LX2)
      1 FORMAT(2I4,/(10F7.3))
        CALL SPDECI(X,LX2,3.0,LX3,IERROR)
        PRINT1,IERROR,LX3,(X(K),K=0,LX3)
        STOP
        END
$ RUN SPA1005
   0   24
  2.000   2.367   2.801   3.349   4.000   4.687   5.307   5.762   6.000   6.038
  5.964   5.909   6.000   6.314   6.837   7.461   8.000   8.246   8.027   7.264
  6.000   4.396   2.693   1.156   0.000
   0    8
  2.000   3.349   5.307   6.038   6.000   7.461   8.027   4.396   0.000
FORTRAN STOP
```

In general, when J is not a power of 2 and whether or not the time-step ratio (R) is even a ratio of integers, one can almost obtain a frequency-preserving interpolation by using a scheme like the one suggested in Fig. 10.16. When

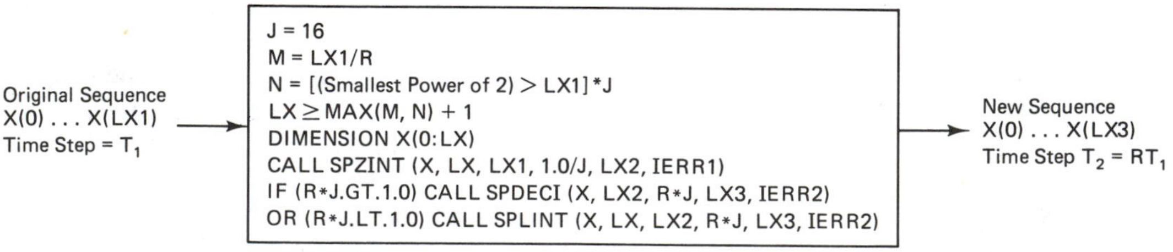

```
J = 16
M = LX1/R
N = [(Smallest Power of 2) > LX1]*J
LX ≥ MAX(M, N) + 1
DIMENSION X(0:LX)
CALL SPZINT (X, LX, LX1, 1.0/J, LX2, IERR1)
IF (R*J.GT.1.0) CALL SPDECI (X, LX2, R*J, LX3, IERR2)
OR (R*J.LT.1.0) CALL SPLINT (X, LX, LX2, R*J, LX3, IERR2)
```

Original Sequence
X(0) . . . X(LX1)
Time Step = T₁

New Sequence
X(0) . . . X(LX3)
Time Step T₂ = RT₁

Figure 10.16. Interpolation or decimation with any time-step ratio, $T_2/T_1 = R$. This version very nearly preserves the original spectrum.

$I = RJ$ is not an integer, the call to SPDECI or SPLINT involves a linear interpolation but, with $J = 16$, the high frequencies in $[x_k]$ still have at least 16 samples per cycle, so, as illustrated in Fig. 10.7, linear interpolation is quite accurate. For greater accuracy we can increase J to 32, 64, or a higher power of 2.

An example using the scheme in Fig. 10.16 is given in SPA1006. In the DO 1 loop the data sequence $[x_k]$, the same sequence used in Fig. 10.10(a), is generated. Then the DFT is taken using SPDFTR and, in the DO 2 loop, the DFT amplitude is scaled by 8.0 and stored in S1, which is printed below the program. We note that S1 is exact to six places with a single nonzero component at $m = 4$. Next, the scheme in Fig. 10.16 is applied to $[x_k]$ using the ratio $R = 0.05$, which would not be a possible ratio in Fig. 10.12. The DFT amplitude of 240 samples (three complete cycles) of $[x_k]$ is then scaled by 120.0 and stored in S2, the first nine components of which are also printed below the program. Here we see that the component with $m = 3$, corresponding with three cycles of $[x_k]$, which should be 1.0, is off in the fourth decimal place and that the neighboring components are close to zero. Thus, interpolation using Fig. 10.16 with $J = 16$ in this case produces an interpolated sequence with a spectrum that is accurate to at least three places, enough for most purposes.

```
        PROGRAM SPA1006
C-SPECTRA OF ORIGINAL (S1) AND INTERPOLATED (S2) SINUSOIDS.
        DIMENSION X(0:513),S1(0:17),S2(0:241)
        PI=4.*ATAN(1.)
        DO 1 K=0,15
          X(K)=SIN(2.*PI*K/4.+0.5)
      1 CONTINUE
        CALL SPDFTR(X,S1,16)
        DO 2 M=0,8
          S1(M)=SQRT(S1(2*M)**2+S1(2*M+1)**2)/8.0
      2 CONTINUE
        CALL SPZINT(X,513,15,1./16.,LX2,IERROR1)
        CALL SPLINT(X,513,LX2,0.05*16,LX3,IERROR2)
        PRINT 3,IERROR1,IERROR2,LX2,LX3
      3 FORMAT('IERROR1,IERROR2,LX2,LX3:',4I5/)
        CALL SPDFTR(X,S2,240)
        DO 4 M=0,120
          S2(M)=SQRT(S2(2*M)**2+S2(2*M+1)**2)/120.0
      4 CONTINUE
        PRINT 5,(M,S1(M),S2(M),M=0,8)
      5 FORMAT('  M SPECTRUM 1 SPECTRUM 2'/(I3,2F11.7))
        STOP
        END
$ RUN SPA1006
IERROR1,IERROR2,LX2,LX3:    0    0  240  300

   M SPECTRUM 1 SPECTRUM 2
   0  0.0000001  0.0000001
   1  0.0000002  0.0000001
   2  0.0000001  0.0000003
   3  0.0000001  0.9992293
   4  1.0000000  0.0000001
   5  0.0000003  0.0000002
   6  0.0000002  0.0000001
   7  0.0000007  0.0000001
   8  0.0000002  0.0000001
FORTRAN STOP
```

10.5 Summary of Calling Sequences

In this chapter we have discussed three routines for decimation and interpolation. These have similar calling sequences, and in this final section we summarize the operation of the routines as follows:

    CALL SPDECI (X, LX, RATIO, LX2, IERROR)
    CALL SPLINT (X, LX, LX1, RATIO, LX2, IERROR)
    CALL SPZINT (X, LX, LX1, RATIO, LX2, IERROR)

Variable	Definition	Input/Output	Remarks
X	Data vector	I,O	Real data
LX	Last index of X(0:LX)	I	Minimum value of LX is different for each routine
LX1	Last index of original sequence	I	No restrictions as long as LX is large enough
RATIO	Time-step ratio, T_2/T_1	I	Range of value differs for each routine
LX2	Last index of output sequence	O	Samples beyond X(LX2) are set to zero
IERROR	Error indicator	O	Indicates invalid value of LX or RATIO

The symbols I and O indicate whether the variables are inputs specified before calling the routine or outputs computed by the routine.

10.6 Exercises

10.1. Assume that the time step is $T_1 = 0.01$ s in the following sequence:
$$[x_k] = [0, 2, 4, 4, 5, 3, 1, -1, 1, 0]$$
Decimate this sequence so that the time step is $T_2 = 0.02$ s. Check the operation of SPDECI in this exercise.

10.2. Use SPDECI to change the time step to $T_2 = 0.015$ s in $[x_k]$ in exercise 10.1. Print the decimated sequence.

10.3. Use SPDECI to change the time step to $T_2 = 0.018$ s in $[x_k]$ in exercise 10.1. Print the decimated sequence.

10.4. Let $T_1 = 0.01$ s in the following sequence:
$$[x_k] = [0, 2, 4, 8, 6, 2, 1, -1, -2, -1, 0]$$
Check the operation of SPLINT by interpolating linearly to obtain $T_2 = 0.005$ s.

10.5. In the sequence in exercise 10.4, interpolate linearly to change the time step to $T_2 = 0.004$ s. Print the new sequence.

10.6. Use SPLINT on the sequence in exercise 4 to change the time step to $T_2 = 0.003$ s.

Print the new sequence. In such a sequence with arbitrary length, which elements will coincide exactly with elements in the original sequence?

10.7. A decaying sinusoidal function is described in equation (3.18). This function and its spectrum are plotted in Fig. 3.4. The plots are based on 1024 samples with sampling rate = 1000 Hz. In Fig. 3.4, we see that the spectrum is concentrated around a peak at about 5.9 Hz. To visualize zero-order-hold accuracy at various sampling frequencies, plot the zero-order-hold version of the waveform with sampling rate equal to 5, 10, 40, and 80 times the frequency of the peak. Compare the result with Fig. 10.6.

10.8. Using first-order instead of zero-order hold, repeat exercise 10.7 at sampling rates of 4, 5, 8, and 10 times the frequency of the peak. Compare your result with Fig. 10.7.

10.9. In this exercise we generate samples of the waveforms in Fig. 10.8. First, compute 512 samples of

$$x(t) = 1.5e^{-5t}\sin(4\pi t)$$

using time step $T_1 = 0.002$ and beginning at $t = 0$. Then compute the spectrum using SPFFTR, truncate the spectrum above 12.5 Hz, and obtain $[x_k]$ using SPIFTR. Second, decimate $[x_k]$ just obtained to produce $[u_k]$ with time step $T_2 = 0.04$ s, noting that the sampling rate is now 25 Hz, or twice the highest frequency in $[x_k]$. Then use SPZINT to produce a third sequence $[v_k]$, which is also limited to 12.5 Hz but has a time step $T_3 = 0.0025$. Plot $[x_k]$, $[u_k]$, and $[v_k]$ for $0 \leq t \leq 1$ and compare with Fig. 10.8(a)–(c).

10.10. In this exercise we examine the DFT of a zero-filled sequence. First, compute 32 samples of $x(t)$ in exercise 10.9, using time step $T = 0.04$ and starting at $t = 0$. Then insert 15 zeros between each pair of samples and after the final sample; then compute the FFT and plot the FFT amplitude. Explain briefly the form of the plot.

10.11. Prove the statement made in connection with Fig. 10.9 and equation (10.11) that half of the spectral repetitions with a zero-filled sequence are duplications of the original spectrum from $m = 0$ to $m = L/2$ and the other half are reflections of this same original spectrum.

10.12. Repeat exercise 10.4 using both SPLINT and SPZINT, and print the interpolated sequences. Then compute and print the amplitude spectra of both sequences. Compare the two spectra.

10.13. Repeat exercise 10.5 using SPLINT and SPZINT. Then compute and print the two amplitude spectra. Comment briefly on the difference between the two spectra.

10.14. In the program SPA1003, explain how SPZINT computes the quantity LX2.

10.15. Execute a version of SPA1003 that produces an interpolated sequence with 16 elements. Then compute and print the amplitude spectrum of the interpolated sequence. Comment briefly on the spectrum.

10.16. Write a program similar to SPA1004 to generate plots similar to Fig. 10.13 but with x_k changed from $\sin(2\pi k/4 + 0.5)$ to $\sin(2\pi k/5 + 0.5)$. Comment on why SPZINT does not produce a perfect sine wave in this case.

10.17. Suppose we have an impulse sequence given by

$$[x_k] = [\ldots\ 0\ 0\ 1\ 0\ 0\ \ldots]$$

Suppose the sequence is very long in both directions and has only the one nonzero sample at $k = 0$ and that the time step is T. What continuous function $x(t)$ has such a sample set and is limited to frequencies below $1/2T$? (*Hint:* To rephrase the question, what continuous function has a spectrum similar to the DFT of $[x_k]$ at frequencies below $1/2T$ and is equal to $[x_k]$ at the sample points $t = 0, \pm T, \pm 2T$, and so on?)

10.18. Demonstrate the answer to exercise 10.17 by using SPZINT with the unit impulse sequence. Start with a sequence of 25 elements having the impulse at its center. Reduce the time step enough to produce a smooth curve, and plot the interpolated result. Also, compute and plot the amplitude spectrum of the interpolated result.

10.19. Using the scheme in Fig. 10.15, reduce the time step in $[x_k]$ of exercise 10.4 from $T_1 = 10$ ms to $T_2 = 3.75$ ms. Print the interpolated sequence.

10.20. Using the scheme in Fig. 10.15, reduce the time step in $[x_k]$ of exercise 10.4 to $T_2 = 0.9375$ ms. Plot the original and interpolated sequences, with straight lines between sample points.

10.21. Run a program similar to SPA1006, but with J in Fig. 10.16 increased to 32. As in SPA1006, use an integral number of cycles of the sinusoid when finding the amplitude spectrum, and note the increased accuracy in the spectrum of the interpolated sequence.

10.22. Using the SPRAND routine with zero mean, without scaling, and with ISEED = 121, generate a random sequence with 500 samples. Then, using the procedure in Fig. 10.16, interpolate to reduce the time step by the ratio $R = 0.2$. Then compute and plot the complete power spectrum of the interpolated sequence. Make the plot in decibels versus frequency index.

10.23. Do exercise 10.22, but with J in Fig. 10.16 increased from 16 to 32.

10.24. Generate 50 samples of the function

$$x(t) = 1.5e^{-5t}\sin(4\pi t)$$

starting at $t = 0$, with time step $T_1 = 40$ ms. Using the procedure in Fig. 10.16, reduce the time step to $T_2 = 1.0$ ms. Plot the interpolated sequence over the interval $0 \leq t \leq 1.0$ s. Then, using 500 samples to obtain a smooth waveform, plot $x(t)$ in the same coordinates, and observe the difference in the plots. Comment briefly on the sources of any differences between the two plots.

10.7 References

[A] Oppenheim, A. V., and A. S. Willsky, *Signals and Systems* (Englewood Cliffs, N.J.: Prentice-Hall, Inc., 1983), Section 8.1.

[B] Tretter, S. A., *Introduction to Discrete-Time Signal Processing* (New York: John Wiley and Sons, 1976), Section 3.5.

[C] Stearns, S. D., *Digital Signal Analysis* (Rochelle Park, N.J.: Hayden, 1975), Chapter 4.

[D] Crochiere, R. E., and L. R. Rabiner, *Multirate Digital Signal Processing* (Englewood Cliffs, N.J.: Prentice-Hall, Inc., 1983), Chapter 2, especially Section 2.3.

[E] Crochiere, R. E., "A General Program to Perform Sampling Rate Conversion of Data by Rational Ratios," in *Programs for Digital Signal Processing* (New York: IEEE Press, 1979), p. 8.2.

chapter **11**

Least-Squares Design and Modeling

11.1 Introduction

The problem of fitting a linear function to a set of data points arises frequently in digital signal processing. One of the applications is data smoothing (other than filtering). It is often more meaningful to characterize experimental results via a smooth function, which is, in some sense, the best fit to the given data set. Data smoothing is also useful in experiments where it is necessary to differentiate the recorded signal [A]. Additional important applications exist in the areas of system identification and deconvolution [F, H, J].

While a number of different error minimization criteria have been suggested, we restrict this discussion to the least-squares approximation [A, B, C]. This technique, which was originated by Gauss, is the most widely used [A]. Simply stated, the least-squares principle involves selecting the function that minimizes the sum of the squared errors. Given two N-point data sets $[x_k]$ and $[y_k]$, the total squared error in this approximation can be expressed as

$$S = \sum_{k=0}^{N-1} [y_k - f(x_k)]^2 \tag{11.1}$$

where $f(x)$ is some function of x_k whose coefficients are adjusted to minimize S [A].

In the next two sections we describe routines that compute the least-squares polynomials for data smoothing. Later in this chapter we discuss some specific tasks involving system identification and deconvolution. Due to the special structures that arise in these applications, different algorithms are available to solve the least-squares problem. In particular, we describe routines for Levinson's algorithm and Durbin's algorithm.

227

11.2 Curve Fitting via the Normal Equations

In this section we consider least-squares approximation via the Lth-order polynomial

$$y = f(x) = b_0 + b_1 x + b_2 x^2 + \cdots + b_L x^L \tag{11.2}$$

Our goal is to find the set of coefficients $[b_n]$ that minimizes the sum

$$S = \sum_{k=0}^{N-1} [y_k - (b_0 + b_1 x_k + \cdots + b_L x_k^L)]^2 \tag{11.3}$$

where $[x_k]$ and $[y_k]$ are N-point data sequences and L, the order of the polynomial, is less than N, the number of data points. In applications described in this text, x, the independent variable, typically corresponds to time, whereas y, the dependent variable, denotes the signal amplitude.

It is well known that the least-squares polynomial approximation in (11.2) is obtained by solving a set of $L + 1$ equations to find the $L + 1$ coefficients of b_n [A, C]. These equations, called the *normal equations*, may be expressed in matrix form as

$$\mathbf{Ab} = \mathbf{c} \tag{11.4}$$

where $\mathbf{A}$ is the $L + 1$ by $L + 1$ symmetric matrix

$$\mathbf{A} = \begin{bmatrix} a_0 & a_1 & \cdots & a_L \\ a_1 & a_2 & \cdots & a_{L+1} \\ \cdot & \cdot & \cdots & \cdot \\ \cdot & \cdot & \cdots & \cdot \\ a_L & a_{L+1} & \cdots & a_{2L} \end{bmatrix} \quad \text{with } a_n = \sum_{k=0}^{N-1} x_k^n$$

and $L + 1$ by 1 vectors $\mathbf{b}$ and $\mathbf{c}$ are given by

$$\mathbf{b} = [b_0 \, b_1 \ldots b_L]^T = \text{polynomial coefficient vector}$$

$$\mathbf{c} = [c_0 \, c_1 \ldots c_L]^T \quad \text{with entries } c_n = \sum_{k=0}^{N-1} y_k x_k^n$$

Note that $\mathbf{A}$ and $\mathbf{c}$ are completely specified by the data sets $[x_k]$ and $[y_k]$.

The solution of (11.4) yields a unique set of $[b_n]$ coefficients, which minimizes the squared error in (11.3) [A]. Unfortunately, except for very low-order polynomials, the normal equations in (11.4) are typically ill-conditioned. Thus, although the b_n coefficients are uniquely defined, in practice it is often difficult to compute them.

A standard technique for solving a matrix equation in the form of (11.4) is to invert the A-matrix and compute the solution vector $\mathbf{b} = \mathbf{A}^{-1}\mathbf{c}$. This approach is feasible only if $\mathbf{A}$ is nonsingular, that is, $\mathbf{A}^{-1}$ exists, and $\mathbf{A}$ is well-conditioned [P]. Since $\mathbf{A}$ in (11.4) is positive definite, we know that it is nonsingular [P]. Even so, some caution should be exercised in attempting this solution to the least-squares problem since, as just mentioned, $\mathbf{A}$ is often ill-conditioned.

Gaussian elimination is probably the most frequently used technique for solving a set of simultaneous linear equations [A, B, D, P]. This involves a successive reduction in the size of the problem. Multiples of one equation are subtracted from other equations in the system in such a way that one unknown is eliminated at each stage of the procedure, until finally there is one equation in one unknown that may be solved immediately. All other unknowns are then found by working the system of equations in reverse order. Unfortunately, this technique also encounters difficulties if the system of equations is singular or ill-conditioned, that is, nearly singular [P]. If this is the case, small changes in the elements of $\mathbf{A}$ and $\mathbf{c}$ may result in large errors in the solution vector $\mathbf{b}$. Even ordinary rounding errors are often enough to distort the solution and make it useless.

More sophisticated techniques, which are beyond the scope of this text, are also available for the solution of matrix equations such as (11.4) [P]. Since this is actually a task in numerical analysis, subroutines which enable solution of the $\mathbf{Ab} = \mathbf{c}$ system of equations are generally provided in computer math libraries.

To provide a simple approach and yet avoid some of the numerical difficulties encountered in solving for the least-squares polynomial in (11.2)–(11.4), the task is separated into two steps. The first subroutine in this section will formulate the $\mathbf{A}$-matrix and the $\mathbf{c}$-vector from data sequences $[x_k]$ and $[y_k]$. The second subroutine, which is included because it is simple and convenient, will solve for the $\mathbf{b}$-vector via Gaussian elimination. It is, however, recommended that math library routines be used to accomplish this second step whenever available (especially for high-order polynomials).

The calling sequence and argument definitions for subroutine SPNORM, which initializes $\mathbf{A}$ and $\mathbf{c}$ in the normal equations in (11.4), are provided next.

CALL SPNORM(X,Y,N,L,A,C,IERROR)

X(0:N − 1)	= data array containing independent variable values
Y(0:N − 1)	= data array containing dependent variable values
N	= number of data points in X and Y
L	= order of polynomial in (11.2); L < N
A(0:L,0:L)	= symmetric A-matrix in (11.4)—returned
C(0:L)	= vector c in (11.4)—returned
IERROR	= 0 no errors detected
	1 invalid number of points—N $\leqslant$ 0
	2 invalid polynomial order—L $\leqslant$ 0 or L $\geqslant$ N

In program SPA1101, SPNORM was used to set up the normal equations for a third-order polynomial fit to data sequences $[x_k]$ and $[y_k]$. The data samples were generated by evaluating the amplitude of a cubic equation at 51 points. The $\mathbf{A}$-matrix and $\mathbf{c}$-vector computed by SPNORM are shown below the program listing.

```
            PROGRAM SPA1101
C-DEMONSTRATE THE USE OF SPNORM
            DIMENSION X(0:50),Y(0:50),A(0:3,0:3),B(0:3),C(0:3)
            DATA N/51/,L/3/
            DO 1 K=0,N-1
              X(K)=K/50.
              Y(K)=10.*(X(K)**3)-3.*(X(K)**2)-5.*X(K)-1.
      1     CONTINUE
            CALL SPNORM(X,Y,N,L,A,C,IERROR)
            IF(IERROR.NE.0)PRINT *,' SPNORM ERROR = ',IERROR
            PRINT 100,((A(I,J),J=0,3),C(I),I=0,3)
    100     FORMAT(19X,'A-MATRIX',24X,'C',/(4F10.2,5X,F10.2))
            END

   $ RUN SPA1101
                        A-MATRIX                            C
         51.00      25.50      17.17      13.01          -99.96
         25.50      17.17      13.01      10.51          -45.30
         17.17      13.01      10.51       8.84          -25.30
         13.01      10.51       8.84       7.65          -15.53
   $
```

In this example, since the $[y_k]$ sequence consisted of *exact*, that is, noise-free, samples from a cubic equation, solution of the third-order normal equations should yield the same polynomial coefficients that were used to generate the data. In order to demonstrate that this is the case, we use subroutine SPGAUS to solve for the **b**-vector. The calling sequence for SPGAUS is

CALL SPGAUS(A,C,L,B,IERROR)

A(0:L,0:L)	=	**A**-matrix in **Ab** = **c** system of equations
C(0:L)	=	vector **c** in **Ab** = **c**
L	=	order of system of equations
B(0:L)	=	solution vector **b**—returned
IERROR	= 0	no errors detected
	1	invalid order—L $\leq$ 0
	2	ill-conditioned—pivot $< 1.E-10$

The SPGAUS subroutine uses Gaussian elimination to find the solution vector **b** that corresponds to the general **Ab** = **c** system of linear equations. The fact that **A** in (11.4) is a symmetric matrix is not used by SPGAUS to reduce the computational burden. As a result, the use of SPGAUS is not restricted to solving normal equations (or to solving other symmetric systems). This routine should, however, be used cautiously, since it is a simple implementation of Gaussian elimination. No row exchanges are performed in order to avoid zero or very small pivots [A, D, P]. These conditions will result in an IERROR=2 return with no solution. This problem should not, however, be encountered in practical signal processing computations with positive definite matrices [P].

SPA1102 utilizes SPGAUS to solve the third-order system of normal equations formulated in SPA1101. The computed coefficients are listed below the program. Note that they are identical to the coefficients of the cubic equation used to generate the $[y_k]$ sequence.

```
            PROGRAM SPA1102
C-DEMONSTRATE THE USE OF SPGAUS
            DIMENSION X(0:50),Y(0:50),A(0:3,0:3),B(0:3),C(0:3)
            DATA N/51/,L/3/
            DO 1 K=0,N-1
              X(K)=K/50.
              Y(K)=10.*(X(K)**3)-3.*(X(K)**2)-5.*X(K)-1.
     1      CONTINUE
            CALL SPNORM(X,Y,N,L,A,C,IERROR)
            IF(IERROR.NE.0) PRINT *,' SPNORM ERROR = ',IERROR
            CALL SPGAUS(A,C,L,B,IERROR)
            IF(IERROR.NE.0)PRINT *,' SPGAUS ERROR = ',IERROR
            PRINT 100,(I,B(I),I=0,3)
     100    FORMAT(' POLYNOMIAL COEFFICIENTS',/(' B(',I1,') = ',F7.2))
            END


            $ RUN SPA1102
            POLYNOMIAL COEFFICIENTS
            B(0) =    -1.00
            B(1) =    -5.00
            B(2) =    -3.00
            B(3) =    10.00
            $
```

In SPA1102, the normal equations were used to find the least-squares polynomial in a case where both the samples and the order of the system were exact. These conditions are not often satisfied in practice. In a typical experiment, the recorded data, that is, the $[y_k]$ sequence, is noisy, and the order of the underlying polynomial can only be estimated. Hence, SPA1103 is provided to demonstrate least-squares polynomial fitting via SPNORM and SPGAUS for a more realistic case.

In SPA1103, the $[y_k]$ sequence is generated using samples from the cubic equation in SPA1102 with additive zero-mean noise. A fifth-order least-squares polynomial is computed in the program. Thus, neither the data samples nor the order of the normal equations is ideal. It is not evident from the magnitudes of the computed coefficients that the underlying polynomial used to generate $[y_k]$ was a cubic equation. It is, however, apparent from the results shown in Fig. 11.1 that the curve defined by the b_n coefficients fits the data samples. In this plot, the asterisks denote the noisy y_k samples, and the solid curve illustrates the computed fifth-order least-squares polynomial.

```
            PROGRAM SPA1103
C-DEMONSTRATE LEAST-SQUARES DATA FITTING VIA SPNORM AND SPGAUS
            DIMENSION X(0:50),Y(0:50),A(0:5,0:5),B(0:5),C(0:5),POLY(0:500)
            DATA N/51/,L/5/,ISEED/12357/
            DO 1 K=0,N-1
              X(K)=K/50.
              Y(K)=10.*(X(K)**3)-3.*(X(K)**2)-5.*X(K)-1.
     +               + .4*(SPRAND(ISEED)-.5)
     1      CONTINUE
            CALL SPNORM(X,Y,N,L,A,C,IERROR)
            IF(IERROR.NE.0) PRINT *,' SPNORM ERROR = ',IERROR
            CALL SPGAUS(A,C,L,B,IERROR)
            IF(IERROR.NE.0)PRINT *,' SPGAUS ERROR = ',IERROR
            DO 2 K=0,500
              PX=K/500.
              POLY(K)=B(0)+B(1)*PX+B(2)*(PX**2)+B(3)*(PX**3)+B(4)*(PX**4)+
     +               B(5)*(PX**5)
     2      CONTINUE
            CALL MCPLOT(0.,.02,Y,51,9,12,2,1,1,1,5.,.3.6)
            CALL MCPLOT(0.,.002,POLY,501,0,0,2,2,1,1,5.,.3.6)
            PRINT 100,(I,B(I),I=0,5)
     100    FORMAT(' POLYNOMIAL COEFFICIENTS',/('   B(',I1,') = ',F7.2))
            END
```

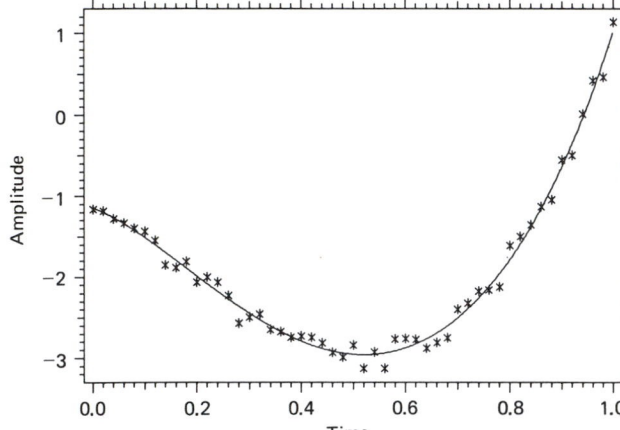

```
$ RUN SPA1103
POLYNOMIAL COEFFICIENTS
    B(0) =    -1.14
    B(1) =    -2.55
    B(2) =   -14.18
    B(3) =    33.49
    B(4) =   -23.09
    B(5) =     8.51
$
```

Figure 11.1. SPA1103: Fifth-order least-squares polynomial computed via SPNORM and SPGAUS.

We have alluded to the difficulties that arise in attempting to solve a set of normal equations that are ill-conditioned or of high order. In these cases, the simple SPGAUS routine may not generate an accurate solution. At this point we provide no alternative beyond suggesting the use of numerical analysis routines from a computer math library. In Section 11.3 we do, however, consider a different technique for least-squares curve fitting.

11.3 Curve Fitting via Orthogonal Polynomials

Here we take a different approach to the problem of finding a curve that provides the least-squares Lth-order fit to a given data sequence. Rather than using the single polynomial in (11.2) as the linear function that minimizes S in (11.1), we use a set of polynomials that are orthogonal over the range of the data sequence [A]. This technique has a number of advantages. The polynomial coefficients can be computed in a single step without having to solve a potentially ill-conditioned system of linear equations. In addition, due to the orthogonality of the polynomial functions, the relative magnitudes of the coefficients will indicate the order of the underlying polynomial even when the system is overspecified [A].

The orthogonal polynomial method discussed next has two limitations. The first is that data samples must occur at regular intervals. This was not essential in the approach described in Section 11.2, since the $[x_k]$ sequence was used explicitly in formulating the normal equations. The second constraint is that the order of the least-squares fit must be less than the number of data samples in $[y_k]$.

The linear function used to minimize the squared error in (11.1) now has the form

$$y = f(x) = b_0 P_0(t) + b_1 P_1(t) + \cdots + b_L P_L(t) \tag{11.5}$$

where

$$t = \frac{x - x_0}{dx}$$

and $[P_n(t)]$ is a family of orthogonal polynomials. The argument t in (11.5) denotes a dummy variable that adjusts the polynomials to compensate for a nonzero starting time x_0 and a nonunity sample interval dx. The polynomials are defined by

$$P_n(t) = \sum_{i=0}^{n} (-1)^i \binom{n}{i} \binom{n+i}{i} \frac{t^{(i)}}{(N-1)^{(i)}}, \qquad n = 0, 1 \ldots, L \tag{11.6}$$

with the following notation being introduced for convenience [A]:

$$u^{(i)} = u(u - 1)(u - 2) \cdots (u - i + 1)$$

$$\binom{m}{i} = \frac{m!}{i!(m - i)!}$$

In (11.6), N specifies the total number of data samples in the $[y_k]$ sequence.

The interpolating polynomials in (11.5) and (11.6) vary in order from 0 to L. By evaluating (11.6) with $n = 0, 1, 2, 3$, we see that the first four polynomials can be written:

$$P_0(t) = 1$$

$$P_1(t) = 1 - \frac{2t}{N - 1}$$

$$P_2(t) = 1 - \frac{6t}{N - 1} + \frac{6t(t - 1)}{(N - 1)(N - 2)} \tag{11.7}$$

$$P_3(t) = 1 - \frac{12t}{N - 1} + \frac{30t(t - 1)}{(N - 1)(N - 2)} - \frac{20t(t - 1)(t - 2)}{(N - 1)(N - 2)(N - 3)}$$

Note that each $P_n(t)$ polynomial could be expressed in the form of (11.2) by combining coefficients of like powers of t. The $[b_n]$ coefficients corresponding to (11.5) that satisfy the least-squares criteria are given by [A]

$$b_n = \frac{\displaystyle\sum_{k=0}^{N-1} y_k P_n(k)}{\displaystyle\sum_{k=0}^{N-1} P_n^2(k)}, \qquad n = 0, 1, \ldots, L \tag{11.8}$$

Due to the orthogonality of the functions in (11.5), these coefficients can be computed exactly using $[y_k]$ and the polynomial definition in (11.6). Thus, we avoid the pitfalls encountered in computing with ill-conditioned systems. In addition, due to the orthogonality of the P_n polynomials, the b_n coefficients are uncoupled [A].

Subroutine SPORTH generates the b_n coefficients defined in (11.8) for an Lth-order fit to the N-point sequence $[y_k]$. The calling sequence is

CALL SPORTH(Y,N,L,ORTHB,WORK,IERROR)

Y(0:N − 1)	= dependent variable data—evenly spaced samples
N	= number of data samples
L	= order of least-squares fit
ORTHB(0:L)	= polynomial coefficients in (11.5)—returned
WORK(0:L,0:L)	= work array used internally
IERROR	= 0 no errors detected
	1 either N or $L \leqslant 0$
	2 $N < L$—not enough points

To generate the curve corresponding to the ORTHB coefficients returned by SPORTH, (11.5) must be evaluated. This, in turn, requires an evaluation of (11.6) for each of the L polynomials at each point on the curve. This is an extremely computationally intensive task, and for this reason the orthogonal polynomial coefficients are of limited use.

Recall from (11.7) that each of the $P_n(t)$ polynomials can be expressed in the form of (11.2). This suggests that (11.5) could be written as

$$y = \bar{b}_0 + \bar{b}_1 t + \cdots + \bar{b}_L t^L \qquad (11.9)$$

where $[\bar{b}_n]$ is a new set of coefficients generated by combining like powers of t in (11.5). This form has several advantages, the primary one being that it is simpler to generate points on a curve via (11.9) than by using (11.5). Equation (11.9) is also a more familiar polynomial function. In applications where curve differentiation is needed, the form of (11.9) is preferable due to its simplicity.

Thus, the only problem that remains is the conversion from the $[b_n]$ coefficients in (11.5) to the $[\bar{b}_n]$ coefficients in (11.9). Equation (11.5) can be expressed in matrix form as follows:

$$y = \mathbf{b}^T \mathbf{P} \mathbf{t} = [b_0\ b_1\ \ldots\ b_L] \begin{bmatrix} 1 & 0 & 0 & . & . & 0 \\ 1 & p_{11} & 0 & . & . & 0 \\ 1 & p_{21} & p_{22} & . & . & 0 \\ . & . & . & . & . & . \\ . & . & . & . & . & . \\ 1 & p_{L1} & . & . & . & p_{LL} \end{bmatrix} \begin{bmatrix} 1 \\ t^{(1)} \\ t^{(2)} \\ . \\ . \\ t^{(L)} \end{bmatrix} \qquad (11.10)$$

where

$$p_{ij} = (-1)^j \binom{i}{j} \binom{i+j}{j} \frac{1}{(N-1)^{(j)}}$$

Our goal is to convert (11.10) into the form of (11.9), which may be written as

$$y = \bar{\mathbf{b}}^T \tilde{\mathbf{t}} = [\bar{b}_0 \ \bar{b}_1 \ ... \ \bar{b}_L] \begin{bmatrix} 1 \\ t \\ \vdots \\ t^L \end{bmatrix} \tag{11.11}$$

This is accomplished by using Stirling's numbers of the first kind to expand the $t^{(i)}$ polynomials [A]. We write

$$t^{(i)} = S_1^{(i)} t + S_2^{(i)} t^2 + \cdots + S_i^{(i)} t^i \tag{11.12}$$

where Stirling's numbers are defined recursively as

$$S_i^{(n+1)} = S_{i-1}^{(n)} - n S_i^{(n)}$$

with $\hspace{8cm}$ (11.13)

$$S_1^{(n+1)} = -n S_1^{(n)} \quad \text{and} \quad S_{n+1}^{(n+1)} = S_n^{(n)}$$

We can now express the t-vector in (11.10) as

$$\mathbf{t} = \mathbf{S} \tilde{\mathbf{t}} \tag{11.14}$$

where

$$\mathbf{S} = \begin{bmatrix} 1 & 0 & 0 & . & . & 0 \\ 0 & 1 & 0 & . & . & 0 \\ 0 & s_{21} & 1 & . & . & 0 \\ . & . & . & . & . & . \\ . & . & . & . & . & . \\ 0 & s_{L1} & s_{L2} & . & . & 1 \end{bmatrix} \quad \text{with } s_{ij} = S_j^{(i)}$$

and Stirling's numbers are as defined in (11.13). By combining (11.10) through (11.14), we can write

$$y = \mathbf{b}^T \mathbf{P} \mathbf{S} \tilde{\mathbf{t}} \tag{11.15}$$

and therefore

$$\bar{\mathbf{b}} = \mathbf{b}^T \mathbf{P} \mathbf{S} \tag{11.16}$$

Thus, the desired $[\bar{b}_n]$ coefficients can be generated directly from the orthogonal polynomial coefficients $[b_n]$ via the matrix multiplications in (11.16).

The key feature of the curve-fitting technique outlined in this section is that no matrix inversions or iterative elimination procedures are needed. All values are computed exactly (within the precision of the computer). In addition, by first computing the orthogonal polynomial coefficients, we gain some insight into the order of the underlying system.

The primary use of subroutine SPORTH as a separate module is for order determination. If the relative magnitudes of the coefficients returned by SPORTH indicate that the order has been overspecified, L can be reduced before proceeding with the coefficient conversion in (11.16).

A more complete subroutine, SPPOLY, performs the entire task, returning $\bar{b}_n$ coefficients as well as the b_n values from SPORTH. The calling sequence is

CALL SPPOLY(DX,Y,N,L,B,ORTHB,WORK,IERROR)

DX	= sample interval, $x_k - x_{k-1}$ (constant)
Y(0:N−1)	= array of N equally spaced samples
N	= number of data samples in Y
L	= order of least-squares fit
B(0:L)	= coefficients: $b_0 + b_1 x + \cdots + b_L x^L$—returned
ORTHB(0:L)	= SPORTH coefficients in (11.5)—returned
WORK(0:L,0:L)	= work array used internally
IERROR	= 0 no errors detected
	1 N or L zero or negative
	2 N < L—too few points for least-squares fit
	3 DX ≤ 0—invalid sample interval

This subroutine first uses SPORTH to generate the b_n coefficients in (11.5). These values are then converted to the $\bar{\mathbf{b}}$-vector in (11.9) via (11.16). Finally, this vector is scaled to compensate for the sample interval DX. Thus, the returned B-vector is in the form of (11.2) rather than in terms of the dummy variable t, as in (11.9). In SPPOLY we have assumed a starting time of $x_0 = 0.0$ for convenience.

Test program SPA1104 repeats the task performed in SPA1102, once again fitting a third-order polynomial to samples generated via a cubic equation. Since the 51 samples are distributed over the interval from 0 to 1, the sample interval is DX = $\frac{1}{50}$. The program and its results are provided next. Note that the B coefficients are again identical to those in the cubic equation used to generate $[y_k]$.

```
        PROGRAM SPA1104
C-DEMONSTRATE DATA FITTING VIA ORTHOGONAL POLYNOMIALS
        DIMENSION Y(0:50),WORK(0:3,0:3),ORTHB(0:3),B(0:3)
        DATA N/51/,L/3/,DX/.02/
        DO 1 K=0,N-1
          PX=K/50.
          Y(K)=10.*(PX**3)-3.*(PX**2)-5.*PX-1.
     1  CONTINUE
        CALL SPPOLY(DX,Y,N,L,B,ORTHB,WORK,IERROR)
        IF(IERROR.NE.0)PRINT *,' SPPOLY ERROR = ',IERROR
        PRINT 100,(I,ORTHB(I),I,B(I),I=0,3)
   100  FORMAT(10X,' POLYNOMIAL COEFFICIENTS',/,(' ORTHB(',I1,') ='
       +      ,F8.2,10X,'B(',I1,') = ',F8.2))
        END

    $ RUN SPA1104
            POLYNOMIAL COEFFICIENTS
        ORTHB(0) =    -1.96      B(0) =     -1.00
        ORTHB(1) =    -0.53      B(1) =     -5.00
        ORTHB(2) =     1.96      B(2) =     -3.00
        ORTHB(3) =    -0.47      B(3) =     10.00
    $
```

In the next example, SPA1105, we again compute the fifth-order least-squares fit to noisy samples from the cubic equation. The results from this program should be compared to those obtained from SPA1103. Note that the relative magnitudes of the last two ORTHB coefficients are very small. This suggests that the order of the least-squares fit was overspecified. In comparing the B-values to those obtained via the solution of the normal equations in SPA1103, we see that they differ slightly. This illustrates that whereas the $[b_n]$ coefficients for an Lth-order least-squares polynomial fit are uniquely defined, actual computed values may vary due to computational procedures. Comparing the plot in Fig. 11.2 with the plot from SPA1103 in Fig. 11.1, we see that the curve generated via the SPA1105 coefficients provides an equally good fit to the $[y_k]$ data.

```
              PROGRAM SPA1105
C-DEMONSTRATE LEAST-SQUARES DATA FITTING VIA ORTHOGONAL POLYNOMIALS
              DIMENSION Y(0:50),WORK(0:5,0:5),ORTHB(0:5),B(0:5),PY(0:500)
              DATA N/51/,L/5/,ISEED/12357/,DX/.02/
              DO 1 K=0,N-1
                PX=K/50.
                Y(K)=10.*(PX**3)-3.*(PX**2)-5.*PX-1. + .4*(SPRAND(ISEED)-.5)
       1      CONTINUE
              CALL SPPOLY(DX,Y,N,L,B,ORTHB,WORK,IERROR)
              IF(IERROR.NE.0)PRINT *,' SPPOLY ERROR = ',IERROR
              DO 2 K=0,500
                PX=K/500.
                PY(K)=B(0)+B(1)*PX+B(2)*(PX**2)+B(3)*(PX**3)+B(4)*(PX**4)+
       +                B(5)*(PX**5)
       2      CONTINUE
              CALL MCPLOT(0.,.02,Y,51,9,12,2,1,1,1,5.,3.6)
              CALL MCPLOT(0.,.002,PY,501,0,0,2,2,1,1,5.,3.6)
              PRINT 100,(I,ORTHB(I),I,B(I),I=0,5)
      100     FORMAT(10X,' POLYNOMIAL COEFFICIENTS',/,(' ORTHB(',I1,') = ',
       +                F8.2,10X,'B(',I1,') = ',F8.2))
              END

              $ RUN SPA1105
                       POLYNOMIAL COEFFICIENTS
              ORTHB(0) =      -1.93        B(0) =      -1.14
              ORTHB(1) =      -0.54        B(1) =      -2.63
              ORTHB(2) =       1.90        B(2) =     -13.62
              ORTHB(3) =      -0.52        B(3) =      32.04
              ORTHB(4) =      -0.02        B(4) =     -21.49
              ORTHB(5) =      -0.03        B(5) =       7.89
              $
```

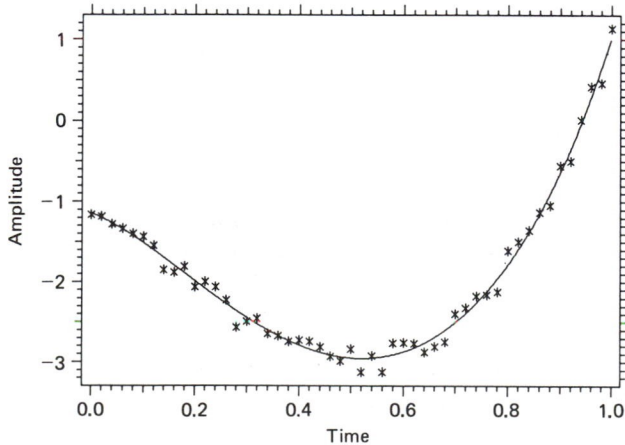

Figure 11.2. SPA1105: Fifth-order least-squares polynomial fit computed via SPPOLY.

Since the results from SPA1105 suggest that the polynomial order was larger than necessary, this task was repeated in SPA1106 with L = 3. A program listing and its results are provided. As expected, the ORTHB coefficients are identical to the first four coefficients from the previous example. The B-coefficients, however, change drastically due to the order reduction. Nevertheless, the plot in Fig. 11.3 indicates that the fit obtained via the third-order system is comparable to the fifth-order polynomial in SPA1105.

```
            PROGRAM SPA1106
C-DEMONSTRATE LEAST-SQUARES DATA FITTING VIA ORTHOGONAL POLYNOMIALS
            DIMENSION Y(0:50),WORK(0:3,0:3),ORTHB(0:3),B(0:3),PY(0:500)
            DATA N/51/,L/3/,ISEED/12357/,DX/.02/
            DO 1 K=0,N-1
            PX=K/50.
            Y(K)=10.*(PX**3)-3.*(PX**2)-5.*PX-1. + .4*(SPRAND(ISEED)-.5)
        1   CONTINUE
            CALL SPPOLY(DX,Y,N,L,B,ORTHB,WORK,IERROR)
            IF(IERROR.NE.0)PRINT *,' SPPOLY ERROR = ',IERROR
            DO 2 K=0,500
            PX=K/500.
            PY(K)=B(0)+B(1)*PX+B(2)*(PX**2)+B(3)*(PX**3)
        2   CONTINUE
            CALL MCPLOT(0.,.02,Y,51,9,12,2,1,1,1,5.,3.6)
            CALL MCPLOT(0.,.002,PY,501,0,0,2,2,1,1,5.,3.6)
            PRINT 100,(I,ORTHB(I),I,B(I),I=0,3)
      100   FORMAT(10X,' POLYNOMIAL COEFFICIENTS',/,(' ORTHB(',I1,') = ',
            +      F8.2,10X,'B(',I1,') = ',F8.2))
            END

        $ RUN SPA1106
                  POLYNOMIAL COEFFICIENTS
            ORTHB(0) =    -1.93       B(0) =    -1.09
            ORTHB(1) =    -0.54       B(1) =    -4.01
            ORTHB(2) =     1.90       B(2) =    -4.91
            ORTHB(3) =    -0.52       B(3) =    11.04
        $
```

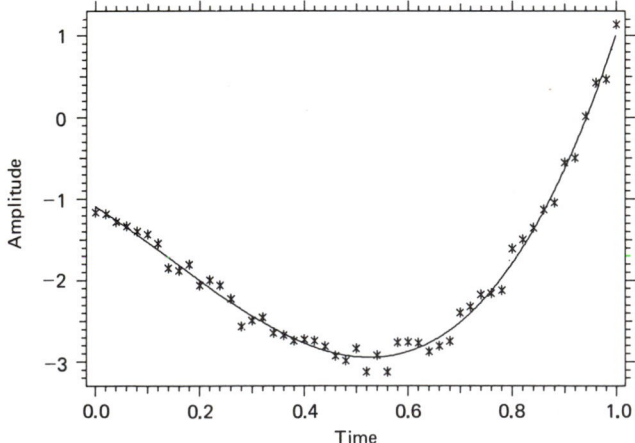

Figure 11.3. SPA1106: Third-order least-squares polynomial fit computed via SPPOLY.

The SPPOLY routine described in this section provides an alternate technique for generating coefficients for the general Lth-order least-squares polynomial to fit a given data sequence. Due to the use of orthogonal polynomials for intermediate steps, the results from this routine are generally more informative and more accurate than those obtained via the solution of the normal equations as described in Section 11.2.

11.4 Levinson's Algorithm

In Section 11.2 we saw that the task of finding the Lth-order least-squares polynomial to fit an arbitrary data sequence could be accomplished by solving the system of linear equations $\mathbf{Ab} = \mathbf{c}$. Although the A-matrix was symmetric, the solution techniques described so far have not utilized this symmetry to improve computational efficiency.

It is often advantageous, or even essential, to exploit special structures in the system of equations $\mathbf{Ab} = \mathbf{c}$ in order to reduce the computational burden of finding the solution. In particular, applications involving system identification, deconvolution, linear prediction, or spectral estimation often require the solution of Toeplitz systems of equations [E, F, G].

In this section we describe a subroutine that uses Levinson's algorithm to solve the matrix equation

$$\mathbf{Ab} = \mathbf{c} \tag{11.17}$$

where $\mathbf{A}$ is both symmetric and Toeplitz, that is,

$$\mathbf{A} = \begin{bmatrix} a_0 & a_1 & . & . & . & a_L \\ a_1 & a_0 & . & . & . & a_{L-1} \\ . & . & & . & . & . \\ . & . & & . & . & . \\ . & . & & . & . & . \\ a_L & a_{L-1} & . & . & . & a_0 \end{bmatrix} \tag{11.18}$$

and the vectors $\mathbf{b}$ and $\mathbf{c}$ are $L + 1$ by 1 with entries $b_0, \ldots, b_L$ and $c_0, \ldots, c_L$, respectively. Note that the A-matrix is completely defined by the elements in any row (or equivalently, in any column).

Levinson's algorithm is an iterative procedure that uses bordering techniques and solves a series of truncated problems. Beginning in the upper left-hand corner of the A-matrix, the solution is initiated by evaluating the equation $a_0 b_0 = c_0$. The first iteration then solves the 2×2 system

$$\begin{bmatrix} a_0 & a_1 \\ a_1 & a_0 \end{bmatrix} \begin{bmatrix} b_0 \\ b_1 \end{bmatrix} = \begin{bmatrix} c_0 \\ c_1 \end{bmatrix} \tag{11.19}$$

At each succeeding step, the size of the problem is incremented by appending the next row and column of $\mathbf{A}$. Thus, the Lth iteration yields the desired solution vector $\mathbf{b}$. The recursions of this algorithm fail only if one of the principal submatrices is singular. This condition results in division by zero and often indicates that the order L has been overspecified [E, H, K, L, M].

Subroutine SPLEVS uses Levinson's algorithm to find the solution of an Lth-order symmetric, Toeplitz system of equations. The calling sequence is

CALL SPLEVS(AVECT,C,L,B,WK,IERROR)

AVECT(0:L)   = first row of symmetric, Toeplitz **A**-matrix
C(0:L)   = vector **c** in **Ab** = **c** matrix equation (11.17)
L   = order of system of equations (11.17) and (11.18)
B(0:L)   = solution vector—returned
WK(0:L)   = work vector used internally
IERROR   = 0   no errors detected
               1   order L < 0
               2   approximate divide by zero ($< 1.E - 10$)

Note that it is not necessary to store the entire $(L + 1)$ by $(L + 1)$ **A**-matrix, since it is completely specified by its first row. The IERROR = 2 return is provided in an attempt to avoid potential mathematical overflow problems. If this condition is detected, the iterative procedure is halted, and the current solution vector is returned.

Before demonstrating the use of SPLEVS, we describe an application in which the necessary symmetric, Toeplitz structure arises.

Consider the general digital system illustrated in Fig. 11.4. Given two N-point data sequences $[x_k]$ and $[y_k]$, our goal is to find the Lth-order finite impulse response (FIR) filter $B(z)$, which produces from $[x_k]$ an estimate $[\hat{y}_k]$ that minimizes the sum of the squared errors S:

$$S = \sum_{k=0}^{N-1} e_k^2 = \sum_{k=0}^{N-1} (y_k - \hat{y}_k)^2$$

$$= \sum_{k=0}^{N-1} [y_k - (b_0 x_k + \cdots + b_L x_{k-L})]^2$$

(11.20)

If we assume that $[x_k]$ and $[y_k]$ are stationary sequences, the optimal (i.e., least-squares) solution is obtained by solving the discrete-time Wiener-Hopf equations [H, N]

$$\begin{bmatrix} r_{xx}(0) & r_{xx}(1) & . & . & . & r_{xx}(L) \\ r_{xx}(1) & r_{xx}(0) & . & . & . & r_{xx}(L-1) \\ . & . & & . & . & . \\ . & . & & . & . & . \\ . & . & & . & . & . \\ r_{xx}(L) & r_{xx}(L-1) & . & . & . & r_{xx}(0) \end{bmatrix} \begin{bmatrix} b_0 \\ b_1 \\ . \\ . \\ . \\ b_L \end{bmatrix} = \begin{bmatrix} r_{yx}(0) \\ r_{yx}(1) \\ . \\ . \\ . \\ r_{yx}(L) \end{bmatrix}$$

(11.21)

where $r_{uv}(n)$ is the total product defined in equation (9.6). Note that (11.21) has

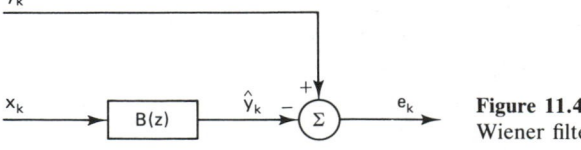

Figure 11.4.   Block diagram for Wiener filtering problem.

the form of (11.18), where the **A**-matrix is formulated via the autocorrelation of $[x_k]$ and the vector **c** is generated via cross correlation, again as in (9.6). The example described above is a special case of the general Wiener filtering problem, since in practice $[x_k]$ and $[y_k]$ are not both available. Note that given only $[x_k]$ and the cross correlation of x_k and y_k, the problem can still be formulated as in (11.21).

Since the autocorrelation matrix in (11.21) is both symmetric and Toeplitz, Levinson's algorithm can be used to solve for the optimal **b**-vector. In SPA1107, we consider the simplified Wiener filtering problem for a first-order system with

$$x_k = 2 \sin\left(\frac{2\pi k}{10}\right), \qquad y_k = 4 \cos\left(\frac{2\pi k}{10}\right) \tag{11.22}$$

In this case, the scaled correlation functions can be evaluated in closed form, yielding

$$a_n = 2 \cos\left(\frac{2\pi n}{10}\right), \qquad c_n = -4 \sin\left(\frac{2\pi n}{10}\right) \tag{11.23}$$

For the data sequences in (11.22), two filter coefficients in $B(z)$ are sufficient to drive the error signal, e_k, to zero [N]. Thus, the filtered x_k sequence, denoted by $\hat{y}_k$ in Fig. 11.4, will be equal to the y_k sequence. This is demonstrated by program SPA1107, which is listed together with its results.

```
            PROGRAM SPA1107
C-DEMONSTRATE THE USE OF LEVINSON'S ALGORITHM
            DIMENSION AVECT(0:1),C(0:1),B(0:1),WK(0:1)
            DIMENSION X(0:50),Y(0:50),PX(0:1)
            DATA AVECT/2.,1.618/,C/0.,-2.351/,L/1/,PY/0/,PX/2*0./,A/0./
            PI=4.*ATAN(1.)
            CALL SPLEVS(AVECT,C,L,B,WK,IERROR)
            IF(IERROR.NE.0)PRINT *,' SPLEVS ERROR = ',IERROR
            DO 1 K=0,50
              X(K)=2.*SIN(2.*PI*K/10.)
              Y(K)=4.*COS(2.*PI*K/10.)
        1   CONTINUE
            CALL SPFILT(B,A,L,1,X,51,PX,PY,IERROR)
            CALL MCPLOT(0.,1.,Y,51,1,0,2,1,2,1,5.,3.6)
            CALL MCPLOT(0.,1.,X,51,9,12,2,2,2,1,5.,3.6)
            PRINT 100,(I,B(I),I=0,1)
      100   FORMAT(' FILTER COEFFICIENTS',/,(' B(',I1,') = ',F10.3))
            END

                  $ RUN SPA1107
                  FILTER COEFFICIENTS
                  B(0) =       2.752
                  B(1) =      -3.402
                  $
```

In the plot in Fig. 11.5, the dashed line results from linear interpolation between the y_k samples, and the asterisks denote $\hat{y}_k$ samples. Note that except for the first sample, the two sequences are identical, as predicted. The initial error is due to the start-up transient in the filtering process.

Since only two filter coefficients were needed to drive the error, e_k, to zero in the previous example, we might suppose that numerical problems would arise if a higher-order least-squares solution is attempted. This point is demonstrated

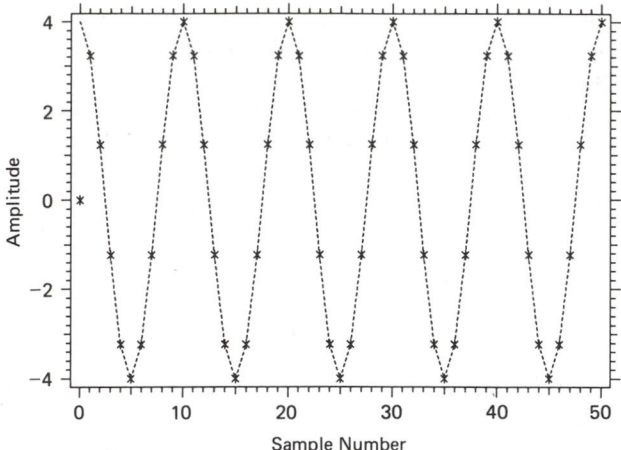

Figure 11.5. SPA1107: Data sequence y_k and the least-squares estimate $\hat{y}_k$ computed via SPLEVS with $L = 1$.

by the results of program SPA1108, in which the previous example is repeated with L = 6. In SPA1108, the procedure was halted after two iterations when an approximate divide-by-zero condition was detected in SPLEVS (IERROR = 2). Thus, coefficients 3 through 6 are still equal to their initial values of zero. Even though the process was not completed, it is evident from the results shown in Fig. 11.6 that the returned **b** vector is an optimal solution. It should be noted that in practical situations we cannot obtain the ideal noiseless data measurements that were assumed to be available in this example. Since noisy data generally masks singularities that result in the divide-by-zero condition, the iterative process will typically continue and a higher-order least-squares solution will be returned.

```
            PROGRAM SPA1108
C-DEMONSTRATE THE USE OF LEVINSON'S ALGORITHM
            DIMENSION AVECT(0:6),C(0:6),B(0:6),WK(0:6)
            DIMENSION X(0:50),Y(0:50),PX(0:6)
            DATA L/6/,PY/0/,PX/7*0./,A/0./
            PI=4.*ATAN(1.)
            DO 1 LL=0,L
              AVECT(LL)=2.*COS(2.*PI*LL/10.)
              C(LL)=-4.*SIN(2.*PI*LL/10.)
      1     CONTINUE
            CALL SPLEVS(AVECT,C,L,B,WK,IERROR)
            IF(IERROR.NE.0)THEN
              PRINT *,' SPLEVS ERROR = ',IERROR
              PAUSE
            ENDIF
            DO 2 K=0,50
              X(K)=2.*SIN(2.*PI*K/10.)
              Y(K)=4.*COS(2.*PI*K/10.)
      2     CONTINUE
            CALL SPFILT(B,A,L,1,X,51,PX,PY,IERROR)
            CALL MCPLOT(0.,1.,Y,51,1,0,2,1,2,1,5.,3.6)
            CALL MCPLOT(0.,1.,X,51,9,12,2,2,2,1,5.,3.6)
            PRINT 100,(I,B(I),I=0,L)
    100     FORMAT(' FILTER COEFFICIENTS',/,(' B(',I1,') = ',F10.3))
            END
```

```
$ RUN SPA1108
  SPLEVS ERROR =              2
FORTRAN PAUSE

FILTER COEFFICIENTS
B(0) =        0.680
B(1) =       -0.049
B(2) =       -2.073
B(3) =        0.000
B(4) =        0.000
B(5) =        0.000
B(6) =        0.000
$
```

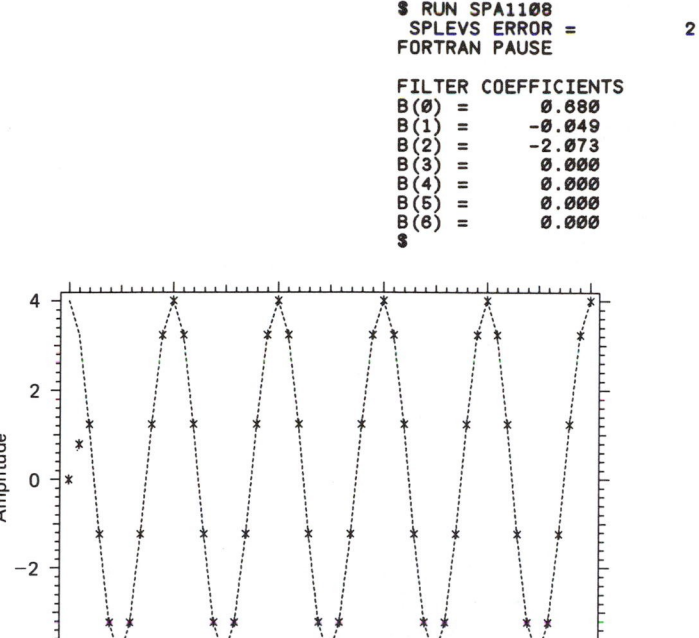

Figure 11.6. SPA1108: Data sequence y_k and the least-squares estimate $\hat{y}_k$ computed via SPLEVS with $L = 6$.

Recall from the earlier discussion of Levinson's algorithm that a truncated problem is solved at each step. This suggests that the **b**-vector returned in SPA1108 is actually the solution that would result from specifying L = 2. This fact is verified by program SPA1109, which is listed next.

```
        PROGRAM SPA1109
C-DEMONSTRATE THE USE OF LEVINSON'S ALGORITHM
        DIMENSION AVECT(0:2),C(0:2),B(0:2),WK(0:2)
        DATA L/2/
        PI=4.*ATAN(1.)
        DO 1 LL=0,L
          AVECT(LL)=2.*COS(2.*PI*LL/10.)
          C(LL)=-4.*SIN(2.*PI*LL/10.)
    1   CONTINUE
        CALL SPLEVS(AVECT,C,L,B,WK,IERROR)
        IF(IERROR.NE.0)PRINT *,' SPLEVS ERROR = ',IERROR
        PRINT 100,(I,B(I),I=0,L)
  100   FORMAT(' FILTER COEFFICIENTS',/,(' B(',I1,') = ',F10.3))
        END
```

```
$ RUN SPA1109
FILTER COEFFICIENTS
B(0) =        0.680
B(1) =       -0.049
B(2) =       -2.073
$
```

The specific relationship between x_k, y_k, and e_k in Fig. 11.4, which results in the symmetric, Toeplitz system of equations in (11.21), occurs in a number of applications in digital signal processing. Thus, Levinson's algorithm is widely applicable.

Consider the problem of system identification, or modeling, illustrated in Fig. 11.7. In this task, the error is minimized when $B(z)$ is the optimal Lth-order FIR model of the unknown system. A very large L is typically required if $B(z)$ is to model an IIR (infinite impulse response) system (that is, a system with poles) accurately. We have restricted the model to be an FIR filter, since if $B(z)$ is an IIR structure, then a set of nonlinear algebraic equations must be solved, and the techniques described herein are not applicable.

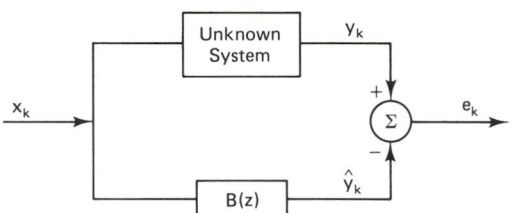

Figure 11.7. Basic structure for modeling, or system identification.

Additional applications in which Levinson's algorithm can be utilized occur in the areas of inverse modeling, equalization, and deconvolution [H, J, N]. The basic structure for these applications is illustrated in Fig. 11.8. The delay $z^{-\Delta}$ is often used to improve the least-squares fit by eliminating the need for $B(z)$ to function as a predictor [N].

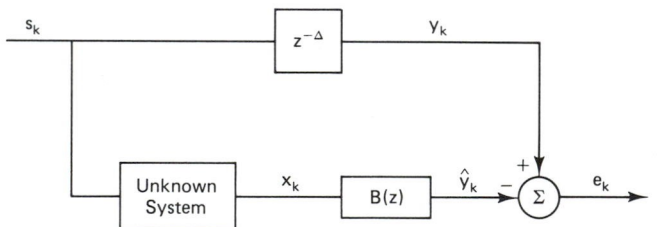

Figure 11.8 Basic structure for inverse modeling, equalization, or deconvolution.

11.5 Durbin's Algorithm

Durbin's algorithm may be used in place of Levinson's algorithm in least-squares prediction, which is a specific subset of the applications described in Section 11.4. In our applications of Durbin's algorithm, the system of equations must have the form

$$\begin{bmatrix} a_0 & a_1 & . & . & . & a_L \\ a_1 & a_0 & . & . & . & a_{L-1} \\ . & . & . & . & . & . \\ . & . & . & . & . & . \\ . & . & . & . & . & . \\ a_L & a_{L-1} & . & . & . & a_0 \end{bmatrix} \begin{bmatrix} b_0 \\ b_1 \\ . \\ . \\ . \\ b_L \end{bmatrix} = \begin{bmatrix} a_1 \\ a_2 \\ . \\ . \\ . \\ a_{L+1} \end{bmatrix} \tag{11.24}$$

Once again using the notation $\mathbf{Ab} = \mathbf{c}$ to define the matrix equation, we note that the $\mathbf{A}$-matrix is both symmetric and Toeplitz, as was required for Levinson's algorithm. In addition, Durbin's algorithm requires the vector $\mathbf{c}$ to be composed of elements from the $\mathbf{A}$-matrix with one additional value, a_{L+1}, appended, as shown in (11.24).

Durbin's algorithm is similar to Levinson's algorithm in that it is again an iterative procedure that solves a series of truncated problems. Note that any system of equations that has the form of (11.24) could also be solved using Levinson's algorithm. When the additional structure is present, however, Durbin's algorithm provides a more efficient means for computing the $\mathbf{b}$-vector.

Subroutine SPDURB, which is described next, generates the solution to a system of equations in the form of (11.24) via Durbin's algorithm. The calling sequence is

CALL SPDURB(AAVECT,L,B,IERROR)

$$\begin{aligned}
\text{AAVECT(0:L}+1) &= \text{augmented } \mathbf{A} \text{ vector } [a_0, a_1, ..., a_{L+1}] \\
\text{L} &= \text{order of system of equations} \\
\text{B(0:L)} &= \text{solution vector—returned} \\
\text{IERROR} &= 0 \quad \text{no errors detected} \\
&\quad\ 1 \quad \text{invalid order } L < 0 \\
&\quad\ 2 \quad \text{approximate divide by zero } (< 1.\text{E}-10)
\end{aligned}$$

As in SPLEVS, the IERROR=2 return is provided in an attempt to avoid numerical problems. Detection of this condition results in termination of the iterative process and return of the current solution vector.

In digital signal processing, Durbin's algorithm can be applied to tasks involving linear prediction. To illustrate the use of SPDURB, we consider the simple one-step predictor shown in Fig. 11.9. Here we see that the predictor $B(z)$ operates on past samples $y_{k-1}, y_{k-2}, ..., y_{k-L-1}$ to predict the current sample y_k.

Once again assuming that all signals are stationary, the system of equations defining the least-squares solution for this task can be written in the form of (11.21). Since the filter input x_k is simply a delayed version of the y_k signal, we know that $r_{xx}(n) = r_{yy}(n)$. In addition, we see that $r_{yx}(n) = r_{yy}(n + 1)$, so the $\mathbf{c}$-vector in $\mathbf{Ab} = \mathbf{c}$ has entries $r_{yy}(1)$ through $r_{yy}(L + 1)$ as specified in (11.24).

Program SPA1110 demonstrates the use of SPDURB for finding the optimal

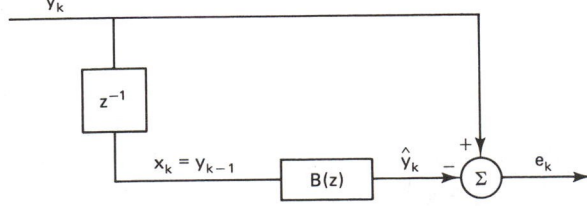

Figure 11.9. Basic structure for one-step prediction.

first-order one-step predictor of $y_k = \sin(2\pi k/10)$. Once again using the closed-form solution in (11.23) for the correlation function, we have AAVECT entries $a_n = 0.5 \cos(2\pi n/10)$. The program listing and its results are provided, together with a plot in Fig. 11.10. Again, the latter illustrates perfect performance by the predictor after sample number $k = 1$.

```
                PROGRAM SPA1110
     C-DEMONSTRATE THE USE OF DURBIN'S ALGORITHM
                DIMENSION AAVECT(0:2),B(0:1),X(0:50),Y(0:50),PX(0:1)
                DATA L/1/,PX/2*0./,A/0./,PY/0./
                PI=4.*ATAN(1.)
                DO 1 LL=0,L+1
                   AAVECT(LL)=.5*COS(2.*PI*LL/10.)
          1     CONTINUE
                CALL SPDURB(AAVECT,L,B,IERROR)
                IF(IERROR.NE.0)THEN
                   PRINT *,' DURBINS ALGORITHM ERROR = ',IERROR
                   PAUSE
                ENDIF
                DO 2 K=0,50
                   Y(K)=SIN(2.*PI*K/10.)
                   X(K)=SIN(2.*PI*(K-1)/10.)
          2     CONTINUE
                CALL SPFILT(B,A,L,1,X,51,PX,PY,IERROR)
                CALL MCPLOT(0.,1.,Y,51,1,0,2,1,2,1,5.,3.6)
                CALL MCPLOT(0.,1.,X,51,9,12,2,2,2,1,5.,3.6)
                PRINT 100,(I,B(I),I=0,L)
        100     FORMAT(' FILTER COEFFICIENTS ',/,(' B(',I1,') =',F10.3))
                END


                        $ RUN SPA1110
                        FILTER COEFFICIENTS
                        B(0) =        1.618
                        B(1) =       -1.000
                        $
```

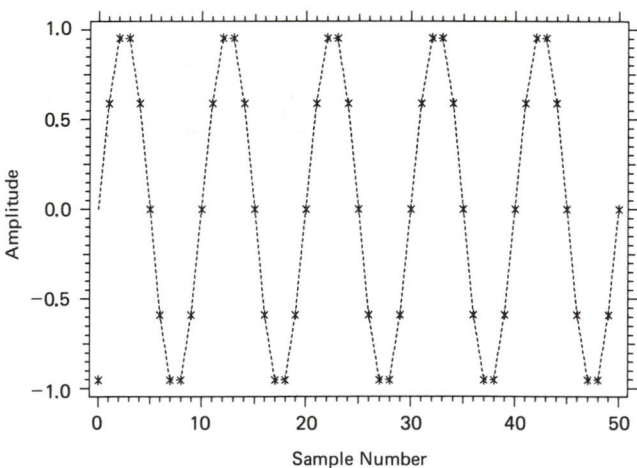

Figure 11.10. SPA1110: Demonstration of one-step prediction via SPDURB with $L = 1$.

Since Durbin's algorithm, like Levinson's algorithm, may encounter numerical problems when the order of the system is overspecified, this task was repeated with L = 10. Note that, as stated previously, this condition is typically masked by noisy data samples in practical situations. A listing of SPA1111 and its results are provided. The plot in Fig. 11.11 shows that the computed $B(z)$ again successfully predicts the y_k sequence. Note that in this case, the approximate divide-by-zero condition was detected after four iterations. From the B-coefficients returned, we see that the optimal tenth-order predictor consists essentially of a one-half period delay and inversion of the sinusoid.

```
                  PROGRAM SPA1111
       C-DEMONSTRATE THE USE OF DURBIN'S ALGORITHM
                  DIMENSION AAVECT(0:11),B(0:10),X(0:50),Y(0:50),PX(0:10)
                  DATA L/10/,PX/11*0./,A/0./,PY/0./
                  PI=4.*ATAN(1.)
                  DO 1 LL=0,L+1
                    AAVECT(LL)=.5*COS(2.*PI*LL/10.)
           1      CONTINUE
                  CALL SPDURB(AAVECT,L,B,IERROR)
                  IF(IERROR.NE.0)THEN
                    PRINT *,' DURBINS ALGORITHM ERROR = ',IERROR
                    PAUSE
                  ENDIF
                  DO 2 K=0,50
                    Y(K)=SIN(2.*PI*K/10.)
                    X(K)=SIN(2.*PI*(K-1)/10.)
           2      CONTINUE
                  CALL SPFILT(B,A,L,1,X,51,PX,PY,IERROR)
                  CALL MCPLOT(0.,1.,Y,51,1,0,2,1,2,1,5.,3.6)
                  CALL MCPLOT(0.,1.,X,51,9,12,2,2,2,1,5.,3.6)
                  PRINT 100,(I,B(I),I=0,L)
         100      FORMAT(' FILTER COEFFICIENTS ',/,(' B(',I2,') =',F10.3))
                  END

                  $ RUN SPA1111
                   DURBINS ALGORITHM ERROR =              2
                  FORTRAN PAUSE

                  FILTER COEFFICIENTS
                  B( 0) =      -0.007
                  B( 1) =       0.004
                  B( 2) =       0.004
                  B( 3) =      -0.007
                  B( 4) =      -1.000
                  B( 5) =       0.000
                  B( 6) =       0.000
                  B( 7) =       0.000
                  B( 8) =       0.000
                  B( 9) =       0.000
                  B(10) =       0.000
                  $
```

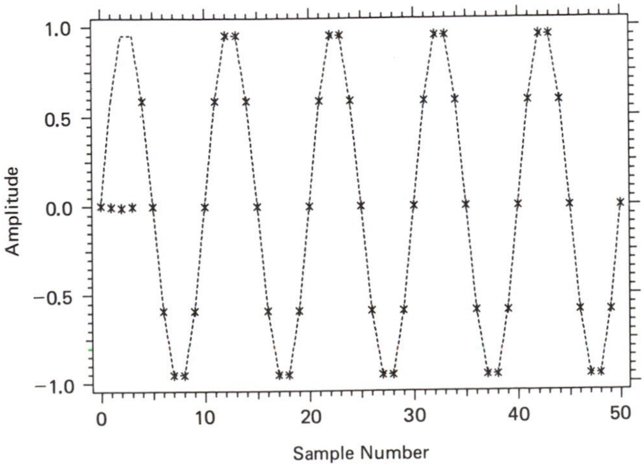

Figure 11.11. SPA1111: Demonstration of one-step prediction via SPDURB with $L = 10$.

The simple one-step predictor structure in Fig. 11.9 is often used for noise reduction in speech and seismic signal processing. Consider, for example, a case where the y_k signal is a sinusoid corrupted by white noise. The least-squares solution will result in a bandpass filter $B(z)$, which predicts the sinusoid and rejects most of the noise, because white noise is uncorrelated from one sample to the next and therefore cannot be predicted. Thus, any noise that passes through the filter to $\hat{y}_k$ will increase the squared error. As a result, $\hat{y}_k$ is actually an enhanced version of the noisy sinusoid y_k. This task is often called *line enhancement*.

As a final point on least-squares design, we note that when $B(z)$ is optimized via one of these algorithms, the minimum total-squared error, that is, the minimum of S in (11.20), may be estimated as

$$S_{\min} \approx r_{yy}(0) - \sum_{n=0}^{L} b_n r_{yx}(n) \tag{11.24}$$

11.6 Summary of Calling Sequences

In this chapter we have discussed routines for least-squares error approximation. This summary defines the calling sequences and argument lists.

```
CALL SPNORM(X,Y,N,L,A,C,IERROR)
CALL SPGAUS(A,C,L,B,IERROR)
CALL SPORTH(Y,N,L,ORTHB,WORK,IERROR)
CALL SPPOLY(DX,Y,N,L,B,ORTHB,WORK,IERROR)
```

Variable	Definition	Routines			
		SPNORM	SPGAUS	SPORTH	SPPOLY
X(0:N−1)	x_k data sequence	I	—	—	—
Y(0:N−1)	y_k data sequence	I	—	I	I
N	Number of data points	I	—	I	I
L	Order of system	I	I	I	I
A(0:L,0:L)	Matrix **A** in **Ab** = **c**	O	I	—	—
B(0:L)	Solution vector	—	O	—	O
C(0:L)	Vector **c** in **Ab** = **c**	O	I	—	—
ORTHB(0:L)	Orthogonal coefficient vector	—	—	O	O
DX	Sample spacing	—	—	—	I
WORK(0:L,0:L)	Work matrix	—	—	X	X
IERROR	Error indicator	O	O	O	O

CALL SPLEVS(AVECT,C,L,B,WK,IERROR)
CALL SPDURB(AAVECT,L,B,IERROR)

Variable	Definition	Routines	
		SPLEVS	SPDURB
AVECT(0:L)	First row of **A**	I	—
AAVECT(0:L+1)	Augmented **A**-vector	—	I
B(0:L)	Solution vector	O	O
C(0:L)	Vector **c** in **Ab** = **c**	I	—
L	Order of system	I	I
WK(0:L)	Work vector	X	—
IERROR	Error indicator	O	O

In this summary, I indicates that the variable is an input to the routine, O indicates that it is returned from the routine, and — means that it is not used. The symbol X denotes work space that must be allocated but not necessarily initialized by the calling program.

11.7 Exercises

11.1. Generate 51-point data sequences $[x_k]$ and $[y_k]$ using the function $y_k = (\frac{1}{8})(35x_k^4 - 30x_k^2 + 3) + 0.1r_k$, where r_k is a sample of zero-mean, unit power density white noise (use the SPRAND function from Chapter 4) and $x_k = k/50$ for $0 \le k \le 50$. Use SPNORM and SPGAUS to find the fifth-order least-squares polynomial. List the coefficients and plot $[y_k]$ together with the polynomial (use at least 500 points so that the curve is relatively smooth).

11.2. Repeat exercise 11.1, this time using the orthogonal polynomials described in Section 11.3. Instead of the $[x_k]$ sequence, use DX = $\frac{1}{50}$. List both sets of coefficients and provide the data plot described above.

11.3. Consider the third-order Legendre polynomial $P_3(x) = 0.5(5x^3 - 3x)$. Form 21-point data sequences by evaluating this function at equally spaced points over $0 \le x \le 1$. Using the techniques described in Section 11.2, compute the least-squares polynomials for $L = 2, 3, 4$. For each case, evaluate the sum of the squared errors using equation (11.3); list the coefficients and the sum S.

11.4. Repeat exercise 11.3 using the SPPOLY routine to generate the coefficients.

11.5. Data sequences are defined by:

$$x_k = [0.0, 0.1, 0.3, 0.4, 0.5, 0.6, 0.7, 0.8, 1.0]$$

$$y_k = 0.5(3x_k^2 - 1)$$

Note that this data has gaps (missing samples) at $x = 0.2, 0.9$. Use SPNORM and SPGAUS to find the second-order least-squares polynomial. Compute the squared error sum S in equation (11.3); plot the results.

11.6. Repeat exercise 11.5, but modify the $[y_k]$ sequence as

$$y_k = 0.5(3x_k^2 - 1) + 0.05r_k$$

Use $r_k = \sqrt{12}(\text{SPRAND(ISEED)} - 0.5)$ to generate the noise samples. Use ISEED = 123.

11.7. With the noisy $[y_k]$ sequence in 11.6 and $[x_k]$ as defined in exercise 11.5, use straight-line interpolation to fill the gaps in the data. Compute the second-order least-squares polynomial first via the normal equations and then by using orthogonal polynomials. Compute the squared error sum S and plot the results for each case.

11.8. Generate a 51-point data sequence that contains nothing but noise, that is, $y_k = r_k$, where r_k is an unaltered sample from the SPRAND function described in Chapter 4. Compute the third-order least-squares polynomial via the normal equations. Evaluate the squared error sum S defined in (11.3) and the value

$$S_1 = \sum_{k=0}^{N-1} (y_k - 0.5)^2$$

List S and S_1; plot $[y_k]$ together with the computed polynomial. Use ISEED = 123.

11.9 Repeat exercise 11.8 using SPPOLY to generate the polynomial coefficients.

11.10. Consider the system identification structure in Fig. 11.7. Assume that the unknown system is described by the FIR filter $H(z) = 1 + 0.2z^{-1} + z^{-2}$. Let the input signal x_k be a 100-point zero-mean, unit power density white noise sequence. Generate the necessary correlation functions using the routines described in Chapter 9. Use Levinson's algorithm to find the second-order least-squares approximation $B(z)$. List the coefficients and plot the error sequence e_k. Use ISEED = 123.

11.11. Repeat exercise 11.10 using a $B(z)$ of order $L = 10$.

11.12. An IIR system is to be modeled using the FIR least-squares approximation. Using the structure in Fig. 11.7, let the unknown system be

$$H(z) = \frac{0.0931 - 0.0975z^{-1}}{1 - 1.8175z^{-1} + 0.9025z^{-2}}$$

Let x_k be a zero-mean, unit power density white noise sequence of 1000 samples. Generate the necessary correlation functions using the routines from Chapter 9. Use Levinson's algorithm to find the 200th-order least-squares approximation. Plot the magnitude response of $H(z)$ and $B(z)$. Use ISEED = 123.

11.13. Consider the inverse modeling structure in Fig. 11.8. Let the unknown system be the all-pole structure

$$H(z) = \frac{1}{1 - 0.6z^{-1}}$$

Let s_k be the white noise sequence described in exercise 11.12. Use a delay of $\Delta = 0$. Generate the first-order least-squares approximation. Plot the magnitude responses for $B(z)$ and the actual system inverse $1/H(z)$.

11.14. Consider the one-step predictor structure in Fig. 11.9. Let $y_k = 2 \sin[2\pi(100)(0.001)k] + r_k$, where r_k is a sample of zero-mean, unit power density white noise. Use 500 samples, and generate the twentieth-order least-squares approximation via Durbin's algorithm. Plot the data sequences $[y_k]$ and $[\hat{y}_k]$ and the magnitude response of $B(z)$.

11.15. Repeat exercise 11.14, but modify y_k by adding a second sinusoidal component: $\sin[2\pi(250)(0.001)k]$.

11.16. Repeat 11.14, but eliminate the sinusoidal component so that the y_k signal is pure noise. Plot only the magnitude response of $B(z)$.

11.8 References

[A] Scheid, F., *Theory and Problems of Numerical Analysis,* Schaum's Outline Series (New York: McGraw-Hill, 1968).

[B] Kreyszig, E., *Advanced Engineering Mathematics* (New York: John Wiley, 1972).

[C] Selby, S., *Standard Mathematical Tables* (CRC Press, 1973).

[D] Lipschutz, S., *Theory and Problems of Linear Algebra,* Schaum's Outline Series (New York: McGraw-Hill, 1968).

[E] Blahut, R., *Fast Algorithms for Digital Signal Processing* (Reading, Mass.: Addison-Wesley, 1985).

[F] Goodwin, G., and R. Payne, *Dynamic System Identification: Experiment Design and Analysis* (New York: Academic Press, 1977).

[G] Haykin, S., *Introduction to Adaptive Filters* (New York: Macmillan, 1984).

[H] Orfanidis, S., *Optimum Signal Processing: An Introduction* (New York: Macmillan, 1985).

[I] Tretter, S., *Introduction to Discrete-Time Signal Processing* (New York: John Wiley, 1976).

[J] Oppenheim, A., and R. Schafer, *Digital Signal Processing* (Englewood Cliffs, N.J.: Prentice-Hall, Inc., 1975).

[K] Carayannis, G., et al., "Fast Recursive Algorithms for a Class of Linear Equations," *IEEE Trans. on ASSP,* ASSP-30, no. 2 (April 1982).

[L] Morf, M., and A. Nehorai, "A Relationship Between the Levinson Algorithm and the Conjugate Direction Method," *IEEE Trans. on ASSP,* ASSP-31, no. 2 (April 1983).

[M] Dunbar, W., "An In-Place Levinson Update Algorithm," *Proc. of the IEEE,* 69, no. 6 (June 1981).

[N] Widrow, B. and S. Stearns, *Adaptive Signal Processing* (Englewood Cliffs, N.J.: Prentice-Hall, Inc., 1985).

[O] Stearns, S., *Digital Signal Analysis* (Rochelle Park, N.J.: Hayden, 1975).

[P] Strang, G., *Linear Algebra and Its Applications* (New York: Academic Press, 1976).

chapter 12

Adaptive Signal Processing

12.1 Introduction

Historically, the basis for adaptive signal processing evolved from techniques developed to enable the adaptive control of time-varying systems. In the 1960s, largely due to the work of Bernard Widrow and his colleagues, it began to be recognized as a separate category in digital signal processing [A–C]. For a number of years, utilization of adaptive signal processing was somewhat limited. In the mid-1970s the topic enjoyed a surge in popularity, which has continued to date.

The gain in popularity of adaptive signal processing is due primarily to advances in digital technology that have increased computing capacities and therefore broadened the scope of digital signal processing as a whole. As a result of the corresponding increase in research efforts, many new applications for adaptive signal processing have been recognized. The field continues to expand due to continuing developments in very-large-scale-integrated circuits (VLSI) and other digital hardware, which have made adaptive techniques feasible in real-time applications.

The key difference between adaptive signal processing methods and the classical signal processing techniques described in earlier chapters is that we are now dealing with time-varying digital systems. Our general system is now as illustrated in Fig. 12.1, where $H_k(z)$ denotes a time-varying transfer function.

The characteristics of the transfer function $H_k(z)$ change, or *adapt*, according to signal conditions. An adaptive algorithm is the equation or set of equations used to adjust the coefficients of $H_k(z)$. To characterize completely the digital system in Fig. 12.1, it would be necessary to analyze its response each time the coefficients are changed. Since we will be dealing with algorithms that update the coefficients during every sample interval, a complete characterization in the conventional sense is impractical. We also note that adaptive signal processing

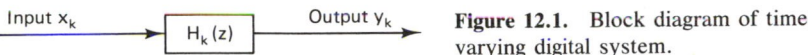

Input x_k → $H_k(z)$ → Output y_k

Figure 12.1. Block diagram of time-varying digital system.

is not appropriate for all applications and, when it is applicable, the adaptive system must be designed very carefully, as illustrated in this chapter.

Without detailing the mechanics of the adaptive process, we can describe a simple interference-canceling task for which adaptive techniques are more useful than classical filtering methods [D, E]. Suppose we have a broadband signal corrupted by power line frequency interference. The classical approach would be to use a notch filter centered at 60 Hz, the nominal line frequency. Since the interfering frequency may in reality deviate from 60 Hz, this filter will attenuate the interference by differing amounts as the frequency varies. A better solution in this case is to use an adaptive filter capable of tracking the line frequency and keeping the notch centered on the incoming interference.

The example just described is one of a large number of applications in which adaptive signal processing is not only appropriate but will achieve results not attainable via classical signal processing techniques. To date, applications have been recognized in such diverse fields as speech analysis, seismic, acoustic, and radar signal processing, and digital filter design. The work has been extended from the single-input time-varying system illustrated in Fig. 12.1 to include adaptive arrays, that is, multiple-input adaptive systems [D, E, G, H, K, P, Y, Z].

Due to the scope of this text, we restrict our discussion to a brief introduction of the topic and a description of the simplest adaptive algorithm, known as the least-mean-squares (LMS) algorithm. We are concerned only with the single-input system shown in Fig. 12.1. A variety of additional algorithms, including many which are designed for specific applications, appear in the literature.

12.2 The Basic Adaptive System

In this presentation we are concerned with adaptive systems whose basic elements are as illustrated in Fig. 12.2. We have the usual input and output sequences $[x_k]$ and $[y_k]$, but the transfer function describing their relationship is now the time-varying $H_k(z)$. We also define a desired response sequence $[d_k]$ and an error

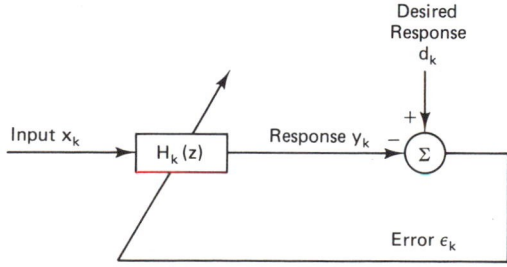

Desired Response d_k

Input x_k → $H_k(z)$ → Response y_k → $+$ $-$ Σ

Error ϵ_k

Figure 12.2. Elements of the basic adaptive system.

sequence $[\varepsilon_k]$, which is the difference between the desired response and the actual response ($\varepsilon_k = d_k - y_k$). The arrow through the $H_k(z)$ box indicates that the error sequence is used to control the adaptation of $H_k(z)$.

The term *desired response* is a misnomer in some respects, but we retain it in order to maintain compatibility with the literature [E, W]. Realistically, if we had access to the true desired response, we would not in general need an adaptive system. In fact, the definition of this desired response is application-dependent and is often the most difficult part of adaptive system specification.

We have said only that the error sequence $[\varepsilon_k]$ is used in some way to control the adaptive process. In particular, our goal is to adapt toward the minimum mean-square-error (MMSE) solution, that is, adjust the coefficients of $H_k(z)$ in order to minimize $E[\varepsilon_k^2]$.

There are two general operating environments to consider. If we are dealing with stationary signals, utilization of an adaptive process is simply a different technique for solving the least-squares problem described in Chapter 11. Since we have not yet restricted the form of the transfer function $H_k(z)$, the solution provided may be more general than that derived via Levinson's algorithm. In the second case, which is of primary interest for real-time applications, we are dealing with nonstationary signals, that is, signals whose statistical properties vary with time. In this environment, the final $H_k(z)$ is not necessarily of interest, since its characteristics are primarily dependent upon the latest signal statistics. In the nonstationary case, examination of the $[y_k]$ or $[\varepsilon_k]$ sequence is generally more informative. Operation in this environment is often referred to as the *tracking mode*. Later in this chapter we provide examples of both stationary and nonstationary operation.

12.3 The LMS Algorithm

The LMS algorithm is the simplest and perhaps the most universally applicable adaptive algorithm in use today [E, W, X]. This algorithm first appeared in the literature in the mid-1960s and has since provided the basis for a great deal of research as well as the evolution of many additional adaptive algorithms [E, F, I, L, M, V].

In this brief presentation we do not provide a detailed derivation or justification of the LMS algorithm. Interested readers should refer to the references at the end of this chapter for additional information.

For the LMS algorithm, the $H_k(z)$ transfer function in Figs. 12.1 and 12.2 is a time-varying FIR filter. The corresponding input-output relationship is similar to (8.1) and is described by

$$y_k = \sum_{n=0}^{L} b_n(k)x_{k-n} \qquad (12.1)$$

where the time-varying character of the filter coefficients is signified by the $b_n(k)$

notation. When the adaptive filter is realized in the direct-form FIR structure, the mean-squared-error performance surface is a quadratic function of the filter coefficients and thus has a single minimum [E]. This is illustrated for the simple one-coefficient case ($L = 0$) in Fig. 12.3, where $b_0(0)$ denotes the initial condition and b_0^* corresponds to the optimal, that is, MMSE, solution. In the general case, which is somewhat more difficult to envision, the surface would be a parabolic function in $(L + 2)$-dimensional space.

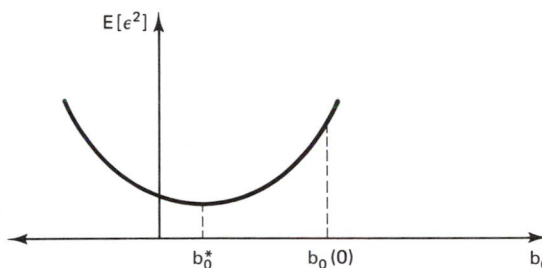

Figure 12.3. MSE performance surface for one-coefficient filter.

The goal of the adaptive process is to adjust the filter coefficients in such a way that they move from any initial condition, $b_n(0)$, toward the MMSE solution, b_n^*. In practical applications we rarely deal with the fixed performance surface just described. In nonstationary environments, the MMSE solution varies as signal conditions change. Thus, the adaptive process must continually adjust the coefficients in order to track the MMSE solution. In adaptive system analysis, however, we typically assume that signal statistics change very slowly, so that signal stationarity may be assumed.

The technique utilized by the LMS algorithm to update filter coefficients is based on the method of steepest descent. This can be described in algorithmic form using vector notation, as follows:

$$\mathbf{B}_{k+1} = \mathbf{B}_k - \mu \nabla_k \tag{12.2}$$

where

$$\text{Coefficient vector:} \quad \mathbf{B}_k = [b_0(k) \ldots b_L(k)]^T$$

$$\text{Gradient vector:} \quad \nabla_k = \frac{\partial E[\varepsilon_k^2]}{\partial \mathbf{B}_k}$$

$$= \left[\frac{\partial E[\varepsilon_k^2]}{\partial b_0(k)} \ldots \frac{\partial E[\varepsilon_k^2]}{\partial b_L(k)} \right]^T$$

and μ is a parameter that controls the rate of convergence. From (12.2) we see that the coefficient updates are proportional to the negative gradient $(-\nabla_k)$ of the performance surface. Thus, when ∇_k is known at each step of the adaptive process, the adjustment always results in a better filter, that is, the MSE decreases from step k to step $k + 1$. In addition, once the MMSE solution is found, the gradient reaches zero so the coefficients remain at their optimal values.

In practice, the algorithm as stated in (12.2) is difficult to implement due to inexact knowledge of the performance surface gradient ∇_k. Various techniques are available for estimating ∇_k [E]. The approach taken in the LMS algorithm is to use a gradient estimate based on the instantaneous squared error,

$$\hat{\nabla}_k = \frac{\partial \varepsilon_k^2}{\partial \mathbf{B}_k} = 2\varepsilon_k \frac{\partial(d_k - y_k)}{\partial \mathbf{B}_k} \tag{12.3}$$

where ε_k, d_k, and y_k are as defined in Fig. 12.2. Since the desired response d_k is independent of the filter coefficients, and the output y_k can be expressed in terms of the filtered input as in (12.1), (12.3) can be written as

$$\hat{\nabla}_k = -2\varepsilon_k \mathbf{X}_k \tag{12.4}$$

where $\mathbf{X}_k$ is a vector of input signal values,

$$\mathbf{X}_k = [x_k\, x_{k-1} \ldots x_{k-L}]^T$$

Combining (12.2) and (12.4), we write the LMS algorithm as follows:

$$\mathbf{B}_{k+1} = \mathbf{B}_k + 2\mu\varepsilon_k \mathbf{X}_k \tag{12.5}$$

where μ is the convergence parameter, ε_k is the error signal from Fig. 12.2, and $\mathbf{X}_k$ is the input signal vector just defined.

From (12.5) we see that given an input signal x_k and a desired signal d_k, our implementation of the LMS adaptive algorithm requires only the selection of the convergence parameter μ. This convergence parameter plays an important role in determining the performance of an adaptive system [E]. Recall that the gradient estimate used in the LMS update is based upon the instantaneous error value ε_k^2 rather than the mean value $E[\varepsilon_k^2]$. Although it has been shown that on the average this form of update moves the coefficients toward the MMSE solution, it is apparent that a single update of the $\mathbf{B}$-vector could contain a considerable error [E]. Thus, a large μ could result in an adaptive process that never converges to the MMSE solution. Conversely, if μ is too small, the coefficient vector adaptation is very slow. The effects of the inaccurate gradient estimate tend to average out in this case, and although the coefficient vector eventually converges if the signals are stationary, it may not converge in nonstationary environments. In other words, if μ is too small, the system may not react rapidly enough to cope with changing signal statistics. To avoid either of these conditions, μ must be neither "too large" nor "too small."

Although a great deal of effort has been devoted to the problem of picking the best value of μ, no universal solution has been found. The guidelines discussed next are simple, generally applicable, and address both stationary and nonstationary cases.

It has been shown that the stable range of μ varies according to the input signal power [E]. If, in place of μ in (12.5), we use the normalized value

$$\mu_n = \frac{\mu}{(L+1)\sigma^2}, \qquad 0 < \mu < 1 \tag{12.6}$$

where $L + 1$ is the number of filter coefficients and σ^2 is the input signal power, we can show that the stable range of μ is always $0 < \mu < 1$. In practice, μ is generally restricted to a small fraction of this stable range in order to smooth the noisy instantaneous gradient estimate in (12.4), as just described.

In some applications, the input signal power is either unknown or is changing with time in a nonstationary environment. In such cases, the σ^2 in (12.6) can be replaced by a time-varying estimate

$$\hat{\sigma}_k^2 = \alpha x_k^2 + (1 - \alpha)\hat{\sigma}_{k-1}^2 \tag{12.7}$$

where x_k is the current input sample and α is a *forgetting factor* in the range $0 < \alpha << 1$. This estimate enables operation in nonstationary environments, since α can be selected to reduce the influence of past input samples. As a general guideline, updating the input signal power via (12.7) amounts to forgetting the past values of x_k exponentially, with a time constant of approximately $1/\alpha$ iterations. The initial value of $\hat{\sigma}^2$ in (12.7) is the user's best a priori estimate of the input signal power.

In the SPNLMS subroutine, we implement the LMS algorithm in the form

$$\mathbf{B}_{k+1} = \mathbf{B}_k + \frac{2\,\mu\,\varepsilon_k \mathbf{X}_k}{(L+1)\hat{\sigma}_k^2}, \qquad 0 < \mu < 1$$
$$\hat{\sigma}_k^2 = \alpha x_k^2 + (1 - \alpha)\hat{\sigma}_{k-1}^2, \qquad 0 \leq \alpha << 1 \tag{12.8}$$

Note that allowing α to equal 0 results in a $\hat{\sigma}_k^2$ that stays at the initial estimate, thus enabling operation with a fixed value of $\hat{\sigma}^2$ if desired. The calling sequence for the SPNLMS subroutine is

CALL SPNLMS(X,N,D,B,L,MU,SIG,AL,PX,IERROR)

$X(0:N-1)$ = data vector (output replaces input)

N   = number of data samples and of adaptive iterations

$D(0:N-1)$ = desired response vector

B(0:L)   = adaptive coefficient vector

L   = order of adaptive system

MU   = REAL convergence parameter μ (0.0 < MU < 1.0)

SIG   = input signal power estimate $\hat{\sigma}_k^2$

AL   = exponential forgetting factor α (0.0 ≤ AL < 1.0)

PX(0:L)   = vector that retains past inputs for block mode

IERROR   = 0   no errors detected
             1   filter order L < 0
             2   convergence parameter MU ≤ 0.0 or MU ≥ 1.0
             3   input signal power estimate SIG ≤ 0
             4   forgetting factor AL < 0.0 or AL ≥ 1.0
             5   response y_k exceeds 1.E10

The coefficient vector, B, must be initialized to zero or some other desired initial condition before calling the subroutine. PX is a vector used internally to retain past values of the input sequence, again enabling block mode operation, as described in Chapter 6. This vector must be initialized (usually to zero) prior to the first call of SPNLMS to specify the initial conditions.

Subroutine SPNLMS is designed to allow several modes of operation. With reference to Fig. 12.2, if the result of interest from the adaptive process is the filtered data vector $[y_k]$, it is available in the array X upon return. If the error sequence $[\varepsilon_k]$ contains the necessary information, it is easily formulated, since $[d_k]$ remains in array D and $[y_k]$ is returned in X. Convergence of the adaptive coefficient vector can easily be observed by specifying $N = 1$ or some other relatively small number. In this case, updated coefficients are returned in B(0:L) after N iterations of the adaptive process. Array PX retains past values so that signal conditions in the adaptive filter are maintained. Note that upon entry, SPNLMS always assumes that the current data sample x_k is in location X(0). This flexible structure should enable the use of this subroutine for nearly any adaptive system application in which the LMS algorithm is used.

In the next two examples, we demonstrate the use of SPNLMS in a process known as adaptive line enhancement. Recall that line enhancement was described briefly in the discussion of the one-step predictor in Chapter 11. The general structure for this application, similar to Fig. 11.9, is illustrated in Fig. 12.4. The desired response in this case contains sinusoidal components with additive broadband noise. Note that the filter input sequence is simply a delayed version of the desired response. The delay, Δ, is selected such that the noise components in d_k and x_k are uncorrelated. Since the MSE is minimized when d_k and y_k match as closely as is possible, during adaptation the filter coefficients are adjusted to pass the sinusoidal components and attenuate the uncorrelated noise.

The adaptive structure in Fig. 12.4 provides information in two forms. During adaptation, the output sequence $[y_k]$ becomes an enhanced version of the original input sequence $[d_k]$. In addition, the magnitude response of the filter develops peaks at the frequency of each of the incoming sinusoids. Thus, the adaptive system functions as a detector as well as an enhancer.

We consider a simple example where d_k contains a single sinusoidal component corrupted by white noise. With white noise, a delay of $\Delta = 1$ is sufficient to

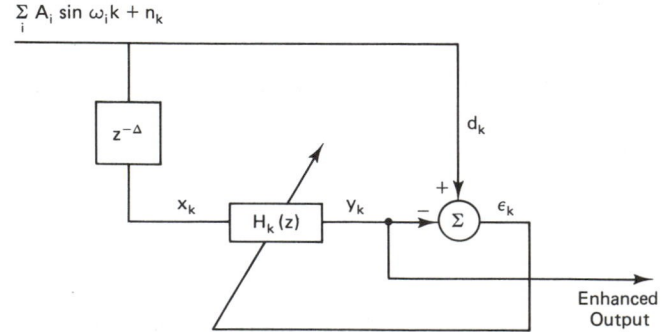

Enhanced
Output

Figure 12.4. General structure for adaptive line enhancement.

decorrelate the noise in d_k and x_k. Our signal definitions are

$$d_k = \sqrt{2}\,\sin\left(\frac{2\pi k}{20}\right) + \sqrt{12}\,(r_k - 0.5)$$

$$x_k = d_{k-1}$$

(12.9)

where r_k is a random number generated via the SPRAND function in Chapter 4. The signal amplitudes in (12.9) result in a signal-to-noise ratio (SNR) equal to 1 and a total input signal power of $\sigma^2 = 2$. The frequency of the sinusoid is 0.05 Hz relative to a 0.5-Hz Nyquist frequency (i.e., the sampling frequency is 1.0 Hz).

In SPA1201 we use the value $\hat{\sigma}^2 = 2.0$, held constant by setting $\alpha = 0.0$, and the convergence parameter in (12.6) is $\mu = 0.10$. The program listing is given, and plots of the desired response and enhanced output sequences are shown in Fig. 12.5.

```
        PROGRAM SPA1201
C-DEMONSTRATE USE OF LMS ALGORITHM FOR LINE-ENHANCEMENT
        DIMENSION X(0:500),D(0:500),B(0:20),PX(0:20)
        REAL MU
        DATA N/501/,MU/.1/,ISEED/12357/,B/21*0./,PX/21*0./
        DATA X(0)/0./,L/20/,SIG/2./,AL/0./
        PI=4.*ATAN(1.)
        DO 1 K=0,N-2
          D(K)=SQRT(2.)*SIN(2.*PI*K/20.)+SQRT(12.)*(SPRAND(ISEED)-.5)
          X(K+1)=D(K)
    1   CONTINUE
        D(500)=SQRT(2.)*SIN(2.*PI*500./20.)+SQRT(12.)*(SPRAND(ISEED)-.5)
        CALL SPNLMS(X,N,D,B,L,MU,SIG,AL,PX,IERROR)
        IF(IERROR.NE.0)THEN
          PRINT *,' SPNLMS ERROR = ',IERROR
          STOP
        ENDIF
        CALL MCPLOT(0.,1.,X,N,0,0,1,1,1,1,5.,3.6)
        CALL MCPLOT(0.,1.,D,N,0,0,1,1,1,1,5.,3.6)
        END
```

In order to demonstrate the effect of changing the convergence parameter, SPA1201 was executed again with the modification $\mu = 0.01$. The resulting enhanced output sequence is shown in Fig. 12.6. Note that the rate of convergence is reduced, but the output is less noisy in this case.

In our next example in program SPA1202, we utilized the signal power estimation in (12.7) by setting $\alpha = 0.01$ and initializing $\hat{\sigma}^2 = 1$. The convergence parameter was again $\mu = 0.01$. The desired response sequence is identical to that in Fig. 12.5. The plots in Fig. 12.7 illustrate the enhanced output sequence and the magnitude response of the adaptive filter upon convergence.

Comparing the response in Fig. 12.7(a) to that in Fig. 12.6, we note that the convergence is slightly faster in Fig. 12.7(a). In SPA1202 the initial rate of convergence was twice that in SPA1201, since in the latter we initialized $\hat{\sigma}^2 = 1$ instead of the actual value, $\sigma^2 = 2$. From the magnitude response in Fig. 12.7(b), we see that the peak filter gain occurs very close to the 0.05-Hz frequency of the sinusoidal component in d_k.

The major purpose of the preceding examples is to illustrate the adjustment of parameters, primarily μ, to obtain better (or worse) performance from an

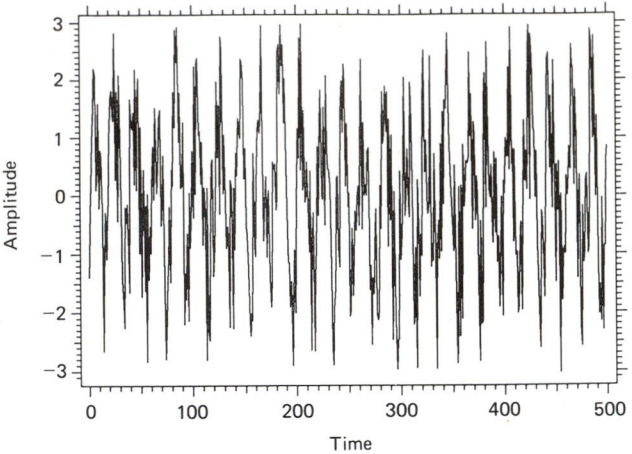

(a) Noisy input sequence, $[d_k]$

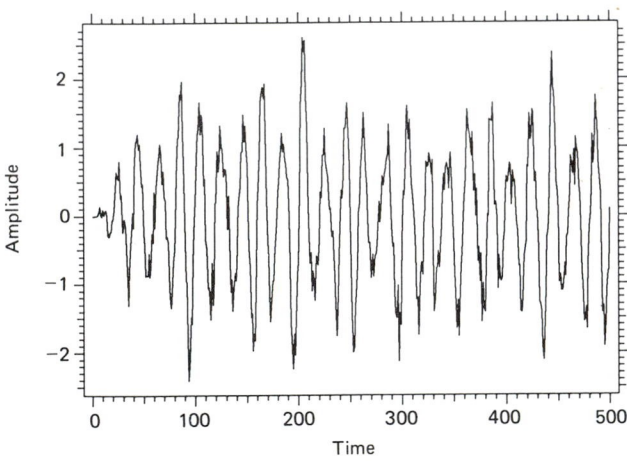

(b) Enhanced output sequence, $[y_k]$

Figure 12.5. SPA1201: Adaptive line enhancement via SPNLMS; $L = 20$, $\mu = 0.1$; $\alpha = 0.0$.

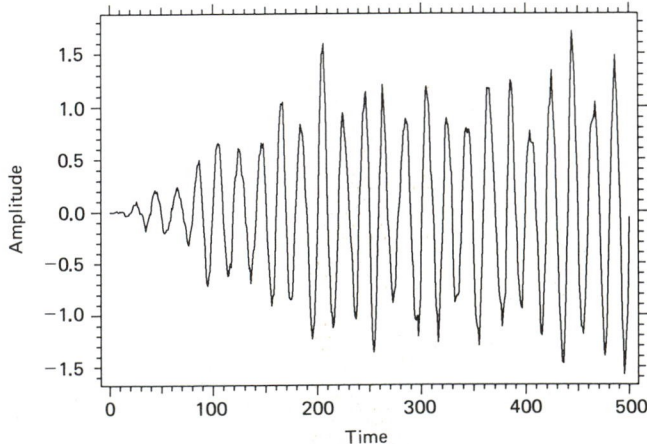

Figure 12.6. SPA1201: Adaptive line enhancement via SPNLMS; $L = 20$, $\mu = 0.01$; $\alpha = 0.0$.

```
      PROGRAM SPA1202
C-DEMONSTRATE USE OF NLMS ALGORITHM FOR LINE-ENHANCEMENT
      DIMENSION X(0:500),D(0:500),B(0:20),PX(0:20),A(1)
      COMPLEX SPGAIN
      REAL MU
      DATA N/501/,MU/.01/,ISEED/12357/,B/21*0./,PX/21*0./
      DATA X(0)/0./,L/20/,SIG/1./,AL/.01/,A(1)/0./
      PI=4.*ATAN(1.)
      DO 1 K=0,N-2
        D(K)=SQRT(2.)*SIN(2.*PI*K/20.)+SQRT(12.)*(SPRAND(ISEED)-.5)
        X(K+1)=D(K)
    1 CONTINUE
      D(500)=SQRT(2.)*SIN(2.*PI*500./20.)+SQRT(12.)*(SPRAND(ISEED)-.5)
      CALL SPNLMS(X,N,D,B,L,MU,SIG,AL,PX,IERROR)
      IF(IERROR.NE.0)THEN
        PRINT *,' SPNLMS ERROR = ',IERROR
        STOP
      ENDIF
      CALL MCPLOT(0.,1.,X,N,0,0,1,1,1,1,5.,3.6)
      DO 2 K=0,500
        X(K)=ABS(SPGAIN(B,A,L,1,K*.5/500.))
    2 CONTINUE
      CALL MCPLOT(0.,.001,X,501,0,0,1,1,3,2,5.,3.6)
      END
```

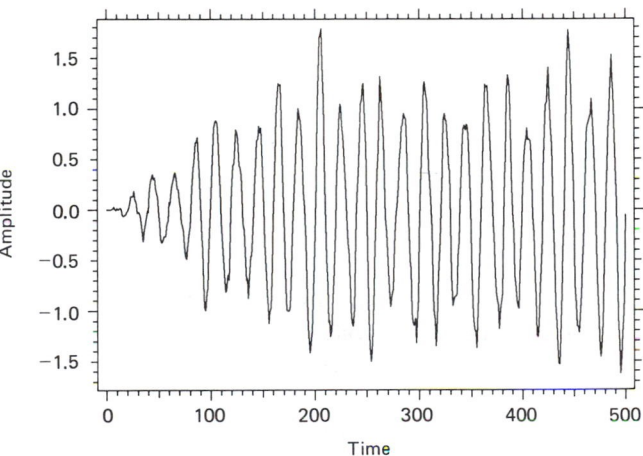

(a) Enhanced output sequence, $[y_k]$

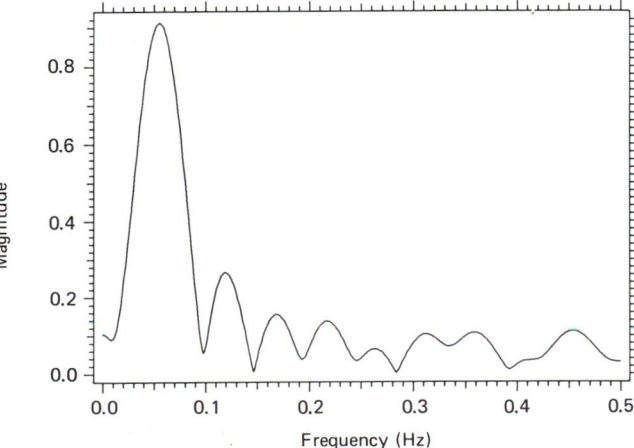

(b) Magnitude response of adaptive filter

Figure 12.7. SPA1202: Adaptive line enhancement via SPNLMS; $L = 20$, $\mu = 0.01$, $\alpha = 0.01$.

adaptive system. Increasing the filter order L generally results in a sharper resonant peak in the magnitude response and a corresponding improvement in the enhanced output signal, but the filter then requires more time to converge. In summary, parameters μ, α, and $\hat{\sigma}^2$ all affect the rate of convergence as well as the accuracy of the converged solution. The effects are less critical when the processing is done off-line, since the experiment can be repeated using a new set of parameters if the solution is not acceptable. Before attempting to perform adaptive signal processing in real time, users should consult the references at the end of this chapter for a more complete discussion of adaptive system performance.

12.4 Miscellaneous Applications

This section is devoted to additional examples of adaptive filtering via the SPNLMS subroutine. Although several specific applications are described, the intent is to give the reader some insight into the potential benefits and inherent dangers of adaptive signal processing.

Adaptive Interference Canceling

The basic structure for adaptive interference canceling is shown in Fig. 12.8. The desired response is composed of a signal component s_k and a noise component n_k, which ideally is uncorrelated with s_k. The filter input is a noise sequence n_k', which is correlated with the noise component in d_k. As in the examples in Section 12.3, minimization of the MSE entails cancelation of the correlated components of d_k and x_k. Thus, the filter converges such that the output y_k is an estimate $\hat{n}_k$ of the noise component in d_k.

Unlike the adaptive line-enhancer output y_k in Fig. 12.4, the adaptive interference canceler output is the error signal ε_k. At this point, a word of caution is in order. The key element that allows this adaptive structure to achieve its goal is the fact that s_k and n_k are uncorrelated, whereas n_k and n_k' are correlated.

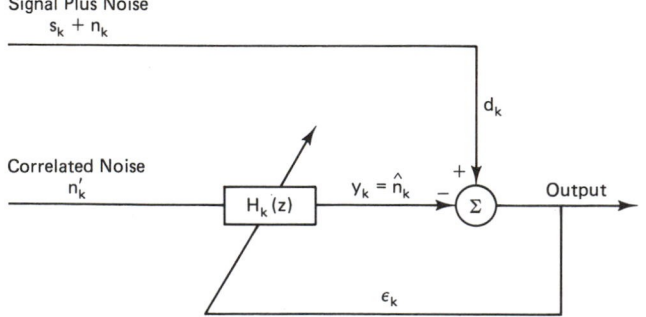

Figure 12.8. Basic structure for adaptive interference canceling.

If these conditions are violated, the adaptive filter may cancel the signal s_k in place of (or in addition to) the noise or could fail to cancel the noise.

To demonstrate the use of this structure, we consider a case involving a broadband signal corrupted by sinusoidal interference. To simplify the discussion, we assume that our signal component is white. In order to demonstrate tracking capabilities, we allow the sinusoidal noise component n_k to increase slowly in amplitude. The correlated noise component n_k' does not drift and differs from n_k in both amplitude and phase. The demonstration is accomplished by SPA1203. The plot in Fig. 12.9 shows the first 250 points of the desired response, that is, the interfering sinusoid plus white noise. The sinusoidal interference, with time-varying amplitude equal to $1.0 + 0.01k$, is evident in this plot.

```
                PROGRAM SPA1203
     C-ADAPTIVE INTERFERENCE CANCELLING
                DIMENSION X(0:500),D(0:500),B(0:4),PX(0:4),SI(0:500)
                REAL MU
                DATA B/5*0./,PX/5*0./,ISEED/12357/
                DATA L/4/,N/501/,MU/.1/,SIG/.005/,AL/0./
                PI=4.*ATAN(1.)
                DO 1 K=0,N-1
                  F=2.*PI*K/20.
                  SI(K)=(1.+.01*K)*SIN(F+PI/3.)
                  D(K)=SQRT(12.)*(SPRAND(ISEED)-.5)+SI(K)
                  X(K)=0.1*SIN(F)
        1       CONTINUE
                CALL MCPLOT(0.,1.,SI,251,0,0,3,1,1,1,5.,3.6)
                CALL MCPLOT(0.,1.,X,251,9,0,3,2,1,1,5.,3.6)
                CALL SPNLMS(X,N,D,B,L,MU,SIG,AL,PX,IERROR)
                IF(IERROR.NE.0)THEN
                  PRINT *,' SPNLMS ERROR = ',IERROR
                  STOP
                ENDIF
                CALL MCPLOT(0.,1.,X,251,1,0,3,3,1,1,5.,3.6)
                CALL MCPLOT(250.,1.,SI(250),251,0,0,2,1,1,1,5.,3.6)
                CALL MCPLOT(250.,1.,X(250),251,1,0,2,2,1,1,5.,3.6)
                CALL MCPLOT(0.,1.,D,251,0,0,1,1,1,1,5.,3.6)
                END
```

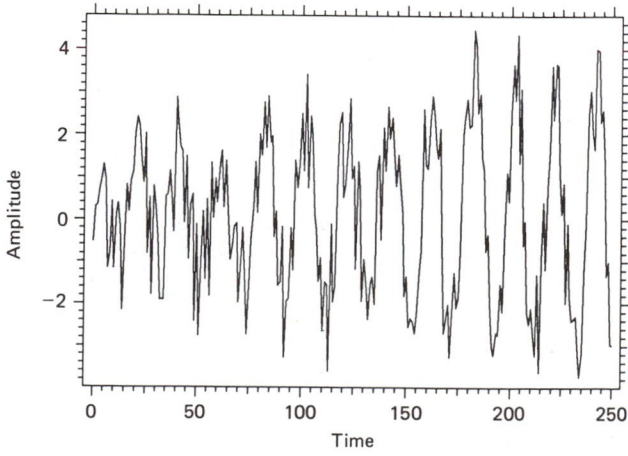

Figure 12.9. SPA1203: Interfering sinusoid plus white noise (desired response) for adaptive interference canceling example.

The plots in Fig. 12.10, also generated by SPA1203, demonstrate the performance of the adaptive interference canceler in this example. Part (a) illustrates the initial adaptation of $H_k(z)$, wherein the correlated noise component n'_k is modified in an attempt to make $\hat{n}_k$ match n_k. The tracking capabilities are demonstrated in Fig. 12.10(b) as the amplitude of $\hat{n}_k$ is adjusted to match the increasing amplitude of n_k.

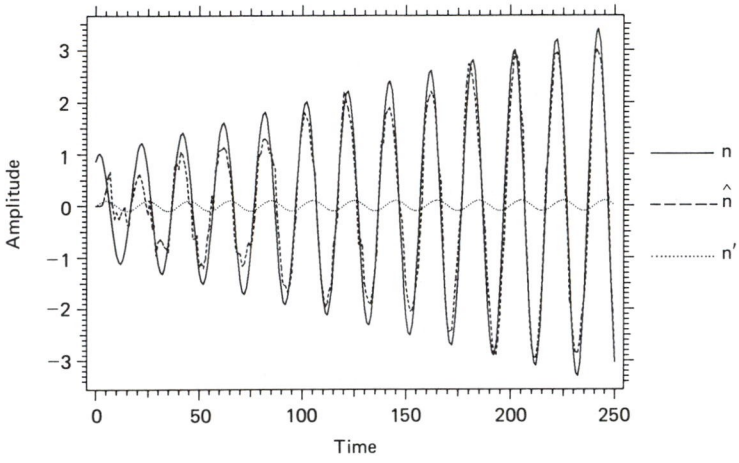

(a) Initial convergence

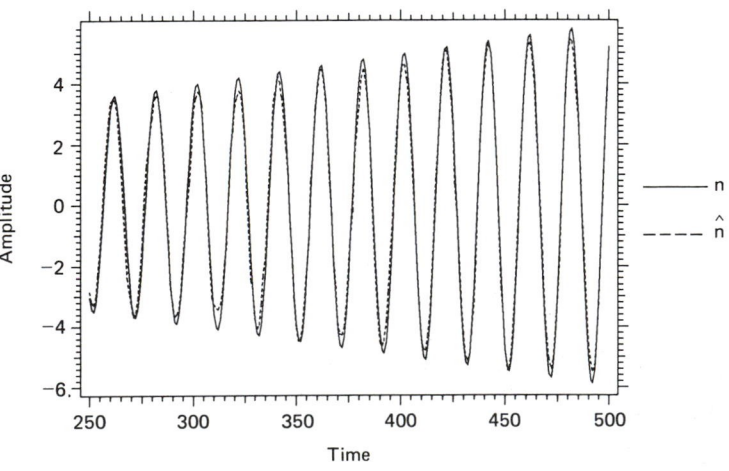

(b) Tracking performance

Figure 12.10. SPA1203: Adaptive interference cancelation via SPNLMS; $L = 4$, $\mu = 0.10$.

From these results, it is evident that the adaptive interference canceler was successful in detecting and matching the interfering sinusoidal component present in the noisy $[d_k]$ sequence. Additional information on adapative interference canceling is available in the literature [D, E].

Adaptive System Identification

In some applications it is advantageous to perform system identification adaptively rather than by using the classical least-squares modeling techniques described in Chapter 11. This is especially true in real-time applications where the characteristics of the unknown system are changing over a period of time. System identification is often an integral part of an adaptive control process. When the model is used to formulate controls, it must reflect the current characteristics of the actual system.

The basic structure for adaptive system identification is illustrated in Fig. 12.11. In this case, the desired response is the response of the unknown system to the input sequence $[x_k]$. Note that if $H_k(z)$ identifies the unknown system exactly, the output error ε_k is zero.

Two conditions must be satisfied if the adaptive filter is to be an accurate model of the unknown system. The first is that the input sequence $[x_k]$ must contain enough information to excite all modes of the system. This is often achieved by using a white noise sequence containing power at all frequencies. Although this is a viable solution for off-line system identification, it is more difficult to utilize in real-time applications, where careful thought must be given to the selection of the input signal. The second condition is that $H_k(z)$ must be of sufficient complexity (i.e., high enough in order) to match the degrees of freedom of the unknown system. For example, if an FIR model is used to identify an IIR system, a very high order $H_k(z)$ is typically required. If the model order is overspecified, the adaptive process may converge more slowly, but the solution will still provide an accurate, though potentially nonunique, model of the system.

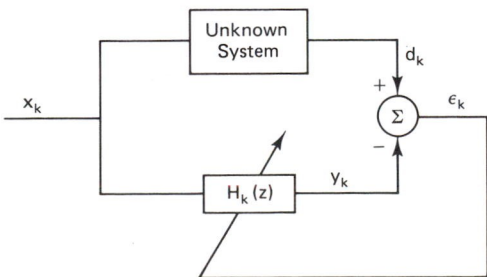

Figure 12.11. Basic structure for adaptive system identification.

An example of adaptive system identification is provided in SPA1204. In this program, we assume that the unknown system is actually a bandpass FIR filter designed via SPFIRD (Chapter 8). The filter is of order 20, whereas the model $H_k(z)$ is of order $L = 30$. The input x_k is a random sequence with unit power density. Results from SPA1204 are provided in two forms. In Fig. 12.12(a), the square of the error sequence is shown. The model converges as this sequence decays to zero. The impulse response of the model together with that of the unknown system is shown in Fig. 12.12(b). It is evident from these curves that the model order was overspecified in this example.

```
      PROGRAM SPA1204
C-ADAPTIVE SYSTEM IDENTIFICATION
      DIMENSION X(0:1000),D(0:1000),B(0:30),PX(0:30)
      DIMENSION SB(0:20),SA(1),SPX(0:20),SPY(1)
      REAL MU
      DATA N/1001/,L/30/,MU/.1/,AL/0./SIG/1./
      DATA B/31*0./,PX/31*0./,SA/0./,SPX/21*0./,SA/0./,ISEED/12357/
      DO 1 K=0,1000
         X(K)=SQRT(12.)*(SPRAND(ISEED)-.5)
    1 CONTINUE
      CALL SPFIRD(20,3,.15,.35,1,SB,IERROR)
         IF(IERROR.NE.0)STOP ' SPFIRD ERROR'
      CALL SPFLTR(SB,SA,20,1,X,N,D,SPX,SPY,IERROR)
         IF(IERROR.NE.0)STOP ' SPFLTR ERROR'
      CALL SPNLMS(X,N,D,B,L,MU,SIG,AL,PX,IERROR)
         IF(IERROR.NE.0)STOP ' SPNLMS ERROR'
      DO 2 K=0,1000
         D(K)=(D(K)-X(K))**2
    2 CONTINUE
      CALL MCPLOT(0.,1.,D,N,0,0,1,1,2,2,5.,3.6)
      CALL MCPLOT(0.,1.,B,31,1,12,2,1,2,1,5.,3.6)
      CALL MCPLOT(0.,1.,SB,21,0,0,2,2,2,1,5.,3.6)
      END
```

Although the model used in SPA1204 was a simple FIR filter, this need not be the case in adaptive system identification in general. If the system to be modeled contains poles as well as zeros, as is typically the case, there are advantages to using an IIR model. The literature contains information regarding a variety of alternate models and algorithms, and also addresses the related topic of adaptive inverse modeling [E, H, I, R, V].

Adaptive Design of Conventional Digital Filters

Although many iterative techniques exist for conventional digital filter design, the procedures using the LMS adaptive algorithm are among the simplest to implement. The basic adaptive system structure for digital filter design is illustrated in Fig. 12.13.

The filter design is specified by the desired amplitude response A_m and phase shift θ_m at each frequency f_m for $m = 1, 2, \ldots M$. Thus, the desired response d_k is formulated as the sum of M sinusoids, each with frequency f_m, amplitude A_m, and phase θ_m. The adaptive filter input is composed of the same M sinusoids with unit amplitude and zero phase shift, so that $H_k(z)$ must acquire the desired gain and phase characteristics in order to drive the error, ε_k, to zero. An example of adaptive filter design is provided in SPA1205.

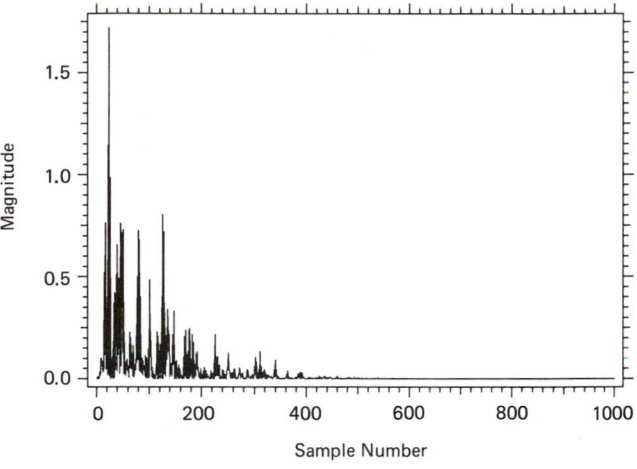

(a) Squared error (ϵ_k^2)

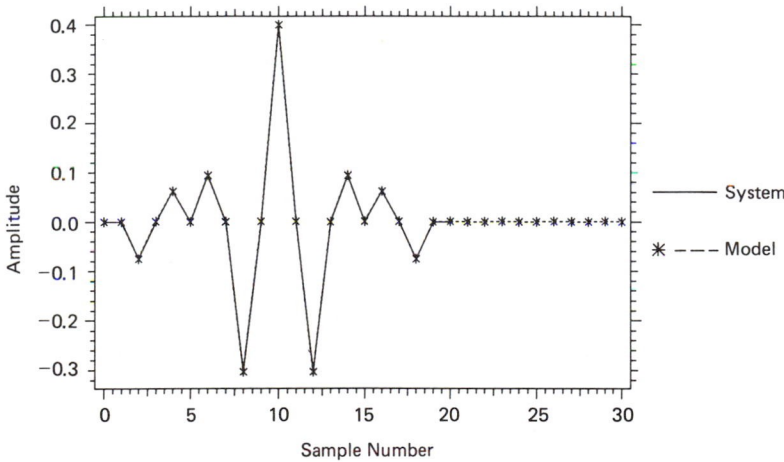

(b) Impulse response of system and model

Figure 12.12.   SPA1204: Adaptive system identification via SPNLMS; $L = 30$, $\mu = 0.10$.

In SPA1205 digital filter characteristics are provided at 41 evenly spaced points ranging from zero to the Nyquist frequency (half the sampling rate). A low-pass filter is specified, with the first 20 gain values of unit amplitude at passband frequencies and the last 20 gain values of zero amplitude at stopband frequencies. The amplitude of the midpoint gain at 0.25 Hz is set to 0.5 to specify a transition band. Linear phase response (delay = $L/2$) is specified.

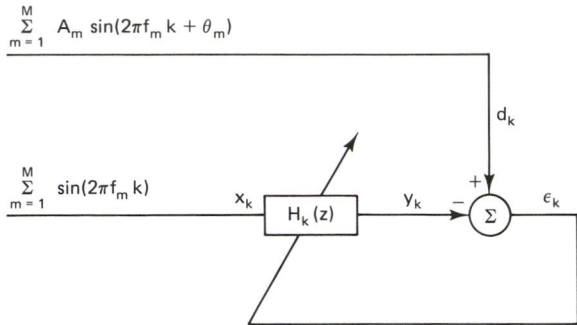

Figure 12.13. Basic structure for adaptive filter design.

The curves in Fig. 12.14(a) show the magnitude response of the converged $H_k(z)$ together with that of the desired filter. In Fig. 12.14(b), the impulse response of $H_k(z)$ is shown. Note the symmetry about the midpoint, which indicates linear phase response.

```
          PROGRAM SPA1205
    C-ADAPTIVE FILTER DESIGN
          DIMENSION X(0:500),D(0:500),B(0:50),PX(0:50),A(1),AMP(0:40)
          REAL MU
          COMPLEX SPGAIN
          DATA L/50/,N/501/,MU/.2/,AL/.0/SIG/.2/
          DATA PX/51*0./,B/51*0./,A/0./,AMP/20*1.,.5,20*0./
          PI=4.*ATAN(1.)
          DO 2 K=0,500
            D(K)=.1*AMP(0)
            X(K)=.1
            DO 1 J=1,40
              D(K)=D(K)+AMP(J)*.1*SIN(2.*PI*.0125*J*(K-L/2.))
              X(K)=X(K)+.1*SIN(2.*PI*.0125*J*K)
    1       CONTINUE
    2     CONTINUE
          CALL SPNLMS(X,N,D,B,L,MU,SIG,AL,PX,IERROR)
          IF(IERROR.NE.0)STOP ' SPNLMS ERROR'
          DO 5 K=0,500
    5     X(K)=ABS(SPGAIN(B,A,L,1,K*.5/500.))
          CALL MCPLOT(0.,.001,X,N,0,0,2,1,3,2,5.,3.6)
          CALL MCPLOT(0.,.0125,AMP,41,1,0,2,2,3,2,5.,3.6)
          CALL MCPLOT(0.,1.,B,51,1,12,1,1,2,1,5.,3.6)
          END
```

In digital filter design, the filter specification samples need not be evenly spaced in frequency. The total number of samples and their spacing will, however, affect the filter design, since the MSE minimization occurs only at those frequencies present in the input sequences. In addition, filter specifications should be consistent with the order of $H_k(z)$ (i.e., a narrow transition band implies a high-order filter, and so on). The adaptive technique provides a great deal of flexibility in digital filter design.

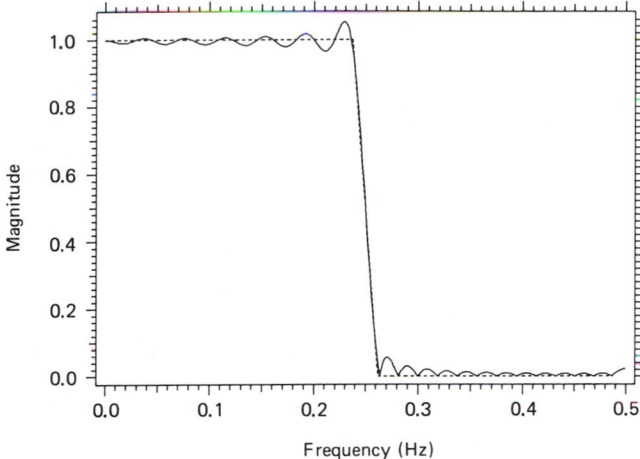

(a) Magnitude responses of $H_k(z)$ and desired filter

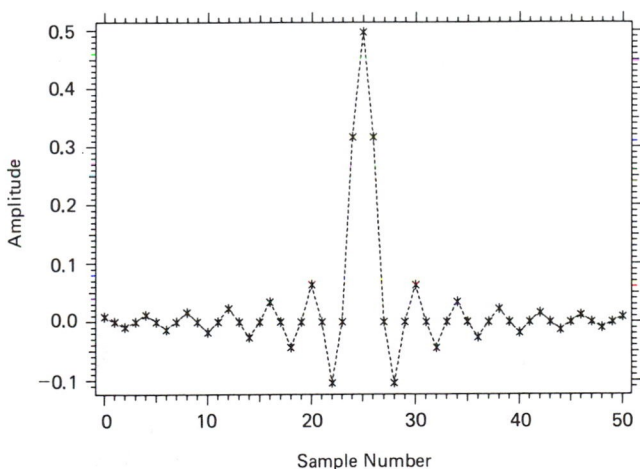

Figure 12.14. SPA1205: Adaptive filter design via SPNLMS; $L = 50$.

(b) Impulse response of $H_k(z)$ after convergence

12.5 Summary of Calling Sequences

Due to the brevity of our discussion of adaptive signal processing, only one subroutine, SPNLMS, was provided in this chapter. We have demonstrated, however, that the LMS algorithm can be utilized in a number of different adaptive system configurations. The calling sequence and argument list are summarized next. As in previous chapters, the symbol I denotes an input to the subroutine and O specifies an output from the subroutine.

CALL SPNLMS(X,N,D,B,L,MU,SIG,AL,PX,IERROR)

Variable	Definition	Routine
		SPNLMS
X(0:N−1)	Data vector (Output replaces Input)	I/O
N	Number of samples and adaptive iterations	I
D(0:N−1)	Desired data sequence	I
B(0:L)	Adaptive filter coefficients	I/O
L	Order of adaptive filter	I
MU	REAL convergence parameter (0.0 < MU < 1.0)	I
SIG	Input signal power estimate (SIG > 0.0)	I
AL	Factor used in SIG update (0.0 ≤ AL << 1.0)	I
PX	Retains past data values for block mode	I/O
IERROR	Return error parameter	O

12.6 Exercises

12.1. Consider the adaptive line enhancer in Fig. 12.4. Generate a 1000-point $[d_k]$ sequence containing two sinusoidal components with additive zero mean, unit power density white noise. The sample interval is $T = 0.001$ s and the sinusoids are specified by: $A_1 = 2, f_1 = 100$ Hz; $A_2 = 0.8, f_2 = 300$ Hz. Call SPNLMS with parameters $L = 40$, $\mu = 0.05$, $\alpha = 0.0$ and $\hat{\sigma}^2$ equal to the input signal power. Plot the magnitude response of the adaptive filter after convergence.

12.2. Use the adaptive line-enhancer structure in Fig. 12.4. The desired response contains three sinusoids with additive zero mean, unit power density white noise. Let $T = 0.01$ s; $A_1 = 1.0, f_1 = 5$ Hz; $A_2 = 0.5, f_2 = 15$ Hz; $A_3 = 1.5, f_3 = 35$ Hz; LMS parameters: $L = 100$, $\mu = 0.01$, $\alpha = 0.0$, $\hat{\sigma}^2 =$ input signal power. Use the SPNLMS subroutine in a block mode operation so that the convergence of $B(z)$ can be observed. Plot the magnitude response of the filter after 100, 200, 500, and 1000 iterations.

12.3. Repeat the task outlined in exercise 12.1, this time using three different filters with orders $L = 10, 20,$ and 50. Plot the magnitude response in each case.

12.4. Consider the adaptive line-enhancer structure in Fig. 12.4. Use a 1000-point non-stationary input sequence containing a sinusoid of unit amplitude and normalized frequency $f = 0.2 + 0.0001k$ Hz-s plus zero mean white noise with power density of 0.5. Use LMS parameters $L = 20$, $\alpha = 0.0$, $\hat{\sigma}^2 =$ signal power. Observe the system performance by plotting the squared error sequence for three cases: $\mu = 0.1, 0.05, 0.01$.

12.5. In the adaptive interference canceler in Fig. 12.8, let

$$d_k = \sqrt{2}\,\sin\left(\frac{2\pi k}{10}\right) + \sin\left(\frac{2\pi k}{15}\right) + 2\sin\left(\frac{2\pi k}{25}\right)$$

$$n'_k = 0.1\sin\left(\frac{2\pi k}{15} + \frac{\pi}{3}\right)$$

Use LMS parameters $L = 10$, $\mu = 0.01$, $\alpha = 0.0$, $\hat{\sigma}^2 =$ signal power. Allow the system to adapt for 500 iterations. Plot the $[\varepsilon_k]$ sequence.

12.6. Repeat exercise 12.5 with

$$n_k' = 0.1 \sin\left(\frac{2\pi k}{10} + \frac{\pi}{4}\right) + 0.15 \sin\left(\frac{2\pi k}{25} + \frac{\pi}{3}\right)$$

12.7. Consider the adaptive interference canceler in Fig. 12.8 with

$$s_k = \frac{4}{\pi} \sum_{n=1,3,5} \frac{1}{n} \sin\left(\frac{2\pi n k}{25}\right) + 0.5 \sin\left(\frac{2\pi 10 k}{25}\right), \qquad k = 0, 1, \ldots, 500$$

$$n_k' = 0.1 \sin\left(\frac{2\pi 10 k}{25} + \frac{\pi}{3}\right)$$

Use LMS parameters $L = 6$, $\mu = 0.01$, $\alpha = 0.01$, $\hat{\sigma}^2 = 1.0$. Plot the error sequence $[\varepsilon_k]$.

12.8. Use the system identification structure in Fig. 12.11. The unknown system is a tenth-order Bessel bandpass filter designed via SPBSSL and SPFBLT (Chapter 7) with band edges at normalized frequencies $f_L = 0.15$ and $f_H = 0.35$ Hz-s (no prescaling is used in SPBSSL). The input sequence is zero mean, unit power density white noise. Use LMS parameters $L = 40$, $\mu = 0.05$, $\alpha = 0.0$, $\hat{\sigma}^2 = 1.0$. Allow the system to adapt for 1000 iterations. Plot the magnitude and phase responses of both the Bessel filter and the FIR model.

12.9. Consider the inverse modeling structure illustrated in Fig. 11.8, and assume that the inverse model is to be formed adaptively, that is, via SPNLMS. Let the unknown system be defined by $H(z) = 2/(1 - 0.6z^{-1})$. Use an input sequence of zero mean, unit power density white noise and delay $\Delta = L/2$. Compute the tenth-order inverse model $B(z)$ using LMS parameters $\mu = 0.05$, $\alpha = 0.0$, and $\hat{\sigma}^2 = 1.0$ and 1000 iterations. Plot the magnitude response of the cascade combination $H(z) \cdot B(z)$.

12.10. Using the basic adaptive system structure in Fig. 12.2, let

$$d_k = \sum_{n=1}^{5} \sin\left(\frac{2\pi k n}{10}\right) \qquad x_k = \sum_{n=1}^{5} 0.5 \sin\left(\frac{2\pi k n}{20}\right)$$

Use LMS parameters $L = 200$, $\mu = 0.1$, $\alpha = 0.0$, and $\hat{\sigma}^2 =$ signal power. After 1000 iterations, plot the magnitude response of the adaptive filter. Explain your result.

12.11. Repeat exercise 12.10 with filters of order $L = 20, 50, 100$.

12.12. Consider the adaptive filter design application illustrated in Fig. 12.13. Design a bandpass filter using specifications:

$$A_m = 0 \qquad f_m = 0, 0.01, 0.02, \ldots, 0.15 \text{ Hz-s}$$

$$A_m = 1 \qquad f_m = 0.2, 0.21, 0.22, \ldots, 0.30 \text{ Hz-s}$$

$$A_m = 0 \qquad f_m = 0.35, 0.36, 0.37, \ldots, 0.50 \text{ Hz-s}$$

Linear phase response

Use LMS parameters $L = 100$, $\mu = 0.1$, $\alpha = 0.0$, $\hat{\sigma}^2 =$ input power. After 1000 iterations, plot the magnitude response of $B(z)$ together with the magnitude specifications. Also plot the impulse response of $B(z)$.

12.13. Repeat exercise 12.12, but specify the passband and stopband frequencies at increments of 0.005 (i.e., specify twice the number of points).

12.14. Design a notch filter using the structure in Fig. 12.13.

$$A_m = 1 \qquad f_m = 0.0, 0.01, 0.02, \ldots, 0.22 \text{ Hz-s}$$

$$A_m = 0 \qquad f_m = 0.25 \text{ Hz-s}$$

$$A_m = 1 \qquad f_m = 0.28, 0.29, 0.30, \ldots, 0.50 \text{ Hz-s}$$

Linear phase response

Use LMS parameters $L = 100$, $\mu = 0.01$, $\alpha = 0.0$, and $\hat{\sigma}^2 =$ input power. After 1000 iterations, plot the magnitude response of $B(z)$ together with the magnitude specifications.

12.7 References

[A] Widrow, B. and M. E. Hoff, Jr., "Adaptive Switching Circuits," 1960 *IRE WESCON Convention Record,* Part 4, pp. 96–104.

[B] Widrow, B., et al., "Adaptive Antenna Systems," *Proc. IEEE,* 55 (December 1967).

[C] Widrow, B., "Adaptive Filters," in Aspects of Network and System Theory (New York: Holt, Rinehart, and Winston, 1970), p. 563.

[D] Widrow, B., et al., "Adaptive Noise Cancelling: Principles and Applications," *Proc. IEEE,* 63, no. 12 (December 1975).

[E] Widrow, B., and S. D. Stearns, *Adaptive Signal Processing* (Englewood Cliffs, N.J.: Prentice-Hall, Inc., 1985).

[F] Ahmed, N., et al., "A Short-Term Sequential Regression Algorithm," *IEEE Trans. ASSP-27,* no. 5 (October 1979).

[G] Clark, G. A. and P. W. Rodgers, "Adaptive Prediction Applied to Seismic Event Detection," *Proc. IEEE,* 69, no. 9 (September 1981).

[H] Cowan, C. and P. Grant, eds., *Adaptive Filters* (Englewood Cliffs, N.J.: Prentice-Hall, Inc., 1985).

[I] David, R. A., "IIR Adaptive Algorithms Based on Gradient Search Techniques," Ph.D. dissertation, Stanford University, August 1981.

[J] Dentino, M., J. McCool, and B. Widrow, "Adaptive Filtering in the Frequency Domain," *Proc. IEEE,* 66 (December 1978).

[K] Etter, D. M. and S. D. Stearns, "Adaptive Estimation of Time Delays in Sampled Data Systems," *IEEE Trans. ASSP-29,* no. 3 (June 1981).

[L] Feintuch, P. L., "An Adaptive Recursive LMS Filter," *Proc. IEEE* (November 1976).

[M] Ferrara, E., "Fast Implementation of LMS Adaptive Filters," *IEEE Trans. on ASSP,* ASSP-28 (August 1980).

[N] Griffiths, L. J., "Rapid Measurement of Digital Instantaneous Frequency," *IEEE Trans. on ASSP,* ASSP-28, no. 2 (April 1975).

[O] Griffiths, L. J., "A Continuously-Adaptive Filter Implemented as a Lattice Structure," *Proc. ICASSP,* ICASSP-77 (May 1977).

[P] Griffiths, L. J., et al., "Adaptive Deconvolution: A New Technique for Processing Time-Varying Seismic Data," *Geophysics* (June 1977).

[Q] Griffiths, L. J., "An Adaptive Lattice Structure for Noise-Cancelling Applications," *Proc. ICASSP,* ICASSP-78 (April 1978).

[R] Treichler, J., et al., "SHARF: An Algorithm for Adapting IIR Digital Filters," *IEEE Trans. ASSP,* ASSP-28, no. 4 (August 1980).

[S] Makhoul, J., "Stable and Efficient Lattice Methods for Linear Prediction," *IEEE Trans. ASSP,* ASSP-25, no. 5 (October 1977).

[T] Griffiths, L. J. and R. S. Medaugh, "A Comparison of Two Fast Linear Predictors," *Proc. ICASSP,* ICASSP-81 (April 1981).

[U] Stearns, S.D. and L. J. Vortman, "Seismic Event Detection Using Adaptive Predictors," *Proc. ICASSP,* ICASSP-81 (March 1981).

[V] White, S. A., "An Adaptive Recursive Digital Filter," *Proc. 9th Asilomar Conf. on Circuits, Systems, and Computers* (November 1975).

[W] Widrow, B., et al., "Stationary and Nonstationary Learning Characteristics of the LMS Adaptive Filter," *Proc. IEEE,* 64, no. 8 (August 1976).

[X] McCool, J. M. and B. Widrow, "A Comparison of Adaptive Algorithms Based on the Methods of Steepest Descent and Random Search," *IEEE Trans. ASSP,* ASSP-24 (September 1976).

[Y] Gooch, R. P., P. F. Titchener, and B. Widrow, "Adaptive Design of Digital Filters," *Proc. ICASSP,* ICASSP-81 (March 1981).

[Z] Zeidler, J. R., et al., "Adaptive Enhancement of Multiple Sinusoids in Uncorrelated Noise," *IEEE Trans. ASSP,* ASSP-26, no. 3 (June 1978).

chapter **13**

Parameter Estimation and Waveform Analysis

13.1 Introduction

In some digital signal processing applications, much of the pertinent information in a signal can be summarized in the form of a few key parameters. These parameters may describe statistical properties or characterize the shape of the waveform in terms of a specific function. The signal rise and/or fall times as well as the maximum and minimum data values may also be of interest.

The routines described in this chapter are aids to parameter estimation and waveform analysis. We first consider the estimation of basic statistical properties by computing time averages. Next we utilize curve-fitting routines to characterize the waveform. Finally, we describe routines that locate the maximum and minimum data values and compute the signal rise and fall times.

13.2 Basic Statistical Parameter Estimation

The primary statistical parameters of interest are the mean and the variance. Although these values are formally defined in terms of ensembles of infinite sequences of random variables, here we utilize a more practical approach [A, B]. We estimate the mean and variance of a data vector by computing the related time averages.

The mean, expectation, or average value of an N-point data sequence $[x_k]$ can be computed as

$$m_x = \frac{1}{N} \sum_{k=0}^{N-1} x_k \tag{13.1}$$

where m_x denotes the mean of $[x_k]$. Function SPMEAN computes the mean of the N-point data sequence, $[x_k]$ via (13.1). The function specification is

FUNCTION SPMEAN(X,N)

$X(0:N-1)$ = data vector
N   = number of samples in data array

If the argument N is invalid (i.e., $N \leq 0$), zero will be returned in place of the desired mean estimate. Before demonstrating the use of SPMEAN, we describe an additional function that computes the variance.

The variance of $[x_k]$ is the most common measure of the squared variation of x_k about its mean value m_x [C]. A related parameter is the *standard deviation*, which is a measure of the deviation of x_k from m_x. The standard deviation of $[x_k]$, σ_x, is equal to the square root of the variance, σ_x^2.

In terms of time averages, the variance of the N-point data sequence $[x_k]$ is defined as

$$\sigma_x^2 = \frac{1}{N-1}\left[\sum_{k=0}^{N-1} x_k^2 - Nm_x^2\right] \tag{13.2}$$

where m_x and σ_x^2 are the mean and variance, respectively [B]. Function SPVARI computes the variance of the data sequence $[x_k]$ via (13.2). The function specification is

FUNCTION SPVARI(X,N)

$X(0:N-1)$ = data sequence
N   = number of points in data array

As with SPMEAN, a value of zero will be returned if argument N is invalid. This will also occur if N is less than 2, since (13.2) cannot be evaluated for that condition.

Test program SPA1301 is provided to illustrate the use of these functions. The data sequences in this example are vectors of uniformly distributed random numbers generated via the SPRAND function. SPRAND, which is described in Chapter 4, generates data with a mean value of $\frac{1}{2}$ and variance equal to $\frac{1}{12}$, or 0.0833333. We evaluated the parameters for five sequences with lengths ranging from N = 10 to N = 100,000. A listing of SPA1301 together with the computed parameters is provided.

From the results of SPA1301, it is apparent that the estimates of the mean and variance of the infinite random sequence tend to be more accurate for the longer data sequences. This is true in general if the data in $X(0:N-1)$ is noisy (i.e., has some randomly distributed components) [A].

In terms of signal characteristics, the mean is often called the DC (i.e., zero frequency) value or bias. For a zero mean signal, the variance is an estimate of the power. If the mean is nonzero, an estimate of the total power in $[x_k]$ is provided by the sum $[\sigma_x^2 + m_x^2]$. See Section 4.2.

```
            PROGRAM SPA1301
      C-COMPUTE MEAN AND VARIANCE OF VECTOR OF RANDOM NUMBERS
            DIMENSION X(0:99999),LEN(5)
            DATA LEN/10,100,1000,10000,100000/,ISEED/12357/
            DO 1 K=0,99999
              X(K)=SPRAND(ISEED)
        1   CONTINUE
            PRINT 100
            DO 2 L=1,5
              AVG=SPMEAN(X,LEN(L))
              VAR=SPVARI(X,LEN(L))
              PRINT 200,LEN(L),AVG,VAR
        2   CONTINUE
      100   FORMAT(' LENGTH',10X,'MEAN',10X,'VARIANCE')
      200   FORMAT(I7,7X,F10.7,7X,F10.7)
            END

      $ RUN SPA1301
        LENGTH          MEAN                VARIANCE
            10        0.4567687           0.0591502
           100        0.5455416           0.0703557
          1000        0.5013196           0.0834480
         10000        0.4972631           0.0833648
        100000        0.4993142           0.0831321
      $
```

13.3 Waveform Analysis via Curve Fitting

In this section we describe software that enables waveform analysis via curve fitting. The basis for this discussion is a technique for fitting a straight line to a set of data points. This method is then extended to provide curve fitting for several basic functions.

The techniques described herein utilize ordered pairs of data values (x_k, y_k) from the corresponding data sequences $[x_k]$ and $[y_k]$. In the applications described in this text, x_k generally is a specific time and y_k is the signal amplitude at that time.

In previous chapters, we generally have defined the time sequence by specifying the starting point and the sampling interval. The amplitude values therefore corresponded to regularly spaced instants in time. Here, by utilizing two distinct sequences, we have relaxed this restriction. For example, suppose we have an irregularly spaced time sequence $[x_k] = [0, T, 2T, 4T, 5T]$, where T is the sample interval in seconds. The corresponding $[y_k]$ sequence specifies the signal amplitude at times 0, T, $2T$, $4T$, and $5T$ seconds. The sample at $3T$ seconds is missing. By fitting a curve to the data provided in $[x_k]$ and $[y_k]$, an estimate of the signal amplitude at $3T$ seconds can be obtained.

Curve-fitting routines are also useful when the measured data is noisy, which is usually the case with physical data. For some experiments, a priori information may suggest that the data waveform should have the characteristics of a specific function. If this is the case, curve fitting will generally provide more informative results than either interpolation or filtering.

The method of least squares, which was introduced in Chapter 11, is the basis for the routines described in this section. Here we restrict the discussion to the simple straight-line fit, which can easily be accomplished without using Levinson's algorithm or other iterative techniques. The general equation for a straight line is

$$y = ax + b \tag{13.3}$$

where a and b are constants.

Utilizing the method of least squares together with data sequences $[x_k]$ and $[y_k]$, estimates of a and b are provided by [B]

$$\hat{a} = \frac{N \sum\limits_{k=0}^{N-1} x_k y_k - \left(\sum\limits_{k=0}^{N-1} x_k \right) \left(\sum\limits_{k=0}^{N-1} y_k \right)}{N \sum\limits_{k=0}^{N-1} x_k^2 - \left(\sum\limits_{k=0}^{N-1} x_k \right)^2} \tag{13.4}$$

$$\hat{b} = m_y - \hat{a} m_x$$

where $\hat{a}$ and $\hat{b}$ are estimates of a and b, respectively, N specifies the length of the data sequences, and m_x and m_y are the means of $[x_k]$ and $[y_k]$ as defined in (13.1).

Subroutine SPLFIT fits a straight line to the N-point data sequences in the arrays $X(0:N-1)$ and $Y(0:N-1)$. The calling sequence is

$$\text{CALL SPLFIT(X,Y,N,A,B,IERROR)}$$

$X(0:N-1)$ = data sequence $[x_k]$
$Y(0:N-1)$ = data sequence $[y_k]$
N   = number of data samples
A   = estimate $\hat{a}$ in (13.4)
B   = estimate $\hat{b}$ in (13.4)
IERROR   = 0   no errors detected
                  1   invalid N; $N \leqslant 0$
                  2   denominator of $\hat{a}$ in (13.4) $< 1.\text{E} - 10$

The IERROR = 2 return is provided in an attempt to avoid arithmetic overflow in case of data sequence problems.

Program SPA1302 is provided to demonstrate the use of SPLFIT. Data was computed as

$$y_k = 0.05 x_k + 0.25 + 0.25(r_k - 0.5)$$

where r_k is a random number generated via the SPRAND function. This equation defines a straight line with parameters $a = 0.05$ and $b = 0.25$ and additive noise varying in amplitude between ± 0.125. In the results plotted in Fig. 13.1, the noisy samples are denoted by the asterisks. The solid line corresponds to the function defined by the SPLFIT estimates $\hat{a}$ and $\hat{b}$.

```
          PROGRAM SPA1302
C-DEMONSTRATE USE OF STRAIGHT LINE FIT ROUTINE
          DIMENSION X(0:50),Y(0:50),PLT(2)
          DATA A/.05/,B/.25/,ISEED/12357/,N/51/
          DO 1 K=0,50
            X(K)=K
            Y(K)=A*X(K)+B +.25*(SPRAND(ISEED)-.5)
    1     CONTINUE
          CALL SPLFIT(X,Y,N,AEST,BEST,IERROR)
          IF(IERROR.NE.0)THEN
            PRINT *,' SPLFIT ERROR =',IERROR
            STOP
          ENDIF
          CALL MCPLOT(0.,1.,Y,51,9,12,2,1,2,1,5.,3.6)
          PLT(1)=BEST
          PLT(2)=AEST*50.+BEST
          CALL MCPLOT(0.,50.,PLT,2,0,0,2,2,2,1,5.,3.6)
          PRINT 100,AEST,BEST
    100   FORMAT(' EQUATION:  Y = ',F5.3,'X + ',F5.3)
          END
```

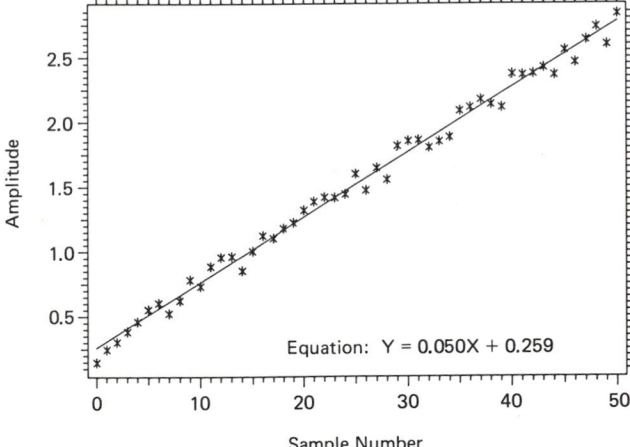

Equation: Y = 0.050X + 0.259

Figure 13.1. SPA1302: Straight-line fitting via SPLFIT.

Note that the $\hat{a}$ and $\hat{b}$ estimates are close to the values $a = 0.05$ and $b = 0.25$ that were used to generate the data. The error in $\hat{b}$ may be attributed to the fact that the mean of the finite additive noise sequence was not precisely zero [A]. Recall from SPA1301 that the mean of the random number sequence generated via SPRAND may deviate considerably from the theoretical value of 0.5 for short $[x_k]$ sequences.

One basic waveform that appears frequently as a recorded signal is the exponential curve

$$y = ab^x \tag{13.5}$$

where a and b are once again constants. This can be expressed in terms of logarithmic functions as

$$\log y = (\log b)x + \log a \tag{13.6}$$

Note that (13.6) has the same general form as the straight-line equation in (13.3).

This suggests that by utilizing the data sequences $[x_k]$ and $[\log y_k]$ and interpreting the results properly, SPLFIT can be used for exponential curve fitting. The SPLFIT estimates $\hat{a}$ and $\hat{b}$ would correspond to $\log b$ and $\log a$, respectively, in (13.6).

To simplify exponential curve fitting, routine SPEXPN is provided to handle reformatting of the data sequences and results such that they are compatible with (13.5). The calling sequence is

$$\text{CALL SPEXPN(X,Y,N,A,B,IERROR)}$$

X(0:N−1)	=	data sequence $[x_k]$
Y(0:N−1)	=	data sequence $[y_k]$
N	=	number of data samples
A	=	estimate for a in (13.5)
B	=	estimate for b in (13.5)
IERROR	= 0	no errors detected
	1	invalid N: $N \le 0$
	2	SPLFIT: IERROR = 2 return
	3	y data ≤ 0

Since the $[y_k]$ sequence in Y(0:N−1) is converted to $[\log y_k]$ internally, all y_k values must be greater than zero. The IERROR = 3 return indicates that this condition has been violated and no parameter estimates have been generated. The SPLFIT subroutine is called internally; thus, the IERROR = 2 return indicates a potential arithmetic overflow.

In test program SPA1303, SPEXPN was utilized to fit an exponential curve to the noisy data generated via

$$y_k = 2.0(0.95)^{x_k} + 0.2(r_k - 0.5)$$

where r_k was again provided by the SPRAND function. Note that the exponential curve is defined by parameters $a = 2.0$ and $b = 0.95$. The results are provided in Fig. 13.2 in the form of a plot showing the noisy sequence samples together with the estimated exponential curve. Note that the parameter estimates $\hat{a} = 2.04$ and $\hat{b} = 0.950$ are nearly the same as the actual values $a = 2.0$ and $b = 0.95$.

An alternate form often used for the exponential function is

$$y = \alpha e^{\beta x} \tag{13.7}$$

where α and β are constants. Note that (13.7) and (13.5) are identical if we define $a = \alpha$ and $b = e^\beta$. Thus, given data sequences $[x_k]$ and $[y_k]$, SPEXPN can be used to estimate α and β as $\hat{\alpha} = \hat{a}$ and $\hat{\beta} = \ln \hat{b}$. This form is especially useful in applications where the signal amplitude decays exponentially, since the time constant β is typically of particular interest.

Our next basic waveform is the power function

$$y = ax^b \tag{13.8}$$

```
                   PROGRAM SPA1303
            C-DEMONSTRATE USE OF EXPONENTIAL FIT ROUTINE
                   DIMENSION X(0:50),Y(0:50),PLT(0:500)
                   DATA A/2./,B/.95/,ISEED/12357/,N/51/
                   DO 1 I=0,50
                     X(I)=I
                     Y(I)=A*(B**I) + .2*(SPRAND(ISEED)-.5)
                1  CONTINUE
                   CALL MCPLOT(0.,1.,Y,51,9,12,2,1,2,1,5.,3.6)
                   CALL SPEXPN(X,Y,N,AEST,BEST,IERROR)
                   IF(IERROR.NE.0)THEN
                     PRINT *,' SPEXPN ERROR = ',IERROR
                     STOP
                   ENDIF
                   DO 2 I=0,500
                     PLT(I)=AEST*(BEST**(I/10.))
                2  CONTINUE
                   CALL MCPLOT(0.,.1,PLT,501,0,0,2,2,2,1,5.,3.6)
                   PRINT 100,AEST,BEST
              100  FORMAT(' EQUATION:  Y = (',F4.2,')(',F5.3,'**X)')
                   END
```

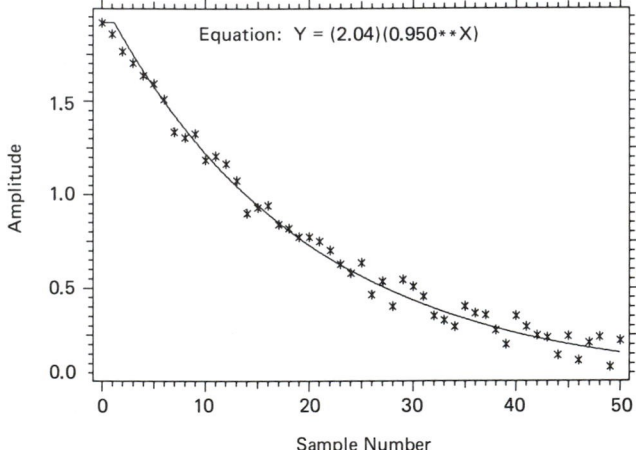

Equation: Y = (2.04)(0.950**X)

Figure 13.2. SPA1303: Exponential curve fitting via SPEXPN.

where a and b are constants. This can be expressed in terms of logarithms as

$$\log y = b(\log x) + \log a \tag{13.9}$$

By defining the data sequences $[\log x_k]$ and $[\log y_k]$, we see that (13.9) has the same general form as (13.3), which describes a straight line. Thus, SPLFIT again generates the basic solution.

Subroutine SPPWRC is provided to handle the intermediate bookkeeping so that A and B are estimates of parameters a and b in (13.8). The calling sequence is

CALL SPPWRC(X,Y,N,A,B,IERROR)

$X(0:N-1)$ = data sequence $[x_k]$

$Y(0:N-1)$ = data sequence $[y_k]$

N = number of data samples

A = estimate of parameter a in (13.8)

B = estimate of parameter b in (13.8)

IERROR = 0   no errors detected
    1   invalid N: N $\leq$ 0
    2   SPLFIT: IERROR = 2 return
    3   X or Y data $\leq$ 0

Both the X and the Y data arrays are converted to logarithmic form internally, so all values must be greater than zero. The IERROR = 3 return indicates that the logarithm of some data sample x_k or y_k could not be computed, and that no parameter estimates were generated.

The use of SPPWRC is demonstrated in program SPA1304. As in the previous examples, we use the SPRAND function and generate a noisy data sequence

$$y_k = 0.5x_k^{0.5} + 0.25(r_k - 0.5)$$

In this example, both a and b are equal to 0.5. Note that $b = 0.5$ corresponds to taking the square root of x_k. The plot in Fig. 13.3 illustrates the results computed by SPA1304. As shown in Fig. 13.3, the parameter estimates computed in SPA1304 via SPPWRC were $\hat{a} = 0.47$ and $\hat{b} = 0.52$. These values are once again near the true values $a = b = 0.5$.

```
          PROGRAM SPA1304
C-DEMONSTRATE USE OF POWER FUNCTION FIT ROUTINE
          DIMENSION X(0:50),Y(0:50),PLT(0:500)
          DATA A/.5/,B/.5/,ISEED/12357/,N/51/
          DO 1 I=0,50
            X(I)=I+1.
            Y(I)=A*(X(I)**B) + .25*(SPRAND(ISEED)-.5)
     1    CONTINUE
          CALL MCPLOT(0.,1.,Y,51,9,12,2,1,2,1,5.,3.6)
          CALL SPPWRC(X,Y,N,AEST,BEST,IERROR)
          IF(IERROR.NE.0)THEN
            PRINT *,' SPPWRC ERROR = ',IERROR
            STOP
          ENDIF
          DO 2 I=0,500
            PLT(I)=AEST*((I/10.+1.)**BEST)
     2    CONTINUE
          CALL MCPLOT(0.,.1,PLT,501,0,0,2,2,1,5.,3.6)
          PRINT 100,AEST,BEST
    100   FORMAT(' EQUATION:  Y = ',F4.2,'*(X**',F4.2,')')
          END
```

The final function for which parameters are estimated via the straight-line fit technique has the general form

$$y = bxe^{ax} \tag{13.9}$$

This is the basic exponential function in (13.7) modified by the x multiplier. In this case, we use natural logarithms and rewrite (13.9) as

$$[\ln y - \ln x] = ax + \ln b \tag{13.10}$$

By utilizing data sequences $[x_k]$ and $[\ln y_k - \ln x_k]$, subroutine SPLFIT can again be used to generate parameter estimates.

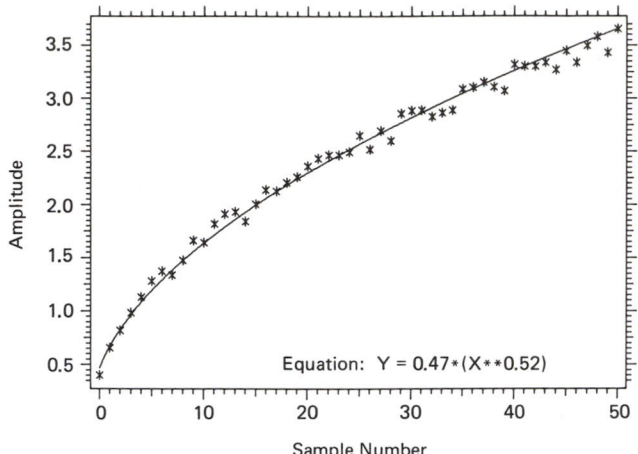

Figure 13.3. SPA1304: Power function fitting via SPPWRC.

Subroutine SPXEXP is provided to facilitate this task. The calling sequence is

CALL SPXEXP(X,Y,N,A,B,IERROR)

$X(0:N-1)$ = data sequence $[x_k]$
$Y(0:N-1)$ = data sequence $[y_k]$
N   = number of data samples
A   = estimate for a in (13.9)
B   = estimate for b in (13.9)
IERROR   = 0   no errors detected
              1   invalid N: $N \leq 0$
              2   SPLFIT: IERROR = 2
              3   X or Y data ≤ 0

Note that all x_k and y_k values must be greater than zero due to the form of (13.10).

Program SPA1305 is provided to demonstrate the use of SPXEXP. The data sequence was generated via

$$y_k = 0.5x_k e^{-0.1xk} + 0.2(r_k - 0.5)$$

where the SPRAND function again provided the r_k values. A plot illustrating the results is shown in Fig. 13.4. Note that in this example, the $\hat{a}$ and $\hat{b}$ estimates are identical to the true values of $a = -0.10$ and $b = 0.50$.

In summary, the curve-fitting routines described in this section provide the least-squared-error fit of some basic functions to the data sequences $[x_k]$ and $[y_k]$. Besides the data itself, the only a priori information necessary is the desired function, which enables us to select the appropriate subroutine.

```
                      PROGRAM SPA1305
          C-DEMONSTRATE USE OF SPXEXP ROUTINE
                      DIMENSION X(0:50),Y(0:50),PLT(0:500)
                      DATA A/-.1/,B/.5/,ISEED/12357/,N/51/
                      DO 1 I=0,50
                        X(I)=I+1.
                        Y(I)=B*X(I)*EXP(A*X(I)) + .2*(SPRAND(ISEED)-.5)
                1     CONTINUE
                      CALL MCPLOT(0.,1.,Y,51,9,12,2,1,2,1,5.,3.6)
                      CALL SPXEXP(X,Y,N,AEST,BEST,IERROR)
                      IF(IERROR.NE.0)THEN
                        PRINT *,' SPXEXP ERROR = ',IERROR
                        STOP
                      ENDIF
                      DO 2 I=0,500
                        PLT(I)=BEST*(I/10.+1.)*EXP(AEST*(I/10.+1.))
                2     CONTINUE
                      CALL MCPLOT(0.,.1,PLT,501,0,0,2,2,2,1,5.,3.6)
                      PRINT 100,BEST,AEST
          100         FORMAT(' EQUATION:  Y = ',F5.2,'*X*EXP(',F5.2,'*X)')
                      END
```

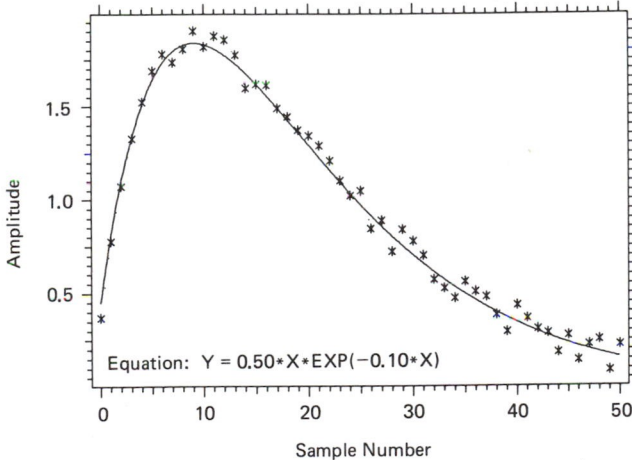

Equation: Y = 0.50*X*EXP(−0.10*X)

Amplitude

Sample Number

Figure 13.4. SPA1305: Exponential curve fitting via SPXEXP.

13.4 Estimation of Rise or Fall Times

The final topic to be considered in this chapter is estimating the rise and fall times of a signal. While these parameters are related to the frequency content, it is sometimes more meaningful to express them as specific time domain values.

To facilitate this task, we first provide a subroutine that may be used to locate the maximum and minimum values of the N-point data sequence in array $X(0:N-1)$. Subroutine SPLMTS returns the array index corresponding to the first occurrence of each extreme as well as the corresponding sample amplitude.

The calling sequence is

CALL SPLMTS(X,N,XMIN,IMIN,XMAX,IMAX)

$X(0:N-1)$ = data sequence $[x_k]$

N   = number of samples in sequence

XMIN   = minimum data value

IMIN   = array index corresponding to XMIN

XMAX   = maximum data value

IMAX   = array index corresponding to XMAX

If argument N is invalid ($N \le 0$), the return arguments are set equal to zero.

In program SPA1306, subroutine SPLMTS is used to locate the amplitude extremes of a sinc function, that is, a function of the form $[\sin \pi t]/\pi t$. A plot of the data sequence is provided in Fig. 13.5. Note that although the maximum value occurs at a unique point, the minimum value occurs at sample numbers 47 and 53. The index returned corresponds to the first occurrence of the minimum data value.

```
                  PROGRAM SPA1306
          C-DEMONSTRATES USE OF SPLMTS SUBROUTINE
                  DIMENSION X(0:100)
                  DATA N/101/
                  PI=4.*ATAN(1.)
                  DO 1 I=0,49
                    X(I)=SIN(2.*PI*.25*(I-50))/(PI*(I-50))
                    X(100-I)=X(I)
                1   CONTINUE
                  X(50)=.5
                  CALL SPLMTS(X,N,XMIN,IMIN,XMAX,IMAX)
                  CALL MCPLOT(0.,1.,X,101,1,12,1,1,2,1,5.,3.6)
                  PRINT 100,IMAX,XMAX,IMIN,XMIN
                100 FORMAT(' MAXIMUM VALUE:  X(',I3,') = ',F5.3,/,
                    +        ' MINIMUM VALUE:  X(',I3,') = ',F5.3)
                  END
                  $ RUN SPA1306
                  MAXIMUM VALUE:  X( 50) = 0.500
                  MINIMUM VALUE:  X( 47) = -.106
                  $
```

In Chapter 1 the signal rise time was defined as the time required for the amplitude to increase from 10% to 90% of its peak value. The fall time is the comparable measure evaluated as the signal decreases in amplitude.

Subroutine SPRFTM will provide the means to evaluate either the rise time or the fall time of the N-point data vector in the array $X(0:N-1)$. The calling sequence is

CALL SPRFTM(IDIR,X,N,T0,T10,T90,T100,IERROR)

IDIR   = direction indicator: 1 = rise time, -1 = fall time

$X(0:N-1)$ = data sequence $[x_k]$

N   = number of data samples in X

T0   = array index corresponding to signal minimum
T10   = time corresponding to 10% amplitude
T90   = time corresponding to 90% amplitude
T100   = array index corresponding to signal global maximum
IERROR   = 0   no errors detected
              1   invalid N: $N \leq 0$
              2   invalid IDIR parameter $\neq \pm 1$
              3   signal maximum = minimum (T0 = T100)

Information computed in SPRFTM is returned in the form of the normalized times T0, T10, T90, and T100. Real values T0 and T100 are the X array indices where the corresponding minimum and maximum were located. In this routine, T100 always corresponds to the global maximum, that is, peak, of the signal. T0 designates the first minimum located either preceding or following the peak (depending upon whether the rise time or the fall time is being measured). Note that T0 does not necessarily correspond to the global minimum. T10 and T90 designate points between T0 and T100 where the signal amplitude has increased by approximately 10% and 90%, respectively. Linear interpolation is used to compute these index values, which are therefore also real. Since the values returned by SPRFTM are in time steps, they can easily be converted to the actual times desired by multiplying by the sample interval T. Thus, the signal rise or fall time is equal to (T90 − T10)T s.

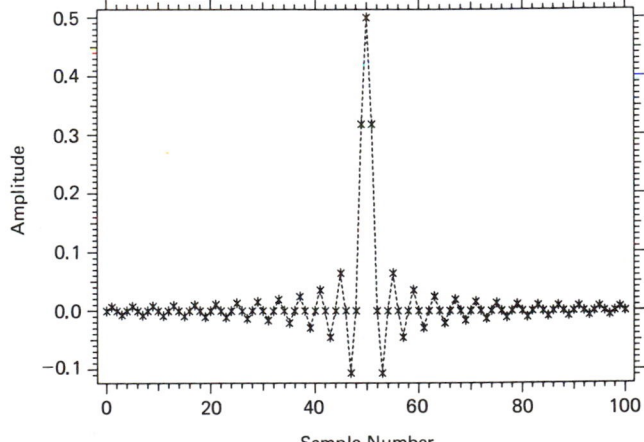

Figure 13.5. SPA1306: Sinc function sequence used to demonstrate the use of subroutine SPLMTS.

To demonstrate the use of SPRFTM, we first consider the exponential function of the form $y = xe^{ax}$, which was used by the SPXEXP routine in the previous section. In program SPA1307 we use SPRFTM to evaluate both the rise time and the fall time of a sequence of samples from this function. A plot of the data vector is provided in Fig. 13.6. On the rising edge, the signal increases in amplitude from 0.0 at T0 = 0 to approximately 3.68 at T100 = 10. Since the amplitude is approximately 0.9 at the end of the first sample interval, linear interpolation results in T10 = 0.10(3.68/0.9) = 0.41. The same technique was used to find T90 = 6.10. On the falling edge, the signal amplitude does not decay to zero; thus the offset must be considered in computing the T10 and T90 values.

```
            PROGRAM SPA1307
      C-DEMONSTRATE USE OF SPRFTM ROUTINE
            DIMENSION X(0:50)
            DATA N/51/
            DO 1 I=0,50
              X(I)=I*EXP(-.1*I)
        1   CONTINUE
            CALL SPRFTM(1,X,N,R0,R10,R90,R100,IERROR)
            IF(IERROR.NE.0)PRINT *,' RISE TIME ERROR=',IERROR
            CALL SPRFTM(-1,X,N,F0,F10,F90,F100,IERROR)
            IF(IERROR.NE.0)PRINT *,' FALL TIME ERROR =',IERROR
            CALL MCPLOT(0.,1.,X,N,1,12,1,1,2,1,5.,3.6)
            PRINT 100
            PRINT 200,R0,R10,R90,R100
            PRINT 300,F0,F10,F90,F100
      100   FORMAT(16X,'MIN',7X,'10%',7X,'90%',7X,'MAX')
      200   FORMAT(' RISE TIME',4X,F6.2,4X,F6.2,4X,F6.2,4X,F6.2)
      300   FORMAT(' FALL TIME',4X,F6.2,4X,F6.2,4X,F6.2,4X,F6.2)
            END
            $ RUN SPA1307
                          MIN        10%        90%        MAX
            RISE TIME     0.00       0.41       6.10      10.00
            FALL TIME    50.00      41.17      15.02      10.00
            $
```

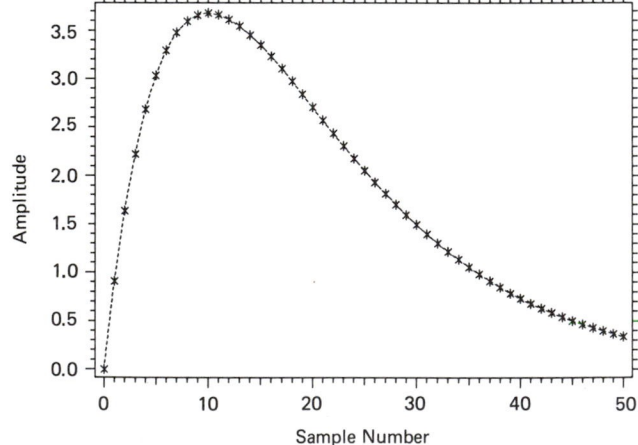

Figure 13.6. SPA1307: Rise and fall times computed via SPRFTM.

The example in program SPA1308 illustrates the use of the SPRFTM routine with a function containing multiple maxima and minima. The signal vector, composed of samples of a shifted sinc function, is shown in Fig. 13.7. Note the symmetry in the computed values of T10 and T90 for the signal rise and fall times. This occurs because the signal itself is symmetric about its midpoint at $k = 25$.

```
            PROGRAM SPA1308
C-DEMONSTRATE USE OF SPRFTM ROUTINE
            DIMENSION X(0:50)
            DATA N/51/
            PI=4.*ATAN(1.)
            DO 1 I=0,24
              X(I)=SIN(2.*PI*.25*(I-25))/(PI*(I-25))
              X(50-I)=X(I)
          1 CONTINUE
            X(25)=.5
            CALL SPRFTM(1,X,N,R0,R10,R90,R100,IERROR)
            IF(IERROR.NE.0)PRINT *,' RISE TIME ERROR=',IERROR
            CALL SPRFTM(-1,X,N,F0,F10,F90,F100,IERROR)
            IF(IERROR.NE.0)PRINT *,' FALL TIME ERROR =',IERROR
            CALL MCPLOT(0.,1.,X,N,1,12,1,1,2,1,5.,3.6)
            PRINT 100
            PRINT 200,R0,R10,R90,R100
            PRINT 300,F0,F10,F90,F100
        100 FORMAT(16X,'MIN',7X,'10%',7X,'90%',7X,'MAX')
        200 FORMAT(' RISE TIME',4X,F6.2,4X,F6.2,4X,F6.2,4X,F6.2)
        300 FORMAT(' FALL TIME',4X,F6.2,4X,F6.2,4X,F6.2,4X,F6.2)
            END
            $ RUN SPA1308
                      MIN        10%        90%        MAX
            RISE TIME  22.00      22.57      24.67      25.00
            FALL TIME  28.00      27.43      25.33      25.00
            $
```

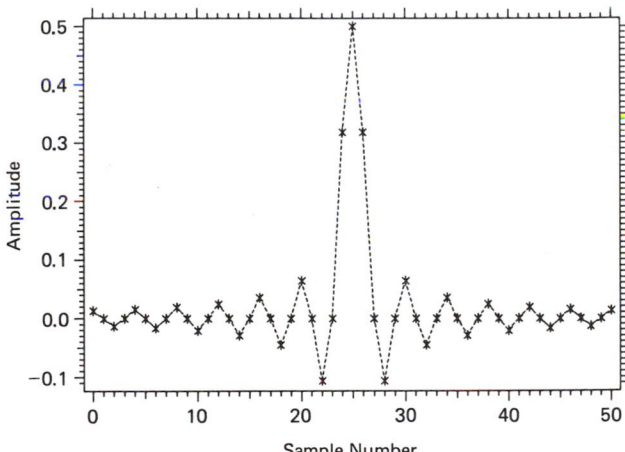

Figure 13.7. SPA1308: Rise and fall time computation via SPRFTM.

13.5 Summary of Calling Sequences

In this chapter we have described a variety of subroutines and functions that enable waveform analysis via parameter estimation. A summary of the modules and their argument lists is provided. As usual, routine inputs and outputs are denoted by I and O, respectively.

```
FUNCTION SPMEAN(X,N)      Output:  mean value of data vector X
FUNCTION SPVARI(X,N)      Output:  variance of data vector X
```

Variable	Definition	Argument type
X(0:N−1)	Data vector $[x_k]$	I
N	Number of data points in X	I

CALL SPLFIT(X,Y,N,A,B,IERROR)   Function: $y = ax + b$
CALL SPEXPN(X,Y,N,A,B,IERROR)   Function: $y = a(b**x)$
CALL SPPWRC(X,Y,N,A,B,IERROR)   Function: $y = a(x**b)$
CALL SPXEXP(X,Y,N,A,B,IERROR)   Function: $y = bx(e**ax)$

Variable	Definition	Argument type
X(0:N−1)	Data vector containing $[x_k]$	I
Y(0:N−1)	Data vector containing $[y_k]$	I
N	Number of data points	I
A	Estimate of a in equations above	O
B	Estimate of b in equations above	O
IERROR	Error-status indicator	O

CALL SPLMTS(X,N,XMIN,IMIN,XMAX,IMAX)

Variable	Definition	Argument type
X(0:N−1)	Data vector containing $[x_k]$	I
N	Number of data points	I
XMIN	Minimum data value	O
IMIN	Array index of XMIN	O
XMAX	Maximum data value	O
IMAX	Array index of XMAX	O

CALL SPRFTM(IDIR,X,N,T0,T10,T90,T100,IERROR)

Variable	Definition	Argument type
IDIR	Direction indicator: 1 = rise time, -1 = fall time	I
X(0:N−1)	Data vector containing $[x_k]$	I
N	Number of data points	I
T0	Time where minimum occurs	O
T10	Time corresponding to 10% amplitude	O
T90	Time corresponding to 90% amplitude	O
T100	Time where global maximum occurs	O
IERROR	Error-status indicator	O

13.6 Exercises

13.1. Compute the mean and variance of a 500-point data sequence $x_k = \sin(2\pi k/20) + r_k$, where r_k is a sample generated via the SPRAND function in Chapter 4. Use these estimates to compute the total power in $[x_k]$.

13.2. A data sequence is defined by $x_k = \sin(2\pi k/20)$. Evaluate the mean and variance for sequences of lengths $N = 20, 30, 40$. Explain the results.

13.3. Use subroutine SPLMTS to find the maximum and minimum of the signal defined in exercise 13.1. Plot the data sequence to verify the results.

13.4. Using the technique outlined in Chapter 4, (4.14), generate a Gaussian random sequence with zero mean and power = 2. Compute estimates for the mean and variance for sequences of lengths $N = 10, 100, 1000, 10000$.

13.5. Generate data sequences: $[x_k] = k$, $k = 0, 1, \ldots, 20$, and $[y_k] = 2.0(0.85)^{x_k} + 0.2(r_k - 0.5)$, where r_k is generated via the SPRAND function. Find the parameter estimates using a curve-fitting routine. Plot the noisy data points together with the curve generated via $\hat{a}$ and $\hat{b}$.

13.6. Repeat the task in exercise 13.5, but omit data points corresponding to $k = 2, 3, 10, 14$. Plot the results.

13.7. Generate data sequences: $[x_k] = k$, $k = 0, 1, \ldots, 50$, and $[y_k] = 3.0e^{-0.1x_k} + 0.3(r_k - 0.5)$, where r_k is generated via the SPRAND function. Find the parameter estimates using a curve fitting routine. Plot the noisy data sequence together with the curve generated via $\hat{a}$ and $\hat{b}$.

13.8. Given the data sequences: $[x_k] = [0, 1, 5]$ and $[y_k] = [0, 1, 25]$. Use SPPWRC to generate estimates for a and b. Plot the resulting power function over the range of $[x_k]$.

13.9. Generate data sequences: $[x_k] = 2k$, $k = 0, 1, \ldots, 30$, and $[y_k] = 2.0x_k e^{-0.1x_k} + (r_k - 0.5)$, where r_k is generated via the SPRAND function. Compute the parameter estimates. Plot the related curve together with the noisy data samples.

13.10. Use the y_k function defined in exercise 13.9, but omit the noise. Evaluate the signal rise and fall times. Plot the data sequence to verify the results.

13.11. Given samples of the function $y_k = 0.5(3x_k^2 - 1)$, where $[x_k]$ is a known sequence, how could SPLFIT be used to generate parameter estimates? Use $[x_k] = [0, 0.1, 0.3, 0.4, 0.5, 0.6, 0.7, 0.8, 1.0]$ and test your solution. Plot the results.

13.12. If the input x_k to a digital filter has mean m_x, then the mean of the output, y_k, is given by $m_y = H(e^{j0})m_x$. To verify this, let $[x_k]$ be a 1000-point sequence generated via SPRAND. Filter this sequence using $H(z) = 2/(1 - 0.5z^{-1})$. Generate estimates for the mean of $[x_k]$ and of the filter output $[y_k]$. Compare these values to the theoretical values given by $m_y = H(e^{j0})m_x$.

13.13. Consider the three-point $[y_k]$ data sequence in exercise 13.8. Evaluate the rise time via SPRFTM. Now generate 100 samples of the function $y = x^2$ at equally spaced points over the interval $0 \le x \le 5$. Evaluate the rise time of this sequence using SPRFTM and compare the results.

13.7 References

[A] Oppenheim, A. V. and R. W. Schafer, *Digital Signal Processing* (Englewood Cliffs, N.J.: Prentice-Hall, Inc., 1975).

[B] Selby, S. M., *CRC Standard Mathematical Tables* (Cleveland, Ohio: CRC Press, 1974).

[C] Stearns, S. D., *Digital Signal Analysis* (Rochelle Park, N.J.: Hayden, 1975).

chapter **14**

Data Windows, Phase Unwrapping, and Unusual Transforms

14.1 Introduction

The signal processing algorithms discussed in this chapter are, at least for the purposes of this book, unrelated short subjects that are included here because they do not fit particularly well in the previous chapters.

Data windows were applied in Chapter 4 in power spectral analysis and in Chapter 8 with finite impulse response (FIR) filters. In this chapter we provide routines for generating several types of data windows and applying them as in the previous chapters.

Phase unwrapping is required generally where the phase spectrum is estimated or when the phase response of a linear system is found, as in Chapter 5. A simple routine for phase unwrapping is included in this chapter, along with an example of unwrapping.

The discrete Fourier transform, which was discussed in Chapter 3 and used throughout this text, is by far the most-used transform in signal processing. Other transforms are, however, useful in certain cases. Among these are the Hilbert, chirp z, and Walsh transforms. These transforms are discussed, along with some routines, at the end of this chapter.

14.2 Data Window Routines

Two data window routines are presented here. The first is a function subprogram that generates a single sample of a selected data window. The second is a subroutine that applies a selected data window to a given data vector. The second routine calls the first during its execution.

The two data window routines are defined as follows:

```
FUNCTION SPWNDO (ITYPE,N,K)
SUBROUTINE SPMASK (X,LX,ITYPE,TSV,IE)
```

ITYPE = type of data window

1—rectangular	4—Hanning
2—tapered rectangular	5—Hamming
3—triangular	6—Blackman

N   = size of window (total number of samples)

K   = sample number, from 0 through $N - 1$

X   = data vector to be masked by window, dimensioned X(0:LX); masking is done in place

LX   = last index in vector X, that is, $LX = N - 1$

TSV   = total squared window value (output)

IE   = error indicator (output); 0 if no error or 1 if ITYPE is out of range

The input variable ITYPE appears in both calling sequences. This variable designates one of six of the most-used data windows, as indicated above. An exact formula for each window type is given in Table 14.1. Note that windows 2, 3, 4, and 6 have endpoints equal to zero; therefore these windows will completely mask out the endpoints of the data vector X; that is, the elements X(0) and X(LX) will be set to zero when SPMASK is called. We also note that the total

TABLE 14.1 DATA-WINDOW FORMULAS

ITYPE	Type	Sample values	Range of k
1	Rectangular	$w_k = 1$	$0 \leqslant k \leqslant N - 1$
2	Tapered rectangular	$w_k = \begin{cases} \frac{1}{2}\left(1 - \cos\dfrac{k\pi}{M+1}\right) \\ 1.0 \\ \frac{1}{2}\left(1 - \cos\dfrac{(N-k-1)\pi}{M+1}\right) \end{cases}$ with $M = \text{INT}[0.1N\text{-}0.2]$	$0 \leqslant k \leqslant M$ $M < k \leqslant N - M - 2$ $N - M - 2 < k \leqslant N - 1$
3	Triangular	$w_k = 1 - \left\| 1 - \dfrac{2k}{N-1} \right\|$	$0 \leqslant k \leqslant N - 1$
4	Hanning	$w_k = \frac{1}{2}\left(1 - \cos\dfrac{2k\pi}{N-1}\right)$	$0 \leqslant k \leqslant N - 1$
5	Hamming	$w_k = 0.54 - 0.46\cos\dfrac{2k\pi}{N-1}$	$0 \leqslant k \leqslant N - 1$
6	Blackman	$w_k = 0.42 - 0.5\cos\dfrac{2k\pi}{N-1}$ $+ 0.08\cos\dfrac{4k\pi}{N-1}$	$0 \leqslant k \leqslant N - 1$

squared window value,

$$TSV = \sum_{k=0}^{N-1} w_k^2 \tag{14.1}$$

where w_k is a window sample, is an output of the subroutine. During execution, the subroutine computes the sum in (14.1), which is often needed for scaling the output.

Our first example of the use of the two routines is in SPA1401. The data vector, which is W(0:255) in this example, is set to 1s in the DO 1 loop, so that after SPMASK is called, W becomes the data window itself, with $N = 256$ samples. The vector X(0:255) is included to provide abscissa samples for the plot routine PXY. The six windows, shown in Fig. 14.1, are plotted in the DO 2 loop. They are plotted in reverse order because the PXY routine uses autoscaling. The TSV array is printed below the listing of SPA1401, and we note the different values of TSV in (14.1) for each window.

```
       PROGRAM SPA1401
C-PLOTS OF SIX DATA WINDOWS.
       DIMENSION X(0:255),W(0:255),TSV(6)
       DO 2 I=6,1,-1
        DO 1 K=0,255
          X(K)=K
          W(K)=1.
     1  CONTINUE
        CALL SPMASK(W,255,I,TSV(I),IE)
        IF(IE.NE.0) STOP
        CALL PXY(X,W,256,1,0,2-I/6+1/I,0,2,.1,.1,.9,.9)
     2 CONTINUE
       PRINT 3,(I,TSV(I),I=1,6)
     3 FORMAT(' TYPE    TSV'/(I4,F12.6))
       STOP
       END
$ RUN SPA1401
 TYPE    TSV
    1   256.000000
    2   222.499969
    3    84.998688
    4    95.624992
    5   101.343437
    6    77.673050
FORTRAN STOP
```

A second example of the use of the data-window routines is given in SPA1402. This example is included to show the effect of our choice of data window on a power spectral estimate with a stationary time series. The DO 1 loop generates 1000 samples of a stationary random signal by passing uniform white noise through a filter given by

$$H(z) = \frac{1}{1 - z^{-1} + 0.8z^{-2}} \tag{14.2}$$

The poles of this filter are near $|z| = 0.9$ on the z-plane, at an angle corresponding with approximately 0.16 Hz-s. Thus the estimated power spectra, shown in Fig. 14.2, are sharply peaked. The six estimated power spectra are computed in SPA1402 in the DO 3 loop. Note that the window type (IWINDO) changes with

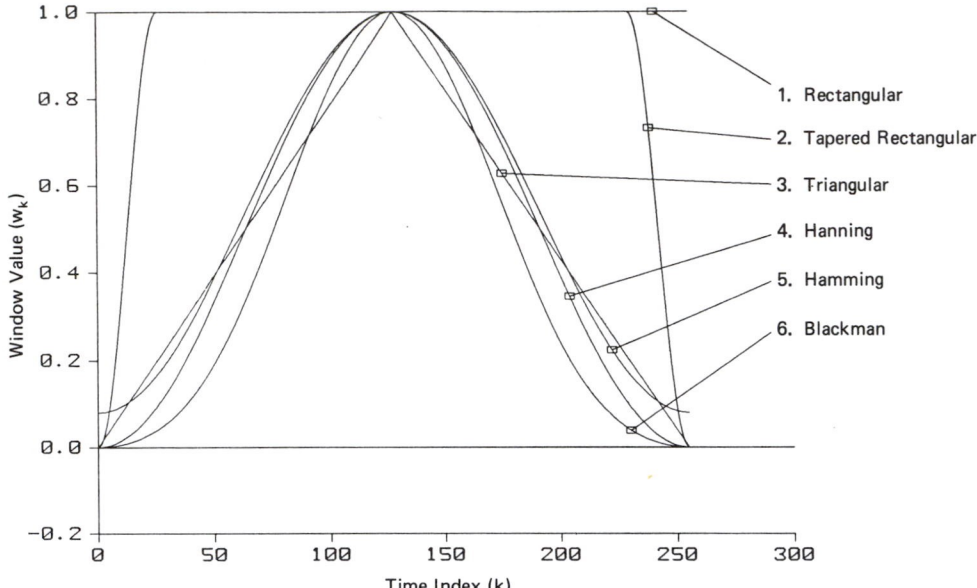

Figure 14.1. Six data windows of length $N = 256$ plotted by the program SPA1401.

each call to SPPOWR. (See Chapter 4, Section 4.2 for the SPPOWR calling sequence description.) Also note that the first data vector element is designated X(0) in the call to SPPOWR. In the DO 3 loop, the window values go in reverse from 6 to 1 only so that the plot routine, HP, will scale the plot shown in Fig. 14.2 correctly.

```
        PROGRAM SPA1402
C-POWER DENSITY SPECTRA USING 6 DIFFERENT DATA WINDOWS.
        DIMENSION X(-2:999),Y(0:16),WK(0:33)
C-GENERATE THE TIME SERIES.
        ISEED=321
        X(-1)=0.
        X(-2)=0.
        DO 1 K=0,999
          X(K)=SQRT(12.)*(SPRAND(ISEED)-.5)+X(K-1)-.8*X(K-2)
      1 CONTINUE
C-COMPUTE AND PLOT THE 6 POWER SPECTRA.
        DO 3 IWINDO=6,1,-1
          CALL SPPOWR(X(0),Y,WK,999,16,IWINDO,.5,NS,IE)
          IF(IE.NE.0) STOP
          DO 2 M=0,16
            Y(M)=10.*ALOG10(Y(M))
      2   CONTINUE
          CALL HP(Y,17,2+1/IWINDO-IWINDO/6,.1,.1,.8,.9)
      3 CONTINUE
        STOP
        END
```

From the results of SPA1402, we conclude that the choice of data window does not have a great effect on spectral estimation with at least some stationary time series. However, the choice does have some effect. In general, the following

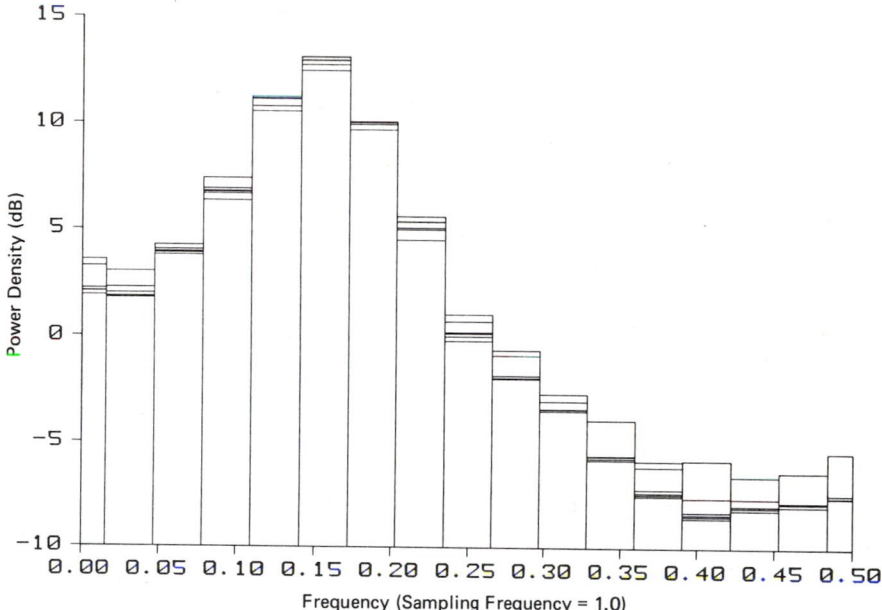

Figure 14.2. Six power density spectra computed by the program example SPA1402. A different data window was used for each spectrum. On the tail of the spectrum from 0.4 to 0.5 Hz-s, the largest values are obtained with the rectangular window.

uses of data windows are recommended:

Long, stationary records	Hamming
(Many blocks, 50% overlap):	Hanning
	Blackman
Nonstationary records:	Tapered rectangular
	Rectangular

The triangular window is often used on correlation estimates before they are transformed to power spectra, as discussed in Section 4.8.

14.3 Phase-Unwrapping Routine

Phase unwrapping is often necessary when the phase characteristic of a linear system is computed using the ATAN2 function, as described in Section 5.6. We saw an example where phase unwrapping was required in Chapter 5, program SPA0506. For another example, we consider the phase shift of the following

FIR filter:

$$H(z) = 1 - 4z^{-1} + 4z^{-2} \tag{14.3}$$

If we view this filter in terms of its pole-zero plot [E] we see a double pole at $z = 0$ and a double zero at $z = 2$ on the z-plane. These produce a phase characteristic that proceeds in the negative direction from 360° (or 0°), around through $-180°$, to $-360°$ (or 0°). If a routine like the example in Chapter 5, which uses the ATAN2 function, is used to compute the phase characteristic of (14.3), the result is as shown in Fig. 14.3(a). The ATAN2 function keeps the phase in the interval $(-180°, +180°)$.

A simple phase-unwrapping routine, called SPUNWR, can be used to correct the jump that occurs in Fig. 14.3(a), or in similar situations. Its calling sequence is

CALL SPUNWR(X,LX,IRD)

X(0:LX) = sequence of phase angles in radians or degrees

LX   = last index in X(0:LX)

IRD   = 1 to indicate that X is in radians; 2 to indicate degrees

The routine simply inserts a 360° (or 2π rad) correction whenever there is a jump in X of more than 180° (or π rad). This simple algorithm suffices for many applications in linear system analysis and in homomorphic signal processing [C], where equivalent unwrapping applications exist. (Specifically, in the latter, one must determine the logarithm of a complex spectrum, and the imaginary or phase part of the complex logarithm contains the same phase ambiguities.) For example, see exercises 14.8 through 14.10.

To unwrap the phase in the previous example of (14.3), we operate with SPUNWR on the computed phase array as in SPA1403 below, where the array contains 201 data points computed in the DO 1 loop. The first call to PXY produces Fig. 14.3(a), where the phase is folded into the range from $-180°$ to $+180°$. Then, after SPUNWR is called, Fig. 14.3(b) is plotted with the second call to PXY. In Fig. 14.3(b) we see the correct phase characteristic of the filter in (14.5).

```
      PROGRAM SPA1403
C-PHASE RESPONSE OF FIR FILTER, H(Z)=1-4*Z**(-1)+4*Z**(-2).
      DIMENSION FREQ(0:200),PHASE(0:200),B(0:2)
      COMPLEX G,SPGAIN
      DATA B/1.,-4.,4./
      PI=4.*ATAN(1.)
      DO 1 M=0,200
        FREQ(M)=0.5*M/200.
        G=SPGAIN(B,0.,2,1,FREQ(M))
        PHASE(M)=ATAN2(AIMAG(G),REAL(G))*180./PI
    1 CONTINUE
      CALL PXY(FREQ,PHASE,201,1,0,0,0,0,.1,.55,.8,.9)
      CALL SPUNWR(PHASE,200,2)
      CALL PXY(FREQ,PHASE,201,1,0,0,0,0,.1,.1,.8,.45)
      STOP
      END
```

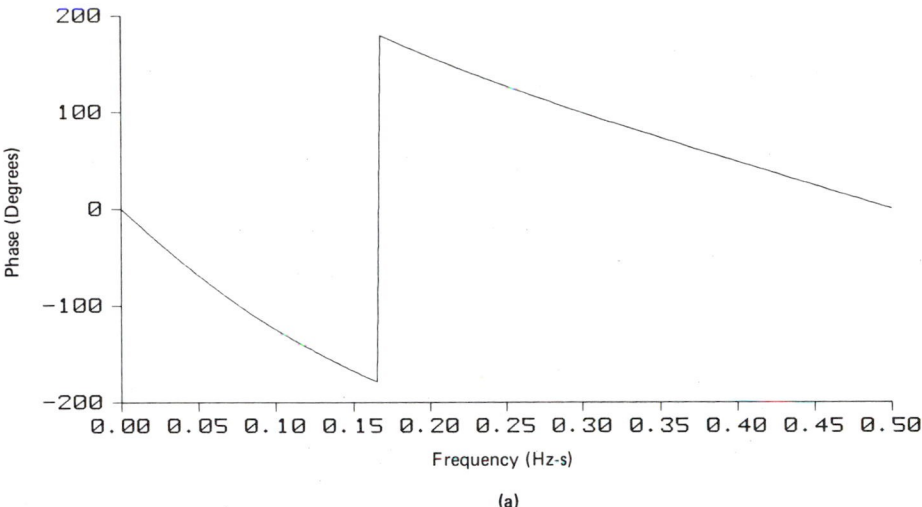

(a)

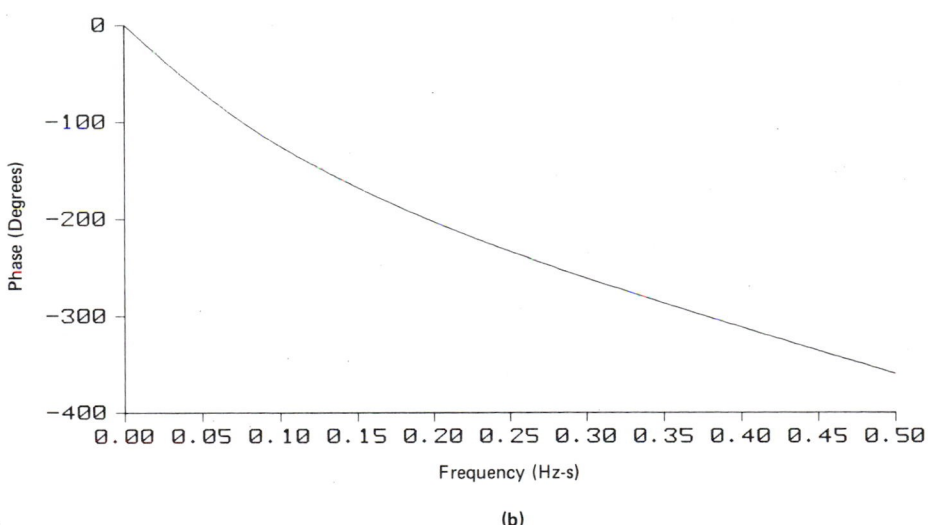

(b)

Figure 14.3.   Phase characteristics computed by the example program SPA1403.
(a) Before unwrapping. (b) After unwrapping.

14.4 A Hilbert Transform Routine

A Hilbert transformer [F] is defined here to be a digital filter that has approximately unit gain and approximately a 90° phase shift at all frequencies. It is useful in creating *analytic* complex signals whose spectra are nonzero only at positive frequencies. Such a process is illustrated in Fig. 14.4, where $HT(z)$ is a Hilbert

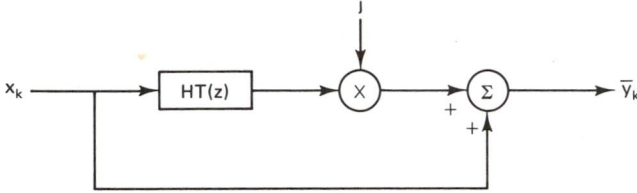

Figure 14.4. Using an ideal Hilbert transformer to transform a real signal x_k into an analytic complex signal $\bar{y}_k$ having a nonzero spectrum only at positive frequencies.

transformer and $[\bar{y}_k]$ is a complex signal whose discrete Fourier transform $[Y_m]$ has negligible magnitude for $m < 0$.

The Hilbert transformer is potentially useful in processing narrow-band signals from sensors in arrays, where quadrature components, shifted 90° in phase, are required [G]. It is also useful in processing and encoding other types of narrow-band signals [H]. If the narrow-band signal is at a single frequency, then the effective Hilbert transformer is simply a delay equal to a quarter of a period at that frequency, but if the signal has spectral content over a range of frequencies, then the effective Hilbert transformer becomes more complicated.

We follow the method described by Oppenheim and Schafer [F] to derive the weights of an FIR Hilbert transformer. Let $H(j\omega)$ be the ideal transfer function of the transformer. Then, to have unit amplitude response and 90° phase lag at all frequencies, we have

$$H(j\omega) = e^{-j\pi/2} = -j; \qquad 0 \leqslant \omega \leqslant \pi \tag{14.4}$$

To have real weights, the transfer function must equal its own conjugate for frequencies from π to 2π, similar to equation (3.9); therefore,

$$H(j\omega) = +j; \qquad \pi < \omega \leqslant 2\pi \tag{14.5}$$

The ideal impulse response $[h_{ik}]$ of $H(j\omega)$ extends infinitely in both directions from the sample h_{i0} and is found as follows as the sampled inverse Fourier transform of $H(j\omega)$:

$$
\begin{aligned}
h_{ik} &= \frac{1}{2\pi} \int_0^{2\pi} H(j\omega) e^{j\omega k}\, d\omega \\
&= \frac{1}{2\pi} \left[-\int_0^{\pi} j e^{j\omega k}\, d\omega + \int_\pi^{2\pi} j e^{j\omega k}\, d\omega \right] \\
&= \begin{cases} 0, & k \text{ even} \\ \dfrac{2}{k\pi}, & k \text{ odd} \end{cases}, \qquad -\infty < k < \infty
\end{aligned}
\tag{14.6}
$$

We note in this result that the sample $h_{i0} = 0$ is included in the category where k is even.

We assume now that our practical FIR Hilbert transformer differs in three ways from the ideal in (14.6). First, it is of finite extent L, where L is odd. Secondly, it is causal, with weights centered around $k = (L - 1)/2$ instead of $k = 0$. Finally, since h_{ik} in (14.6) is truncated, we apply a data window to improve the gain characteristic of the practical transformer. (We recall that this idea was discussed for FIR filter design in Chapter 8.)

The introduction of causality simply adds a delay of $(L - 1)/2$ samples, which is easy to compensate. The truncation and windowing effects are shown first in Fig. 14.5. Here we see the amplitude gain of a Hilbert transformer of length $L = 33$, which should ideally be one at all frequencies. With the rectangular window (that is, no window), we see large deviations from 1. With the Hamming and Blackman windows, the gain is close to 1 over a wide frequency range. As shown in Fig. 14.6, the phase characteristic is nearly perfect for $L = 33$, no matter which window is used. The small perturbations are due to rounding noise in a computer with 32-bit words. Considering Figs. 14.5 and 14.6, we choose the Hamming window, which gives the widest frequency range with a flat gain characteristic.

With the Hamming window, the truncation effects are shown more explicitly in Fig. 14.7, where we see the Hilbert transformer amplitude gain for filters of

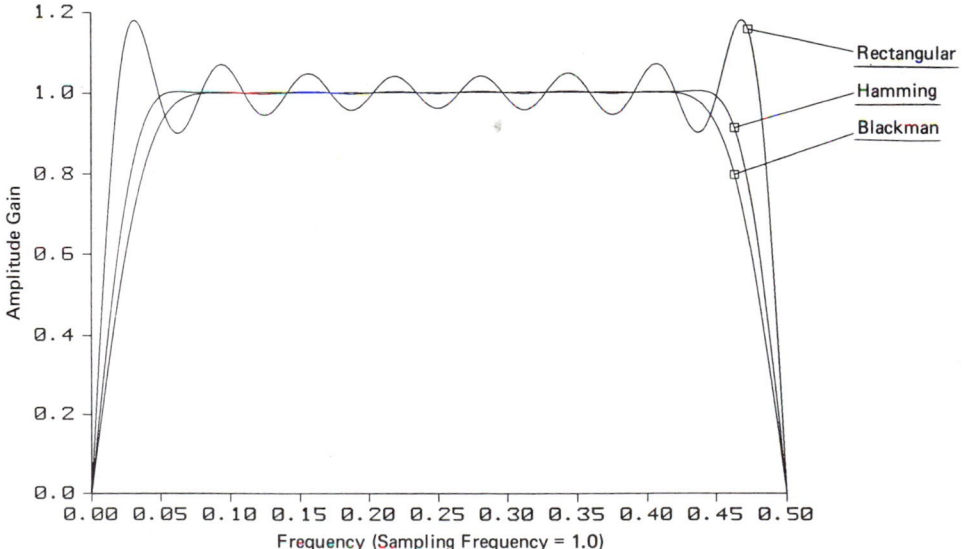

Figure 14.5. Amplitude gain of Hilbert transformers of length $L = 33$ using rectangular, Hamming, and Blackman data windows.

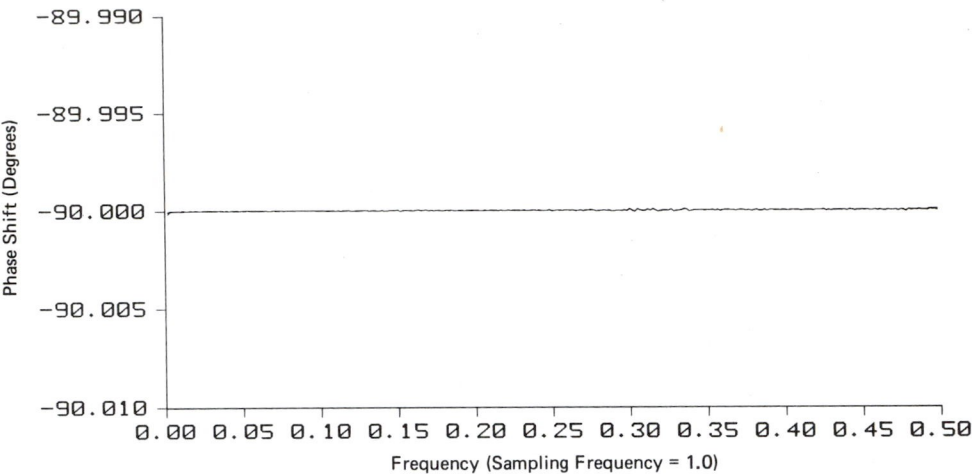

Figure 14.6.   Phase shift of the three Hilbert transformers of length $L = 33$.

length $L = 9$ through 257. The filter length must be chosen with these effects in mind.

A routine has been written to generate a set of Hilbert transformer weights, that is, the FIR filter coefficients for a Hilbert transformer, using the Hamming

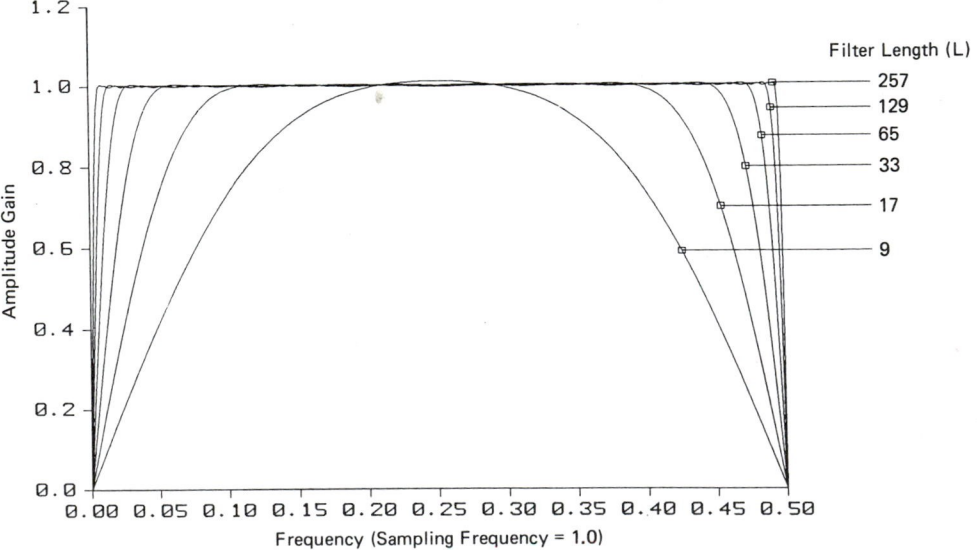

Figure 14.7.   Effects of length (L) on amplitude gain of the Hilbert transformer using the Hamming window.

window. The calling sequence is

$$\text{CALL SPHILB(X,LX)}$$

X(0:LX) = weight vector
LX   = last index of X(0:LX)

When SPHILB is called, the transformer weights are computed and placed in the vector X. The transformer length L is equal to LX + 1 if LX is even, or equal to LX if LX is odd. (If LX is odd the last element, X(LX) is not used.) The computed weights begin with X(0) and are centered around X$((L-1)/2)$, as described earlier. Thus, the transformer introduces a delay of $(L-1)/2$ samples in addition to the 90° phase shift.

An example of the use of SPHILB is seen in SPA1404. The example follows the scheme in Fig. 14.8. A sine wave with 30 samples per period, x_k, is generated and transformed with a Hilbert transformer having $L = 65$ weights. Referring to Fig. 14.7, we see that the transformer has near unity gain at the sine wave frequency, which is $\frac{1}{30}$, or 0.033 Hz-s. The transformed waveform y_k is compared with the delayed input x_{dk}. The delay compensates for the transformer delay of $(L-1)/2 = 32$ samples discussed earlier. The results from SPA1404 are plotted in Fig. 14.9. After the filter transient, which lasts 64 time steps (filter length), has died out, we see that y_k lags x_{dk} by 90°.

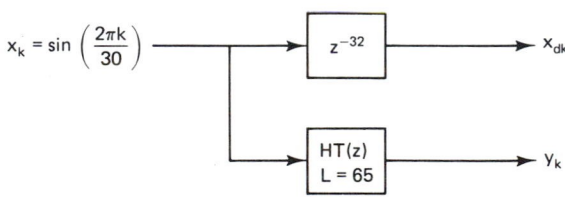

Figure 14.8. Scheme used to generate the signals $[x_{dk}]$ and $[y_k]$ in SPA1404, which are plotted in Fig. 14.9.

```
        PROGRAM SPA1404
C-HILBERT TRANSFORM OF A SINE WAVE WITH 30 SAMPLES/PERIOD.
        DIMENSION T(0:150),X(0:150),XD(0:150),Y(0:150)
        DIMENSION B(0:64),PX(0:64)
        DATA PX/65*0./
        PI=4.*ATAN(1.)
C-GENERATE THE SINE WAVE PLUS A VERSION DELAYED BY 32 SAMPLES.
        DO 1 K=0,150
          T(K)=K
          X(K)=SIN(2.*PI*K/30.)
          XD(K)=SIN(2.*PI*(K-32)/30.)
      1 CONTINUE
C-GENERATE THE HILBERT-TRANSFORMED VERSION USING L=65 WEIGHTS.
        CALL SPHILB(B,64)
        CALL SPFLTR(B,0.,64,1,X,151,Y,PX,PX,IE)
        IF(IE.NE.0) PAUSE 1
C-PLOT THE DELAYED AND TRANSFORMED WAVEFORMS.
        CALL PXY(T,XD,151,1,0,1,0,2,.1,.3,.9,.8)
        CALL PXY(T,Y,151,1,0,3,0,0,0,0,0,0)
        STOP
        END
```

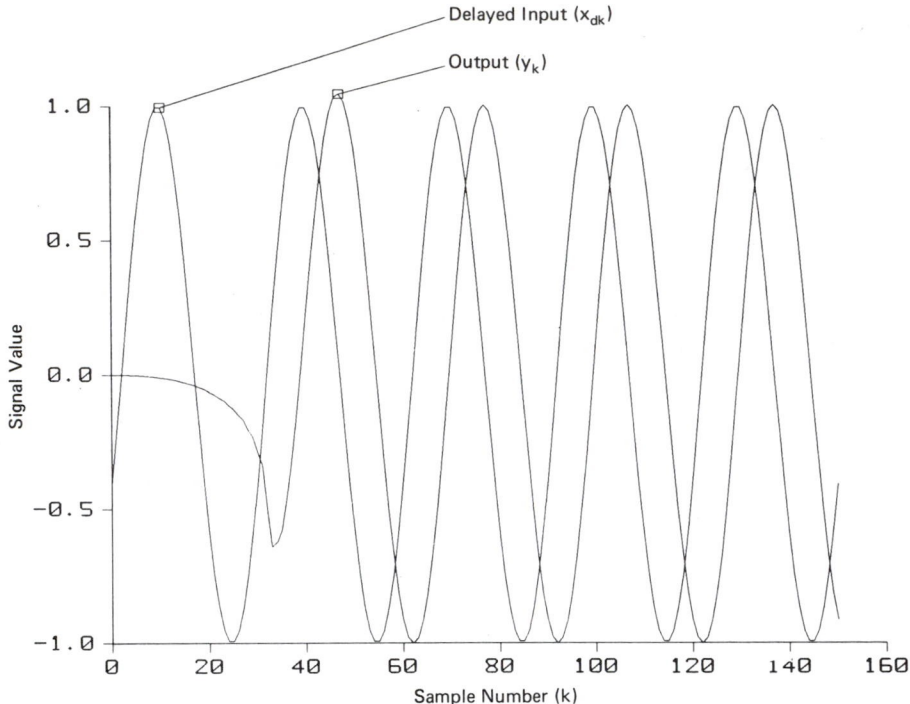

Figure 14.9. Plots of the signals x_{dk} and y_k in Fig. 14.8, showing the 90° phase lag effected by the Hilbert transformer.

14.5 A Chirp *z*-Transform Routine

In this section we introduce a routine, SPCHRP, that implements the chirp *z*-transform [I, J] as a method for computing a "zoom" discrete Fourier transform (DFT), which gives increased detail of the Fourier spectrum over a restricted frequency range. The SPCHRP routine has essentially the same use as the SPCOMP (DFT component) function described in Chapter 3. The advantages of SPCHRP over SPCOMP are that it is generally faster (and much faster if the time series is long) and that the numerical rounding errors tend to be less than those of SPCOMP because SPCHRP uses the FFT. On the other hand, SPCHRP is more complicated to use than SPCOMP and is also less flexible in that the computed DFT components must be evenly spaced over the given frequency range.

The theory of the SPCHRP algorithm is described by Rabiner, Schafer, and Rader [I]. We can understand the basic idea of the algorithm by using the following argument. Suppose we wish to compute the DFT of an *N*-point time series $[x_k]$ over a range of frequencies given by

$$\omega_m = \omega_0 + m\Delta\omega, \qquad m = 0, 1, \ldots, M \tag{14.7}$$

Similar to equation (3.5) in Chapter 3, the $M + 1$ DFT components are given by

$$X_m = \sum_{k=0}^{N-1} x_k e^{-jk\omega_m}, \qquad m = 0, 1, \ldots, M \tag{14.8}$$

Now we define the following complex constants:

$$\begin{aligned} A &\triangleq e^{-j\omega_0} \\ W &\triangleq e^{-j\Delta\omega/2} \end{aligned} \tag{14.9}$$

If we substitute (14.7) into (14.8) and use these definitions, we obtain the following expression for X_m:

$$X_m = \sum_{k=0}^{N-1} x_k A^k W^{2mk}, \qquad m = 0, 1, \ldots, M \tag{14.10}$$

Next we make a substitution due to Bluestein [K], given by

$$2mk = k^2 + m^2 - (m - k)^2 \tag{14.11}$$

and (14.10) becomes

$$X_m = W^{m^2} \sum_{k=0}^{N-1} [x_k A^k W^{k^2}] W^{-(m-k)^2}, \qquad m = 0, 1, \ldots, M \tag{14.12}$$

Here we have a result that can be evaluated as illustrated in Fig. 14.10 using N-point fast Fourier transforms (FFTs). First we multiply the data sequence $[x_k]$ by $A^k W^{k^2}$ to obtain the N-point complex sequence $[y_k]$, in brackets in (14.12). Then $[y_k]$ is convolved with $[W^{k^2}]$ using the product-of-transforms method described in Chapter 9. Finally the convolution $[g_m]$ is multiplied by $[W^{m^2}]$ to obtain $[X_m]$.

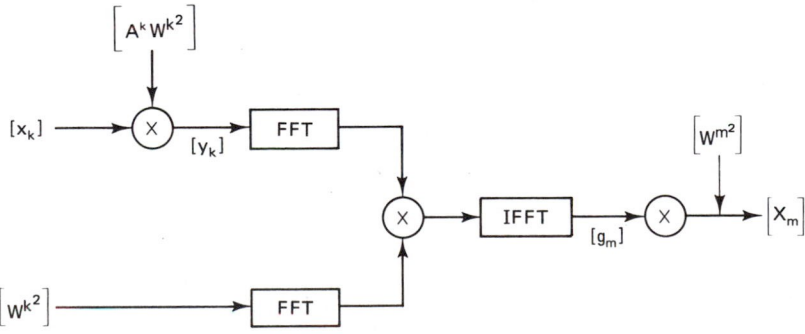

Figure 14.10. Diagram of the chirp z-transform algorithm showing how the $(M + 1)$-point transform $[X_m]$ is obtained from the N-point sequence $[x_k]$.

Thus we see that Fig. 14.10 gives an implementation of the form of the DFT in (14.12) that takes advantage of the speed of the FFT. The term *chirp* comes from the use of W^{k^2} in the algorithm. From (14.9), W^{k^2} is seen to be a complex sinusoid whose frequency $k\Delta w/2$ increases linearly with time k.

The SPCHRP subroutine operates as in Fig. 14.10. The calling sequence for SPCHRP is

CALL SPCHRP(X,LX,LX1,F1,F2,WORK,LX2,IERROR)

X(0:LX)	= *complex* data vector containing (but longer than) the time series $[x_k]$, which may be real or complex, in X(0) through X(LX1)
LX	= last index of X(0:LX); LX + 1 must be a power of 2
LX1	= last index of *N*-point input data sequence, that is, LX1 + 1 = *N*; must be less than LX
F1	= initial frequency (Hz-s) corresponding to $\omega_0/2\pi$ in (14.10)
F2	= final frequency (Hz-s) corresponding with $(\omega_0 + M\Delta\omega)/2\pi$.
WORK(0:LX)	= complex work array, same size as X(0:LX).
LX2	= last frequency index *M* computed by the subroutine as *M* = LX − LX1.
IERROR	= error indicator: 0 if no error is detected; 1 if LX is not greater than LX1 + 1, 2 if F2 is not greater than F1, or 3 if LX + 1 is not a power of 2

In these specifications, we note that the input time series is in X(0:LX1), that is, in the first part of X(0:LX), which is a complex array. The input time series may be either real or complex. Then, after execution, the output spectrum is in X(0:LX2), with LX2 computed as

$$LX2 = LX - LX1 \tag{14.13}$$

Thus the number of complex points in the output DFT depends on LX, which is a power of two minus one and is greater than LX1. The points in the output DFT are distributed over the frequency domain in accordance with (14.7), in which

$$F1 = \frac{\omega_0}{2\pi}$$

$$F2 = \frac{(\omega_0 + M\Delta\omega)}{2\pi} \tag{14.14}$$

$$LX2 = M = LX - LX1$$

An example of the use of SPCHRP is shown in SPA1405. The example is essentially the same as the SPCOMP example in Figs. 3.4 and 3.5. Here we have a real time series* with $N = 924$ points generated and stored in the complex

---

* The time series is shifted in Chapter 3, but shifting does not affect the amplitude spectrum.

array X(0) through X(923). Then SPCHRP is called to compute LX2 + 1 = (2047 − 923) + 1 = 1125 complex DFT values from F1 = 0.0 to F2 = 0.03 Hz-s in accordance with (14.14). Then the DFT amplitude is plotted in Fig. 14.11 using the PXY plot routine. After adjusting the frequency scales, we see that Figs. 3.5 and 14.11 are identical, so SPCHRP and SPCOMP produce the same result in this case.

```
      PROGRAM SPA1405
C-USES THE CHIRP Z-TRANSFORM TO FIND THE SPECTRUM OF
C-EXP(-K/176)*SIN(2*PI*K/160) FROM 0.0 TO 0.03 HZ-S.
      DIMENSION FREQ(0:1124),AMPL(0:1124)
      COMPLEX X(0:2047),WORK(0:2047)
      PI=4.*ATAN(1.)
C-GENERATE THE TIME SERIES.
      DO 1 K=0,923
        X(K)=EXP(-K/176.)*SIN(2.*PI*K/160.)
    1 CONTINUE
C-GENERATE THE DFT IN THE DESIRED FREQUENCY RANGE.
      CALL SPCHRP(X,2047,923,0.0,0.03,WORK,LX2,IE)
      IF(IE.NE.0) STOP
C-PLOT THE AMPLITUDE SPECTRUM.
      DO 2 M=0,LX2
        FREQ(M)=0.0+M*(0.03-0.0)/LX2
        AMPL(M)=ABS(X(M))
    2 CONTINUE
      CALL PXY(FREQ,AMPL,LX2+1,1,0,0,0,2,.1,.1,.9,.9)
      STOP
      END
```

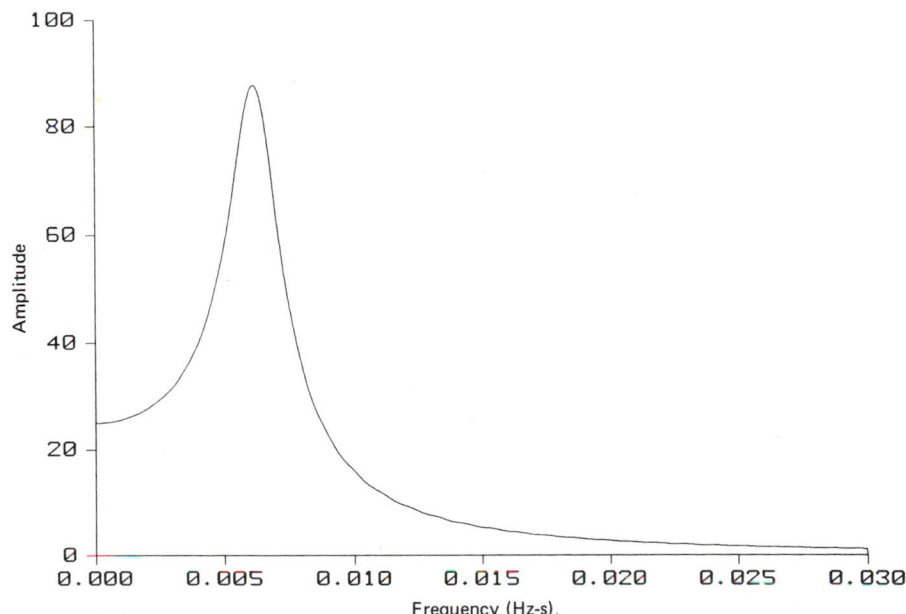

Figure 14.11. Amplitude spectrum of the time series plotted in Fig. 3.4, computed using the SPCHRP routine.

14.6 Walsh Functions and Transforms

Walsh functions [L, M] are binary (± 1) sequences that are mutually orthogonal with respect to a given sequence length. For example, the following two Walsh functions with length $N = 8$ are orthogonal:

$$1 \quad -1 \quad -1 \quad 1 \quad 1 \quad -1 \quad -1 \quad 1$$
$$1 \quad -1 \quad -1 \quad 1 \quad -1 \quad 1 \quad 1 \quad -1 \tag{14.15}$$

We see that the inner product* of the two functions is zero; thus they are mutually orthogonal.

A number of different Walsh transforms exist. The transforms are based on sets of Walsh functions like those in (14.15) in the same way that the DFT is based on sine and cosine functions. Thus, one would convert a fast Fourier transform routine like SPFFTC, which is described in Chapter 3, into a fast Walsh transform routine by substituting Walsh functions for the sine and cosine (or complex exponential) functions found in SPFFTC.

Examples of Walsh transforms may be found in the references. Since a number of different transforms exist under the heading of *Walsh* or *Walsh-Hadamard* transforms and since the use of Walsh transforms is restricted to a few applications, we do not include Walsh transform routines here. However, we include a short routine for generating Walsh functions, so that the reader can generate the functions and use them in transform routines or in any other context.

The routine for generating Walsh functions is a function subprogram called SPWLSH. Like the FFT routines in this book, it is designed to work with data sequences of length N equal to a power of 2. Its calling sequence is

FUNCTION SPWLSH(LOGN,NSEQ,K)

LOGN = log base 2 of the sequence length N
NSEQ = sequency index (0 through $N - 1$).
K   = time index (0 through $N - 1$).

The SPWLSH function generates a single *Walsh-ordered* [L] Walsh coefficient, analogous to a single value of the sine or cosine function used in the DFT. The *sequency index* NSEQ is analogous to the frequency index m used previously. The time index K is the same as used previously.

An example of the use of SPWLSH is given in SPA1406. The program uses SPWLSH to compute all of the Walsh coefficients for a sequence length of $N = 8$ and then prints the results. We note that as the sequency index increases, the number of transitions between ± 1 in the sequence of Walsh coefficients also increases. We also note that the eight Walsh functions (rows) are mutually orthogonal with respect to the eight sample points.

---

* The inner product of $[x_k]$ and $[y_k]$ is the sum $x_0 y_0 + x_1 y_1 + x_2 y_2 + \cdots$.

```
            PROGRAM SPA1406
C-WALSH COEFFICIENTS FOR SEQUENCE LENGTH N=8.
            DIMENSION COEFF(0:7)
            PRINT '('' SEQ. INDEX  --WALSH COEFFICIENTS---------'')'
            DO 2 NSEQ=0,7
              DO 1 K=0,7
                COEFF(K)=SPWLSH(3,NSEQ,K)
        1     CONTINUE
              PRINT '(I6,5X,8F4.0)', NSEQ,COEFF
        2   CONTINUE
            STOP
            END
$ RUN SPA1406
  SEQ. INDEX  --WALSH COEFFICIENTS----------
          0        1.  1.  1.   1.   1.  1.  1.  1.
          1        1.  1.  1.   1.  -1. -1. -1. -1.
          2        1.  1. -1.  -1.  -1. -1.  1.  1.
          3        1.  1. -1.  -1.   1.  1. -1. -1.
          4        1. -1. -1.   1.   1. -1. -1.  1.
          5        1. -1. -1.   1.  -1.  1.  1. -1.
          6        1. -1.  1.  -1.  -1.  1. -1.  1.
          7        1. -1.  1.  -1.   1. -1.  1. -1.
FORTRAN STOP
```

To compute a Walsh-ordered Walsh transform of a data sequence with $N = 8$, we would simply multiply the data vector by the Walsh coefficient matrix printed by SPA1406. Thus, with the use of SPWLSH, the reader could easily write a Walsh transform routine. See, for example, exercises 14.20 through 14.22.

14.7 Summary of Calling Sequences

Several routines, not all related, have been introduced in this chapter. The calling sequences and argument lists are summarized as follows.

FUNCTION SPWNDO (ITYPE,N,K)
CALL SPMASK (X,LX,ITYPE,TSV,IERR)

Variable	Definition	Input/ Output	Remarks
SPWNDO	Data window value	O	Single window value w_k
ITYPE	Type of window	I	1—rectangular
			2—tapered rectangular
			3—triangular
			4—Hanning
			5—Hamming
			6—Blackman
N	Size of window	I	Total number of samples
K	Sample number	I	Position within window
X	Data to be windowed	I,O	Dimensioned X(0:LX)
LX	Last index of X	I	Usually LX = $N - 1$
TSV	Total squared window value	O	Equation (14.1)
IERR	Error indicator	O	0 = no error
			1 = ITYPE out of range

CALL SPUNWR(X,LX,IRD)

Variable	Definition	Input/ Output	Remarks
X	Sequence of phase angles	I,O	Dimensioned X(0:LX)
LX	Last index of X	I	
IRD	Radian/degree indicator	I	1—phase angles in radians
			2—phase angles in degrees

CALL SPHILB(X,LX)

Variable	Definition	Input/ Output	Remarks
X	Hilbert transformer weight vector	O	Dimensioned X(0:LX)
LX	Last index of X	I	Usually LX is even and the number of weights, LX + 1, is odd.

CALL SPCHRP(X,LX,LX1,F1,F2,WORK,LX2,IERR)

Variable	Definition	Input/ Output	Remarks
X	Complex data vector containing (but longer than) the data to be transformed	I,O	Dimensioned COMPLEX X(0:LX)
LX	Last index of X	I	LX = power of 2
LX1	Last index of input data in X(0) through X(LX1)	I	Number of data points = LX1 + 1
			Number of spectral points = M = LX − LX1
F2	Final frequency in Hz-s	I	$(\omega_0 + M\Delta\omega)/2\pi$ in (14.7)
WORK	Complex work array	—	Dimensioned COMPLEX WORK (0:LX)
LX2	Last frequency index	O	LX2 = M = LX − LX1
IERR	Error indicator	O	0 = No error
			1 = LX1 ≥ LX − 1
			2 = F1 ≥ F2
			3 = LX + 1 ≠ power of 2

FUNCTION SPWLSH(LOGN,NSEQ,K)

Variable	Definition	Input/ Output	Remarks
SPWLSH	Walsh function value	O	Single value, 1.0 or −1.0
LOGN	Log base 2 of sequence length	I	Sequence length $N = 2^{LOGN}$
NSEQ	Sequency index	I	$0 \le NSEQ \le N − 1$
K	Time index	I	$0 \le K \le N − 1$

14.8 Exercises

14.1. Make a plot similar to Fig. 14.1 for the Hamming and Hanning windows with $N = 32$ data samples. Show individual window values as points on the plot.

14.2. Repeat exercise 14.1 for the tapered rectangular and Blackman windows.

14.3. Generate the data sequence

$$x_k = \sin\left(\frac{2\pi k}{175}\right), \qquad k = 0, 1, \ldots, 999$$

Using SPMASK, apply the following data windows to $[x_k]$. For each application, plot $[x_k]$ with and without the window applied.
(a) Tapered rectangular
(b) Triangular
(c) Hanning
(d) Hamming
(e) Blackman

14.4. Using SPRAND with ISEED $= 123$, generate a white noise sequence with unit power and with 1000 samples. Then produce a bandlimited sequence $[x_k]$ by filtering the white noise through a two-section low-pass Butterworth filter with cutoff (-3 dB) at 0.2 Hz-s. For each window type below, plot $[x_k]$ unwindowed and then windowed on the same plot.
(a) Tapered rectangular
(b) Triangular
(c) Hanning
(d) Hamming
(e) Blackman

14.5. Using SPWNDO, check the six total squared window values computed and printed by SPA1401.

14.6. Generate the bandlimited sequence $[x_k]$ described in exercise 14.4. Then, using each of the following data windows, compute and plot a power spectrum. To compute each spectrum, use SPPOWR with half-overlapping segments of size 32, that is, with LY $= 16$. Make a plot of decibels against frequency similar to Fig. 14.2.
(a) Rectangular
(b) Tapered rectangular
(c) Triangular
(d) Hanning
(e) Hamming
(f) Blackman

14.7. Repeat exercise 14.6 for the sequence $[x_k]$ in exercise 14.3. Note the effect of different windows on leakage. As in exercise 14.6, use the following windows.
(a) Rectangular
(b) Tapered rectangular
(c) Triangular
(d) Hanning
(e) Hamming
(f) Blackman

14.8. Write a program similar to SPA1403 for the phase response of

$$H(z) = 1 - 5z^{-1} + 8z^{-2} - 4z^{-3}$$

Make a plot similar to Fig. 14.3.

14.9. Using the SPGAIN and ATAN2 functions, compute and plot the phase characteristic of

$$H(z) = \frac{1.0 + 1.98z^{-2} + 0.9801z^{-4}}{1.0 - z^{-1} + 0.99z^{-2}}$$

Make a continuous plot of degrees against frequency using 201 equally spaced points from 0.0 through 0.5 Hz-s. Then process this same phase data using SPUNWR and plot the result. Make both plots on the same set of axes.

14.10. Repeat exercise 14.9 for the phase characteristic of

$$H(z) = \frac{1.0 + z^{-2} + 0.9801z^{-4}}{1.0 - z^{-1} + 0.99z^{-2}}$$

Explain the result in this case using a pole-zero plot.

14.11. Describe the simplest Hilbert transformer for the signal $x_k = \cos(0.2\pi k + \alpha)$.

14.12. Using the SPCOMP function, make a plot similar to Fig. 14.5 for the amplitude gain of a Hilbert transformer of length $L = 33$ but with the tapered rectangular, triangular, Hanning, and Hamming windows applied to the filter weights.

14.13. Write a program similar to SPA1404 to demonstrate the Hilbert transformer routine. Use a cosine wave instead of a sine wave, with 16 samples per cycle. Use Fig. 14.7 to select a minimal transformer size for this application. (Do not use size 65.) Make a plot similar to Fig. 14.9.

14.14. Generate a bandlimited sequence with 1024 samples similar to $[x_k]$ in exercise 14.4. Then, using the scheme in Fig. 14.4, generate an approximately analytic sequence $[\bar{y}_k]$. Use a Hilbert transformer of length 121 and do not forget to account for the delay due to the causality of the transformer. Finally, compute the amplitude spectrum of the sequence $[\bar{y}_k]$ using SPFFTC. Plot the amplitude spectrum over the frequency range $-0.5 \leq \nu \leq 0.5$. (*Hint:* See equation (3.8).)

14.15. Repeat exercise 14.14 using the signal $[x_k]$ in exercise 14.3.

14.16. Generate the sequence

$$x_k = e^{-k/200}\sin\left(\frac{2\pi k}{150}\right), \qquad k = 0, 1, \ldots, 1023$$

Using SPFFTR, compute and plot the amplitude spectrum of the entire sequence. Then, using SPCHRP, compute and plot a detailed amplitude spectrum from 0.004 to 0.010 Hz-s.

14.17. Examine the leakage phenomenon by using SPCHRP to compute the spectrum of the sequence $[x_k]$ in exercise 14.3. Make a smooth plot of the amplitude spectrum from 0.00 to 0.02 Hz-s.

14.18. Check the result of exercise 14.17 by using SPCOMP to make the same plot.

14.19. Write a program to print the Walsh coefficient matrix for a sequence length of $N = 16$.

14.20. Write a Walsh transform routine SPDWTR similar to SPDFTR that uses SPWLSH to generate the Walsh coefficients for each index pair (k, m) in place of $e^{j2\pi km/N}$.

The sequence length N is restricted to be a power of 2. The only arrays in SPDWTR should be input and output vectors X(0:LX) and Y(0:LX), both real.

14.21. Use SPDWTR in exercise 14.20 to compute and plot the *sequency spectrum* of the following sequence:

$$[x_k] = [2 \quad 2 \quad -2 \quad -2 \quad 2 \quad 2 \quad -2 \quad -2]$$

14.22. Use SPDWTR in exercise 14.20 to compute and plot the sequency spectrum of

$$x_k = \sin\left(\frac{2\pi k}{16}\right), \qquad k = 0, 1, ..., 127$$

14.9 References

[A] Nuttall, A. H., and G. C. Carter, "A Generalized Framework for Power Spectral Estimation," *IEEE Trans. on ASSP,* ASSP-28 (June 1980): 334.

[B] Yuen, C. K., "Quadratic Windowing in the Segment Averaging Method for Power Spectrum Computation," *Technometrics* (May 1978): 195.

[C] Oppenheim, A. V., and R. W. Schafer, *Digital Signal Processing* (Englewood Cliffs, N.J.: Prentice-Hall, Inc., 1975), Chapter 10.

[D] Tribolet, J. M., "A New Phase Unwrapping Algorithm," *IEEE Trans. on ASSP,* ASSP-25 (April 1977): 170–177.

[E] Oppenheim, A. V., and R. W. Schafer, *Digital Signal Processing* (Englewood Cliffs, N.J.: Prentice-Hall, Inc., 1975), Chapter 2.

[F] Oppenheim, A. V., and R. W. Schafer, *Digital Signal Processing* (Englewood Cliffs, N.J.: Prentice-Hall, Inc., 1975), Chapter 7.

[G] Widrow, B., and S. D. Stearns, *Adaptive Signal Processing* (Englewood Cliffs, N.J.: Prentice-Hall, Inc., 1985), Chapter 13.

[H] Tretter, S. A., *Introduction to Discrete-Time Signal Processing* (New York: John Wiley, 1976), Section 2.5.

[I] Rabiner, L. R., R. W. Schafer, and C. M. Rader, "The Chirp z-Transform Algorithm and Its Application," *BSTJ,* 48, no. 5 (May-June, 1969): 1249–1292.

[J] Oppenheim, A. V., and R. W. Schafer, *Digital Signal Processing* (Englewood Cliffs, N.J.: Prentice-Hall, Inc., 1975), Section 6.6.

[K] Bluestein, L. I., "A Linear Filtering Approach to the Computation of the Discrete Fourier Transform," *1968 Northeast Electronics Research and Engineering Meeting Record,* 10 (November 1968), 218–219.

[L] Elliott, D. F., and K. R. Rao, *Fast Transforms—Algorithms, Analyses, Applications* (New York: Academic Press, 1982), Chapter 8.

[M] Ahmed, N., and K. R. Rao, *Orthogonal Transforms for Digital Signal Processing* (New York: Springer-Verlag, 1975), Chapters 5 and 6.

Subprogram Listings in Fortran 77

A complete listing of all of the subprograms in this book is included here. Comment statements describing each algorithm and its calling sequence are included in each subprogram. Most of the calling sequences are summarized at the end of each chapter in the text. The few routines not mentioned are used by other routines. The following listing gives a cross reference to the relevant chapter for each subprogram along with the purpose of the subprogram. The listing is followed by the complete subprogram listings.

ALPHABETICAL LISTING OF ROUTINES

Routine	Chapter	Purpose or result
SPBFCT	7,11	Ratio of factorials
SPBILN	7	Bilinear analog to digital transformation
SPBSSL	7	Bessel low-pass analog filter coefficients
SPBWCF	7	Butterworth low-pass analog filter coefficients
SPCBII	7	Chebyshev Type 2 low-pass analog filter coefficients
SPCFLT	6	Cascade filtering of real data in place
SPCHBI	7	Chebyshev Type 1 low-pass analog filter coefficients
SPCHRP	14	Chirp z-transform of a complex sequence
SPCOMP	3	Single component of the DFT of a real sequence
SPCONV	9	Fast convolution of two real data sequences
SPCORR	9	Fast correlation of two real data sequences
SPCROS	4	Cross power spectrum of two real sequences
SPDECI	10	Linear decimation of a real data sequence
SPDFTC	3	DFT of a complex sequence of any length
SPDFTR	3	DFT of a real sequence of any length
SPDURB	11	Durbin's solution of special Toeplitz equations
SPEXPN	13	Least-squares coefficients of $a(b{**}x)$
SPFBLT	7	Conversion from analog to digital filter
SPFFTC	3	FFT of a complex sequence of length = power of 2
SPFFTR	3	FFT of a real sequence of length = power of 2
SPFILT	6	Direct-form filtering of real data in place

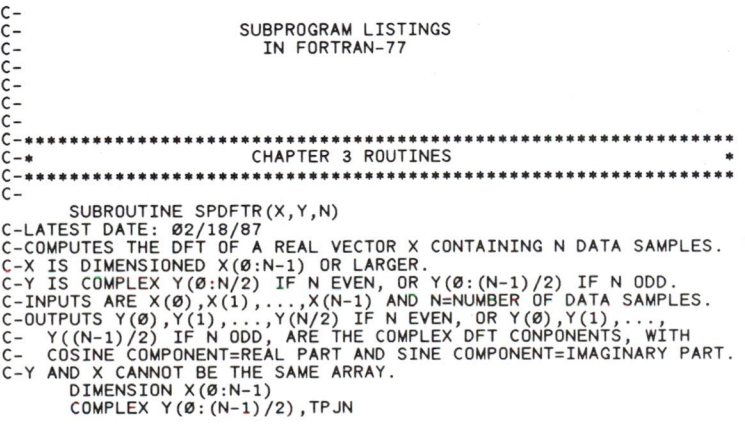

```
C-
C-                      SUBPROGRAM LISTINGS
C-                         IN FORTRAN-77
C-
C-
C-
C-
C-****************************************************************
C-*                    CHAPTER 3 ROUTINES                        *
C-****************************************************************
C-
        SUBROUTINE SPDFTR(X,Y,N)
C-LATEST DATE: 02/18/87
C-COMPUTES THE DFT OF A REAL VECTOR X CONTAINING N DATA SAMPLES.
C-X IS DIMENSIONED X(0:N-1) OR LARGER.
C-Y IS COMPLEX Y(0:N/2) IF N EVEN, OR Y(0:(N-1)/2) IF N ODD.
C-INPUTS ARE X(0),X(1),...,X(N-1) AND N=NUMBER OF DATA SAMPLES.
C-OUTPUTS Y(0),Y(1),...,Y(N/2) IF N EVEN, OR Y(0),Y(1),...,
C-   Y((N-1)/2) IF N ODD, ARE THE COMPLEX DFT CONPONENTS, WITH
C-   COSINE COMPONENT=REAL PART AND SINE COMPONENT=IMAGINARY PART.
C-Y AND X CANNOT BE THE SAME ARRAY.
        DIMENSION X(0:N-1)
        COMPLEX Y(0:(N-1)/2),TPJN
```

```
      TPJN=CMPLX(0.,-8.*ATAN(1.)/N)
      DO 2 M=0,N/2
        Y(M)=X(0)
        DO 1 K=1,N-1
          Y(M)=Y(M)+X(K)*EXP(TPJN*K*M)
    1   CONTINUE
    2 CONTINUE
      RETURN
      END
C-
      SUBROUTINE SPIDTR(Y,X,N)
C-LATEST DATE: 11/12/85
C-SAME AS SPDFTR EXCEPT REVERSED; Y=SPECTRUM, X=OUTPUT TIME SERIES.
C-SPECTRAL DATA IS ASSUMED TO BE IN COMPLEX Y(0) THRU Y(N/2).
C-N=TIME SERIES LENGTH SHOULD BE EVEN.  TIME SERIES (SCALED BY N)
C-IS COMPUTED IN X(0) THRU X(N-1).  COMPLEX Y(0:N/2) AND DIMENSION
C-X(0:N-1) ARE ASSUMED, ALTHOUGH THE ARRAYS MAY BE LARGER.
      DIMENSION X(0:N-1)
      COMPLEX Y(0:N/2),TPJN
      TPJN=CMPLX(0.,8.*ATAN(1.)/N)
      DO 2 K=0,N-1
        X(K)=Y(0)+Y(N/2)*(-1)**K
        DO 1 M=1,N/2-1
          X(K)=X(K)+2.*REAL(Y(M)*EXP(TPJN*K*M))
    1   CONTINUE
    2 CONTINUE
      RETURN
      END
C-
      SUBROUTINE SPDFTC(X,Y,N,ISIGN)
C-LATEST DATE: 02/20/87
C-COMPUTES THE DFT OF A COMPLEX VECTOR X HAVING N COMPLEX SAMPLES.
C-X AND Y ARE SPECIFIED COMPLEX X(0:N-1),Y(0:N-1) OR LARGER.
C-INPUTS ARE X(0),X(1),...,X(N-1)=COMPLEX DATA VECTOR.
C-           N=NUMBER OF DATA SAMPLES.
C-           ISIGN=-1 FOR FORWARD OR +1 FOR INVERSE TRANSFORM.
C-OUTPUTS ARE Y(0),Y(1),...,Y(N-1)=COMPLEX DFT COMPONENTS, WITH
C-  COSINE COMPONENT=REAL PART AND SINE COMPONENT=IMAGINARY PART.
C-Y AND X CANNOT BE THE SAME ARRAY.
      COMPLEX X(0:N-1),Y(0:N-1),TPJN
      TPJN=CMPLX(0.,ISIGN*8.*ATAN(1.)/N)
      DO 2 M=0,N-1
        Y(M)=X(0)
        DO 1 K=1,N-1
          Y(M)=Y(M)+X(K)*EXP(TPJN*K*M)
    1   CONTINUE
    2 CONTINUE
      RETURN
      END
C-
      SUBROUTINE SPFFTR(X,N)
C-LATEST DATE: 11/12/85
C-FFT ROUTINE FOR REAL TIME SERIES (X) WITH N=2**K SAMPLES.
C-COMPUTATION IS IN PLACE, OUTPUT REPLACES INPUT.
C-DIMENSION X(0:N+1) (REAL; NOT COMPLEX) OR LARGER.
C-REAL TIME SERIES (INPUT) IS IN X(0) THRU X(N-1).
C-FIRST OUTPUT FFT COMPONENT IS X(0)=REAL PART; X(1)=IMAGINARY,
C-ETC.  LAST COMPONENT IS X(N)=REAL AND X(N+1)=IMAGINARY PART.
C-IMPORTANT:  N MUST BE AT LEAST 4 AND MUST BE A POWER OF 2.
      COMPLEX X(0:N/2),U,TMP
      TPN=8.*ATAN(1.)/N
      CALL SPFFTC(X,N/2,-1)
      X(N/2)=X(0)
      DO 1 M=0,N/4
        U=CMPLX(SIN(M*TPN),COS(M*TPN))
        TMP=((1.+U)*X(M)+(1.-U)*CONJG(X(N/2-M)))/2.
        X(M)=((1.-U)*X(M)+(1.+U)*CONJG(X(N/2-M)))/2.
        X(N/2-M)=CONJG(TMP)
    1 CONTINUE
      RETURN
      END
C-
```

```
      SUBROUTINE SPIFTR(X,N)
C-LATEST DATE: 02/20/87
C-INVERSE FFT OF THE COMPLEX SPECTRUM OF A REAL TIME SERIES.
C-X AND N ARE THE SAME AS IN SPFFTR.  IMPORTANT: N MUST BE A POWER
C-OF 2 AND X MUST BE DIMENSIONED X(0:N+1) (REAL ARRAY, NOT COMPLEX).
C-THIS ROUTINE TRANSFORMS THE OUTPUT OF SPFFTR BACK INTO THE INPUT,
C-SCALED BY N.  COMPUTATION IS IN PLACE, AS IN SPFFTR.
      COMPLEX X(0:N/2),U1,TMP
      TPN=8.*ATAN(1.)/N
      DO 1 M=0,N/4
        U1=CMPLX(SIN(M*TPN),-COS(M*TPN))
        TMP=(1.+U1)*X(M)+(1.-U1)*CONJG(X(N/2-M))
        X(M)=(1.-U1)*X(M)+(1.+U1)*CONJG(X(N/2-M))
        X(N/2-M)=CONJG(TMP)
    1 CONTINUE
      CALL SPFFTC(X,N/2,1)
      RETURN
      END
C-
      SUBROUTINE SPFFTC(X,N,ISIGN)
C-LATEST DATE: 02/20/87
C-FAST FOURIER TRANSFORM OF N=2**K COMPLEX DATA POINTS USING TIME
C-DECOMPOSITION WITH INPUT BIT REVERSAL.  N MUST BE A POWER OF 2.
C-X MUST BE SPECIFIED COMPLEX X(0:N-1)OR LARGER.
C-INPUT IS N COMPLEX SAMPLES, X(0),X(1),...,X(N-1).
C-COMPUTATION IS IN PLACE, OUTPUT REPLACES INPUT.
C-ISIGN = -1 FOR FORWARD TRANSFORM, +1 FOR INVERSE.
C-X(0) BECOMES THE ZERO TRANSFORM COMPONENT, X(1) THE FIRST,
C-AND SO FORTH.  X(N-1) BECOMES THE LAST COMPONENT.
      COMPLEX X(0:N-1),T
      PISIGN=4*ISIGN*ATAN(1.)
      MR=0
      DO 2 M=1,N-1
        L=N
    1   L=L/2
        IF(MR+L.GE.N) GO TO 1
        MR=MOD(MR,L)+L
        IF(MR.LE.M) GO TO 2
        T=X(M)
        X(M)=X(MR)
        X(MR)=T
    2 CONTINUE
      L=1
    3 IF(L.GE.N) RETURN
      DO 5 M=0,L-1
        DO 4 I=M,N-1,2*L
          T=X(I+L)*EXP(CMPLX(0.,M*PISIGN/FLOAT(L)))
          X(I+L)=X(I)-T
          X(I)=X(I)+T
    4   CONTINUE
    5 CONTINUE
      L=2*L
      GO TO 3
      END
C-
      COMPLEX FUNCTION SPCOMP(X,N,F)
C-LATEST DATE: 02/20/87
C-COMPUTES A SINGLE COMPLEX DFT COMPONENT OF A REAL VECTOR X.
C-X MUST BE SPECIFIED DIMENSION X(0:N-1) OR LARGER.
C-INPUTS ARE X(0),X(1),...,X(N-1) = REAL DATA SEQUENCE,
C-          N=NO. OF DATA SAMPLES.
C-          F=FREQUENCY IN HZ-S; SAMPLING FREQUENCY=1.0.
C-OUTPUT IS COMPLEX SPCOMP=COMPLEX DFT COMPONENT.  NOTE THAT
C-SPCOMP MUST BE DECLARED COMPLEX IN CALLING PROGRAM.
      DIMENSION X(0:N-1)
      RAD=8.*ATAN(1.)*F
      SPCOMP=X(0)
      DO 1 K=1,N-1
        SPCOMP=SPCOMP+X(K)*EXP(CMPLX(0.,-K*RAD))
    1 CONTINUE
      RETURN
      END
```

```
C-*****************************************************************
C-*                    CHAPTER 4 ROUTINES                         *
C-*****************************************************************
C-
        SUBROUTINE SPPOWR(X,Y,WORK,LX,LY,IWINDO,OVRLAP,NSGMTS,IERROR)
C-LATEST DATE: 02/20/87
C-COMPUTES RAW PERIODOGRAM, AVERAGED OVER SEGMENTS OF X(0:LX).
C-X(0),X(1),---,X(LX)=INPUT DATA SEQUENCE.
C-Y(0),Y(1),---,Y(LY)=OUTPUT PERIODOGRAM.  LY MUST BE A POWER OF 2.
C-WORK=WORK ARRAY DIMENSIONED AT LEAST WORK(0:2*LY+1).
C-LX=LAST INDEX IN DATA SEQUENCE AS ABOVE.
C-LY=FREQUENCY INDEX CORRESPONDING TO HALF SAMPLING RATE.  POWER OF 2.
C-SEGMENT LENGTH IS 2*LY.  DATA LENGTH (LX+1) MUST BE AT LEAST THIS BIG.
C-IWINDO=DATA WINDOW TYPE, 1(RECTANGULAR), 2(TAPERED RECTANGULAR),
C-  3(TRIANGULAR), 4(HANNING), 5(HAMMING), OR 6(BLACKMAN). SEE CH. 14.
C-OVRLAP=FRACTION THAT EACH DATA SEGMENT OF SIZE 2*LY OVERLAPS ITS
C-  PREDECESSOR.  MUST BE GREATER THAN OR EQUAL 0 AND LESS THAN 1.
C-NSGMTS=NO. OVERLAPPING SEGMENTS OF X AVERAGED TOGETHER.  OUTPUT.
C-IERROR=0  NO ERROR DETECTED.
C-        1   IWINDO OUT OF RANGE (1-6).
C-        2   LX TOO SMALL, I.E., LESS THAN 2*LY-1.
C-        3   LY NOT A POWER OF 2.
        DIMENSION X(0:LX),Y(0:LY),WORK(0:2*LY+1)
        IERROR=2
        IF(LX+1.LT.2*LY) RETURN
        IERROR=3
        BASE=LY
    1   BASE=BASE/2.
        IF(BASE-2.) 7,2,1
    2   DO 3 M=0,LY
          Y(M)=0.
    3   CONTINUE
        NSHIFT=MIN0(2*LY,MAX0(1,INT(2*LY*(1.-OVRLAP)+.5)))
        NSGMTS=1+(LX+1-2*LY)/NSHIFT
        DO 6 ISEGMT=0,NSGMTS-1
          DO 4 NSAMP=0,2*LY-1
            WORK(NSAMP)=X(NSHIFT*ISEGMT+NSAMP)
    4     CONTINUE
          CALL SPMASK(WORK,2*LY-1,IWINDO,TSV,IERROR)
          IF(IERROR.NE.0) RETURN
          CALL SPFFTR(WORK,2*LY)
          DO 5 M=0,LY
            Y(M)=Y(M)+(WORK(2*M)**2+WORK(2*M+1)**2)/(TSV*NSGMTS)
    5     CONTINUE
    6   CONTINUE
    7   RETURN
        END
C-
        FUNCTION SPRAND(ISEED)
C-LATEST DATE: 11/13/85
C-UNIFORM RANDOM NUMBER FROM 0.0 TO 1.0.
C-INITIALIZE BY SETTING ISEED, THEN LEAVE ISEED ALONE.
        ISEED=2045*ISEED+1
        ISEED=ISEED-(ISEED/1048576)*1048576
        SPRAND=FLOAT(ISEED+1)/1048577.0
        RETURN
        END
C-
        SUBROUTINE SPCROS(X1,X2,Y,WORK,LX,LY,IWINDO,OVRLAP,NSGMTS,IERROR)
C-LATEST DATE: 02/20/87
C-SIMILAR TO SPPOWR, BUT FOR THE AVG. CROSS-SPECTRUM OF 2 SEQUENCES.
C-X1(0) THRU X1(LX) AND X2(0) THRU X2(LX) ARE THE INPUT DATA SEQUENCES.
C-COMPLEX Y(0),Y(1),---,Y(LY)=OUTPUT SPECTRUM.  LY=POWER OF 2.
C-WORK=COMPLEX WORK ARRAY DIMENSIONED AT LEAST WORK(0:2*LY-1).
C-LX=LAST INDEX IN INPUT DATA SEQUENCES AS ABOVE.
C-LY=FREQUENCY INDEX CORRESP. TO HALF SAMPLING RATE=POWER OF 2.
C-SEGMENT LENGTH IS 2*LY.  DATA LENGTH (LX+1) MUST BE AT LEAST THIS BIG.
C-IWINDO=DATA WINDOW TYPE, 1(RECTANGULAR), 2(TAPERED RECTANGULAR),
C-  3(TRIANGULAR), 4(HANNING), 5(HAMMING), OR 6(BLACKMAN).  SEE CH. 14.
C-OVRLAP=FRACTION THAT EACH DATA SEGMENT OF SIZE 2*LY OVERLAPS ITS
C-  PREDECESSOR.  MUST BE GREATER THAN OR EQUAL 0 AND LESS THAN 1.
C-NSGMTS=NO. OVERLAPPING SEGMENTS OF X AVERAGED TOGETHER.  OUTPUT.
C-IERROR=0  NO ERROR DETECTED.
C-        1   IWINDO OUT OF RANGE (1-6).
C-        2   LX TOO SMALL, I.E., LESS THAN 2*LY-1.
```

```
C-        3  LY NOT A POWER OF 2.
          DIMENSION X1(0:LX),X2(0:LX)
          COMPLEX Y(0:LY),WORK(0:2*LY-1)
          IERROR=1
          IF(IWINDO.LT.1.OR.IWINDO.GT.6) RETURN
          IERROR=2
          IF(LX+1.LT.2*LY) RETURN
          IERROR=3
          BASE=LY
     1    BASE=BASE/2.
          IF(BASE-2.) 7,2,1
     2    DO 3 M=0,LY
          Y(M)=(0.,0.)
     3    CONTINUE
          NSHIFT=MIN0(2*LY,MAX0(1,INT(2*LY*(1.-OVRLAP)+.5)))
          NSGMTS=1+(LX+1-2*LY)/NSHIFT
          TSV=0.
          DO 6 ISEGMT=0,NSGMTS-1
            DO 4 NSAMP=0,2*LY-1
              W=SPWNDO(IWINDO,2*LY,NSAMP)
              IF(ISEGMT.EQ.0) TSV=TSV+W*W
              INDEX=NSHIFT*ISEGMT+NSAMP
              WORK(NSAMP)=CMPLX(W*X1(INDEX),W*X2(INDEX))
     4      CONTINUE
          CALL SPFFTC(WORK,2*LY,-1)
          Y(0)=Y(0)+REAL(WORK(0))*AIMAG(WORK(0))/(TSV*NSGMTS)
          DO 5 M=1,LY
            P=AIMAG(WORK(M)*WORK(2*LY-M))
            Q=WORK(M)*CONJG(WORK(M))-WORK(2*LY-M)*CONJG(WORK(2*LY-M))
            Y(M)=Y(M)+CMPLX(0.5*P,0.25*Q)/(TSV*NSGMTS)
     5      CONTINUE
     6    CONTINUE
          IERROR=0
     7    RETURN
          END


C-*********************************************************************
C-*                    CHAPTER 5 ROUTINES                            *
C-*********************************************************************
C-
          COMPLEX FUNCTION SPGAIN(B,A,LB,LA,FREQ)
C-LATEST DATE: 02/20/87
C-THIS FUNCTION COMPUTES THE COMPLEX GAIN OF ANY CAUSAL LINEAR
C-SYSTEM IN DIRECT FORM.  FOR A PARALLEL SYSTEM, JUST TREAT EACH
C-SECTION AS DIRECT AND ADD THE RESULTS.  FOR A CASCADE SYSTEM,
C-TREAT EACH SECTION AS DIRECT AND MULTIPLY THE RESULTS TOGETHER.
C-YOU MUST SPECIFY:  DIMENSION B(0:LB),A(1:LA); COMPLEX GAIN.
C-THE LINEAR SYSTEM IS ASSUMED TO HAVE THE TRANSFER FUNCTION
C-
C-        B(0)+B(1)*Z**(-1)+B(2)*Z**(-2)+...+B(LB)*Z**(-LB)
C- H(Z)=----------------------------------------------------
C-        1.0+A(1)*Z**(-1)+A(2)*Z**(-2)+...+A(LA)*Z**(-LA)
C-
C-FOR AN FIR SYSTEM, USE LA=1 AND A(1)=0.  DO NOT USE LA=0.
C-FREQ=FREQUENCY IN HZ-S, I.E., SAMPLING FREQUENCY=1.0.
C-NOTE: COMPUTATION IS LIMITED BY "SMALL" AND "BIG" BELOW.
          DIMENSION B(0:LB),A(1:LA)
          COMPLEX Z1,BSUM,ASUM
          SMALL=1.E-10
          BIG=1.E10
          Z1=CEXP(CMPLX(0.,-8.*ATAN(1.)*FREQ))
          BSUM=0.
          IF(LB.LE.0) GO TO 2
          DO 1 I=LB,1,-1
            BSUM=(BSUM+B(I))*Z1
     1    CONTINUE
     2    ASUM=0.
          DO 3 I=LA,1,-1
            ASUM=(ASUM+A(I))*Z1
     3    CONTINUE
          IF(ABS(1.0+ASUM).LT.SMALL) SPGAIN=BIG
          IF(ABS(1.0+ASUM).GE.SMALL) SPGAIN=(B(0)+BSUM)/(1.0+ASUM)
          RETURN
```

```
            END
C-
            SUBROUTINE SPRESP(X,Y,LX,LY,B,A,LB,LA)
C-LATEST DATE: 11/13/85
C-TIME DOMAIN RESPONSE OF A CAUSAL LINEAR SYSTEM.
C-USER MUST DIMENSION X(0:LX),Y(0:LY),B(0:LB),A(1:LA)
C-X=USER-SUPPLIED INPUT SIGNAL WITH FIRST SAMPLE AT X(0).
C-THIS SIGNAL IS ASSUMED TO CONTINUE WITH ITS FINAL VALUE,
C- I.E., X(LX+1)=X(LX), ETC.  THUS A UNIT IMPULSE IS (1.,0.)
C- WITH LX=1, A UNIT STEP COULD BE (1.,1.) WITH LX=1, ETC.
C-Y(0) THRU Y(LY)=COMPUTED RESPONSE TO INPUT SIGNAL X.
C-B,A=COEFFICIENTS OF CAUSAL LINEAR SYSTEM WITH
C-
C-        B(0)+B(1)*Z**(-1)+B(2)*Z**(-2)+...+B(LB)*Z**(-LB)
C- H(Z)=-------------------------------------------------
C-        1.0+A(1)*Z**(-1)+A(2)*Z**(-2)+...+A(LA)*Z**(-LA)
C-
C-FOR AN FIR SYSTEM, USE LA=1 AND A(1)=0.  DO NOT USE LA=0.
C-NOTE: ZERO INITIAL CONDITIONS ARE ASSUMED.
            DIMENSION X(0:LX),Y(0:LY),B(0:LB),A(1:LA)
            DO 3 K=0,LY
              SUM=0.0
              DO 1 N=0,LB
                SAMPLE=0.0
                IF(K-N.GE.0) SAMPLE=X(MIN0(K-N,LX))
                SUM=SUM+B(N)*SAMPLE
      1       CONTINUE
              DO 2 N=1,LA
                SAMPLE=0.0
                IF(K-N.GE.0) SAMPLE=Y(K-N)
                SUM=SUM-A(N)*SAMPLE
      2       CONTINUE
              Y(K)=SUM
      3     CONTINUE
            RETURN
            END


C-***************************************************************
C-*                  CHAPTER 6 ROUTINES                         *
C-***************************************************************
C-
            SUBROUTINE SPFILT(B,A,LB,LA,X,N,PX,PY,IERROR)
C-LATEST DATE: 11/13/85
C-FILTERS N-POINT DATA SEQUENCE IN PLACE USING ARRAY X
C-TRANSFER FUNCTION COEFFICIENTS ARE IN ARRAYS B AND A
C-          B(0)+B(1)*Z**(-1)+.......+B(LB)*Z**(-LB)
C-   H(Z) = ---------------------------------------
C-          1+A(1)*Z**(-1)+.......+A(LA)*Z**(-LA)
C-PX SAVES PAST VALUES OF INPUT X
C-PY SAVES PAST VALUES OF OUTPUT Y
C-IERROR=0    NO ERRORS DETECTED
C-      1     FILTER RESPONSE EXCEEDS 1.E10
            DIMENSION X(0:N-1),B(0:LB),A(LA),PX(0:LB),PY(LA)
            IERROR=1
            DO 5 K=0,N-1
              PX(0)=X(K)
              X(K)=0.
              DO 1 L=0,LB
                X(K)=X(K)+B(L)*PX(L)
      1       CONTINUE
              DO 2 L=1,LA
                X(K)=X(K)-A(L)*PY(L)
      2       CONTINUE
              IF(ABS(X(K)).GT.1.E10) RETURN
              DO 3 L=LB,1,-1
                PX(L)=PX(L-1)
      3       CONTINUE
              DO 4 L=LA,2,-1
                PY(L)=PY(L-1)
      4       CONTINUE
              PY(1)=X(K)
```

```
      5 CONTINUE
        IERROR=0
        RETURN
        END
C-
        SUBROUTINE SPFLTR(B,A,LB,LA,X,N,Y,PX,PY,IERROR)
C-LATEST DATE: 11/13/85
C-FILTERS N-POINT DATA SEQUENCE X AND RETURNS OUTPUT IN Y
C-TRANSFER FUNCTION COEFFICIENTS ARE IN ARRAYS B AND A
C-           B(0)+B(1)*Z**(-1)+........+B(LB)*Z**(-LB)
C-     H(Z) = -----------------------------------------
C-             1+A(1)*Z**(-1)+.......+A(LA)*Z**(-LA)
C-PX RETAINS PAST VALUES OF INPUT X
C-PY RETAINS PAST VALUES OF OUTPUT Y
C-IERROR=0     NO ERRORS DETECTED
C-     1       OUTPUT EXCEEDS 1.E10
C-NOTE: OUTPUT ARRAY Y MUST BE INITIALIZED BY CALLING PROGRAM
        DIMENSION X(0:N-1),B(0:LB),A(LA),Y(0:N-1),PX(0:LB),PY(LA)
        IERROR=1
        DO 5 K=0,N-1
         PX(0)=X(K)
         SUM=0.0
         DO 1 L=0,LB
          SUM=SUM+B(L)*PX(L)
      1    CONTINUE
         DO 2 L=1,LA
          SUM=SUM-A(L)*PY(L)
      2    CONTINUE
         IF(ABS(SUM).GT.1.E10) RETURN
         DO 3 L=LB,1,-1
          PX(L)=PX(L-1)
      3    CONTINUE
         DO 4 L=LA,2,-1
          PY(L)=PY(L-1)
      4    CONTINUE
         PY(1)=SUM
         Y(K)=Y(K)+SUM
      5 CONTINUE
        IERROR=0
        RETURN
        END
C-
        SUBROUTINE SPCFLT(B,A,LS,NS,X,N,PX,PY,IERROR)
C-LATEST DATE: 11/13/85
C-FILTERS N-POINT DATA SEQUENCE IN PLACE USING ARRAY X
C-TRANSFER FUNCTION IS COMPOSED OF NS SECTIONS IN CASCADE WITH
C-     MTH STAGE TRANSFER FUNCTION
C-         B(0,M)+B(1,M)*Z**(-1)+......+B(LS,M)*Z**(-LS)
C-     H(Z) = -------------------------------------------
C-             1+A(1,M)*Z**(-1)+......+A(LS,M)*Z**(-LS)
C-PX RETAINS PAST VALUES OF INPUT X
C-PY RETAINS PAST VALUES OF OUTPUT Y
C-IERROR=0     NO ERRORS DETECTED
C-     1 - NS  OUTPUT AT STAGE [IERROR] EXCEEDS 1.E10
        DIMENSION X(0:N-1),B(0:LS,NS),A(LS,NS),PX(0:LS,NS),PY(LS,NS)
        DO 5 M=1,NS
         IERROR=M
         DO 3 K=0,N-1
          PX(0,M)=X(K)
          X(K)=B(0,M)*PX(0,M)
          DO 1 LL=1,LS
           X(K)=X(K)+B(LL,M)*PX(LL,M)-A(LL,M)*PY(LL,M)
      1     CONTINUE
          IF(ABS(X(K)).GT.1.E10) RETURN
          DO 2 LL=LS,2,-1
           PX(LL,M)=PX(LL-1,M)
           PY(LL,M)=PY(LL-1,M)
      2     CONTINUE
          PX(1,M)=PX(0,M)
          PY(1,M)=X(K)
      3    CONTINUE
      5 CONTINUE
        IERROR=0
        RETURN
        END
```

```
C-
        SUBROUTINE SPPFLT(B,A,LS,NS,X,N,Y,PX,PY,IERROR)
C-LATEST DATE: 11/13/85
C-FILTERS N-POINT DATA SEQUENCE X AND RETURNS OUTPUT IN Y
C-TRANSFER FUNCTION IS COMPOSED OF NS SECTIONS IN PARALLEL WITH
C-          MTH SECTION DEFINED BY
C-        B(0,M)+B(1,M)*Z**(-1)+........+B(LS,M)*Z**(-LS)
C-  H(Z) = --------------------------------------------------
C-                1+A(1,M)*Z**(-1)+..........+A(LS,M)*Z**(-LS)
C-PX RETAINS PAST VALUES OF INPUT X
C-PY RETAINS PAST VALUES OF OUTPUT Y
C-IERROR=0      NO ERRORS DETECTED
C-     1 - NS   OUTPUT AT STAGE [IERROR]  EXCEEDS 1.E10
        DIMENSION X(0:N-1),Y(0:N-1),B(0:LS,NS),A(LS,NS)
        DIMENSION PX(0:LS,NS),PY(LS,NS)
        DO 1 K=0,N-1
          Y(K)=0.0
    1   CONTINUE
        DO 5 M=1,NS
          IERROR=M
          DO 4 K=0,N-1
            PX(0,M)=X(K)
            SUM=B(0,M)*PX(0,M)
            DO 2 LL=1,LS
              SUM=SUM+B(LL,M)*PX(LL,M)-A(LL,M)*PY(LL,M)
    2       CONTINUE
            IF(ABS(SUM).GT.1.E10) RETURN
            DO 3 LL=LS,2,-1
              PX(LL,M)=PX(LL-1,M)
              PY(LL,M)=PY(LL-1,M)
    3       CONTINUE
            PX(1,M)=PX(0,M)
            PY(1,M)=SUM
            Y(K)=Y(K)+SUM
    4     CONTINUE
    5   CONTINUE
        IERROR=0
        RETURN
        END
C-
        SUBROUTINE SPLFLT(KAPPA,NU,L,X,N,PAST,IERROR)
C-LATEST DATE: 11/13/85
C-FILTERS N-POINT DATA SEQUENCE IN PLACE USING ARRAY X
C-LATTICE TRANSFER FUNCTION COEFFICIENTS IN REAL ARRAYS KAPPA AND NU
C-PAST RETAINS OLD VALUES TO ENABLE BLOCK MODE FILTERING
C-IERROR=0      NO ERRORS DETECTED
C-     1      OUTPUT EXCEEDS 1.E10
        DIMENSION X(0:N-1),PAST(0:L)
        REAL KAPPA(0:L-1),NU(0:L)
        IERROR=1
        DO 5 K=0,N-1
          SUM=X(K)
          DO 2 LL=L,1,-1
            SUM=SUM-KAPPA(LL-1)*PAST(LL-1)
            PAST(LL)=PAST(LL-1)+KAPPA(LL-1)*SUM
    2     CONTINUE
          PAST(0)=SUM
          X(K)=0.0
          DO 3 LL=0,L
            X(K)=X(K)+NU(LL)*PAST(LL)
    3     CONTINUE
          IF(ABS(X(K)).GT.1.E10) RETURN
    5   CONTINUE
        IERROR=0
        RETURN
        END
C-
        SUBROUTINE SPLTCF(B,A,L,KAPPA,NU,WORK,IERROR)
C-LATEST DATE: 11/13/85
C-CONVERTS TRANSFER FUNCTION COEFFICIENTS FROM DIRECT FORM TO LATTICE
C-DIRECT FORM H(Z) IS DEFINED BY
C-        B(0)+B(1)*Z**(-1)+...........+B(L)*Z**(-L)
C-  H(Z) = ------------------------------------------
C-             1+A(1)*Z**(-1)+..........+A(L)*Z**(-L)
C-LATTICE COEF ARE RETURNED IN REAL ARRAYS KAPPA AND NU
C-IERROR=0      CONVERSION WITH NO ERRORS DETECTED
```

```
C-      1       UNSTABLE H(Z)
C-WORK ARRAY IS USED INTERNALLY.
      DIMENSION B(0:L),A(L),WORK(0:L,2)
      REAL KAPPA(0:L-1),NU(0:L)
      IERROR=1
      WORK(0,1)=1.
      DO 1 LL=1,L
        WORK(LL,1)=A(LL)
    1 CONTINUE
      DO 5 LL=L,1,-1
        DO 2 J=0,LL
          WORK(J,2)=WORK(LL-J,1)
    2   CONTINUE
        KAPPA(LL-1)=WORK(LL,1)
        IF(ABS(KAPPA(LL-1)).GE.1.0) RETURN
        DO 3 J=0,LL
          WORK(J,1)=(WORK(J,1)-KAPPA(LL-1)*WORK(J,2))/
     +            (1.-KAPPA(LL-1)**2)
    3   CONTINUE
        NU(LL)=B(LL)
        DO 4 J=0,LL
          B(J)=B(J)-NU(LL)*WORK(J,2)
    4   CONTINUE
    5 CONTINUE
      NU(0)=B(0)
      IERROR=0
      RETURN
      END




C-*****************************************************************
C-*                   CHAPTER 7 ROUTINES                         *
C-*****************************************************************
C-
      SUBROUTINE SPBILN(D,C,LN,B,A,WORK,IERROR)
C-LATEST DATE: 11/13/85
C-CONVERTS ANALOG H(S) TO DIGITAL H(Z) VIA BILINEAR TRANSFORM
C-      ANALOG TRANSFER FUNCTION          DIGITAL TRANSFER FUNCTION
C-          D(L)*S**L+.....+D(0)              B(0)+......+B(L)*Z**-L
C-   H(S)=---------------------       H(Z)=----------------------
C-          C(L)*S**L+.....+C(0)              1+.......+A(L)*Z**-L
C-H(S) IS ASSUMED TO BE PRE-SCALED AND PRE-WARPED
C-LN SPECIFIES THE LENGTH OF THE COEFFICIENT ARRAYS
C-FILTER ORDER L IS COMPUTED INTERNALLY
C-WORK IS AN INTERNAL ARRAY (2D) SIZED TO MATCH COEF ARRAYS
C-IERROR=0   NO ERRORS DETECTED IN TRANSFORMATION
C-      1    ALL ZERO TRANSFER FUNCTION
C-      2    INVALID TRANSFER FUNCTION; Y(K) COEF=0
      DIMENSION D(0:LN),C(0:LN),B(0:LN),A(LN),WORK(0:LN,0:LN)
      DO 1 I=LN,0,-1
        IF(C(I).NE.0..OR.D(I).NE.0.)GO TO 2
    1 CONTINUE
      IERROR=1
      RETURN
    2 L=I
      DO 5 J=0,L
        WORK(0,J)=1.
    5 CONTINUE
      TMP=1.
      DO 6 I=1,L
        TMP=TMP*FLOAT(L-I+1)/FLOAT(I)
        WORK(I,0)=TMP
    6 CONTINUE
      DO 8 I=1,L
        DO 7 J=1,L
          WORK(I,J)=WORK(I,J-1)-WORK(I-1,J)-WORK(I-1,J-1)
    7   CONTINUE
    8 CONTINUE
      DO 10 I=L,0,-1
        B(I)=0.
        ATMP=0.
```

```
      DO 9 J=Ø,L
        B(I)=B(I)+WORK(I,J)*D(J)
        ATMP=ATMP+WORK(I,J)*C(J)
    9   CONTINUE
      SCALE=ATMP
      IF(I.NE.Ø) A(I)=ATMP
   1Ø CONTINUE
      IERROR=2
      IF(SCALE.EQ.Ø.) RETURN
      B(Ø)=B(Ø)/SCALE
      DO 12 I=1,L
        B(I)=B(I)/SCALE
        A(I)=A(I)/SCALE
   12 CONTINUE
      DO 14 I=L+1,LN
        B(I)=Ø.Ø
        A(I)=Ø.Ø
   14 CONTINUE
      IERROR=Ø
      RETURN
      END
C-
      SUBROUTINE SPFBLT(D,C,LN,IBAND,FLN,FHN,B,A,WORK,IERROR)
C-LATEST DATE: Ø5/19/86
C-CONVERTS NORMALIZED LP ANALOG H(S) TO DIGITAL H(Z)
C-      ANALOG TRANSFER FUNCTION          DIGITAL TRANSFER FUNCTION
C-      D(M)*S**M+.....+D(Ø)                 B(Ø)+.....+B(L)*Z**-L
C-  H(S)=--------------------           H(Z)=---------------------
C-      C(M)*S**M+.....+C(Ø)                 1+......+A(L)*Z**-L
C-FILTER ORDER L IS COMPUTED INTERNALLY
C-IBAND=1   LOWPASS          FLN=NORMALIZED CUTOFF IN HZ-SEC
C-      2   HIGHPASS         FLN=NORMALIZED CUTOFF IN HZ-SEC
C-      3   BANDPASS         FLN=LOW CUTOFF; FHN=HIGH CUTOFF
C-      4   BANDSTOP         FLN=LOW CUTOFF; FHN=HIGH CUTOFF
C-LN SPECIFIES COEFFICIENT ARRAY SIZE
C-WORK(Ø:LN,Ø:LN) IS A WORK ARRAY USED INTERNALLY
C-RETURN IERROR=Ø     NO ERRORS DETECTED
C-              1     ALL ZERO TRANSFER FUNCTION
C-              2     BILIN: INVALID TRANSFER FUNCTION
C-              3     FILTER ORDER EXCEEDS ARRAY SIZE
C-              4     INVALID FILTER TYPE PARAMETER (IBAND)
C-              5     INVALID CUTOFF FREQUENCY
      DIMENSION D(Ø:LN),C(Ø:LN),B(Ø:LN),A(LN),WORK(Ø:LN,Ø:LN)
      PI=4.*ATAN(1.)
      IERROR=Ø
      IF(IBAND.LT.1.OR.IBAND.GT.4) IERROR=4
      IF(FLN.LE.Ø..OR.FLN.GT.Ø.5) IERROR=5
      IF(IBAND.GE.3.AND.FLN.GE.FHN) IERROR=5
      IF(IBAND.GE.3.AND.FHN.GT.Ø.5) IERROR=5
      IF(IERROR.NE.Ø) RETURN
      DO 1 I=LN,Ø,-1
        IF(C(I).NE.Ø..OR.D(I).NE.Ø.) GO TO 2
    1 CONTINUE
      IERROR=1
      RETURN
    2 M=I
      W1=TAN(PI*FLN)
      L=M
      IF(IBAND.LE.2) GO TO 3
      L=2*M
      W2=TAN(PI*FHN)
      W=W2-W1
      WØ2=W1*W2
    3 CONTINUE
      IERROR=3
      IF(L.GT.LN) RETURN
      GO TO (3Ø,2Ø,4Ø,2Ø) IBAND
C-SUBSTITUTION OF 1/S TO GENERATE HIGHPASS (HP,BS)
   2Ø CONTINUE
      DO 25 MM=Ø,M/2
        TMP=D(MM)
        D(MM)=D(M-MM)
        D(M-MM)=TMP
```

```
            TMP=C(MM)
            C(MM)=C(M-MM)
            C(M-MM)=TMP
   25 CONTINUE
        IF(IBAND.EQ.4) GO TO 40
C-SCALING S/W1 FOR LOWPASS,HIGHPASS
   30 CONTINUE
        DO 35 MM=0,M
          D(MM)=D(MM)/(W1**MM)
          C(MM)=C(MM)/(W1**MM)
   35 CONTINUE
        GO TO 100
C-SUBSTITUTION OF (S**2+W0**2)/(W*S)  BANDPASS,BANDSTOP
   40 CONTINUE
        DO 45 LL=0,L
          WORK(LL,0)=0.
          WORK(LL,1)=0.
   45 CONTINUE
        DO 52 MM=0,M
          TMPD=D(MM)*(W**(M-MM))
          TMPC=C(MM)*(W**(M-MM))
          DO 50 K=0,MM
            LS=M+MM-2*K
            TMP=SPBFCT(MM,MM)/(SPBFCT(K,K)*SPBFCT(MM-K,MM-K))
            WORK(LS,0)=WORK(LS,0)+TMPD*(W02**K)*TMP
            WORK(LS,1)=WORK(LS,1)+TMPC*(W02**K)*TMP
   50    CONTINUE
   52 CONTINUE
        DO 55 LL=0,L
          D(LL)=WORK(LL,0)
          C(LL)=WORK(LL,1)
   55 CONTINUE
C-SUBSTITUTE (Z-1)/(Z+1)
  100 CONTINUE
        CALL SPBILN(D,C,LN,B,A,WORK,IERROR)
        RETURN
        END
C-
        SUBROUTINE SPBWCF(L,K,LN,D,C,IERROR)
C-LATEST DATE: 11/13/85
C-GENERATES KTH SECTION COEFFICIENTS FOR LTH ORDER NORMALIZED
C-      LOWPASS BUTTERWORTH FILTER
C-SECOND ORDER SECTIONS: K<=(L+1)/2
C-ODD ORDER L:  FINAL SECTION WILL CONTAIN 1ST ORDER POLE
C-LN DEFINES COEFFICIENT ARRAY SIZE
C-ANALOG COEFFICIENTS ARE RETURNED IN D AND C
C-IERROR=0      NO ERRORS DETECTED
C-       1      INVALID FILTER ORDER L
C-       2      INVALID SECTION NUMBER K
        DIMENSION D(0:LN),C(0:LN)
        PI=4.*ATAN(1.)
        IERROR=0
        IF(L.LE.0) IERROR=1
        IF(K.LE.0.OR.K.GT.INT((L+1)/2)) IERROR=2
        IF(IERROR.NE.0) RETURN
        D(0)=1.
        C(0)=1.
        DO 1 I=1,LN
          D(I)=0.
          C(I)=0.
    1 CONTINUE
        TMP=K-(L+1.)/2.
        IF(TMP.EQ.0.) THEN
          C(1)=1.
        ELSE
          C(1)=(-2.)*COS((2*K+L-1)*PI/(2*L))
          C(2)=1.
        ENDIF
        RETURN
        END
C-
        SUBROUTINE SPCHBI(L,K,LN,EP,D,C,IERROR)
C-LATEST DATE: 11/13/85
C-GENERATES KTH SECTION COEFFICIENTS FOR LTH ORDER NORMALIZED
C-            LOWPASS CHEBYSHEV TYPE I ANALOG FILTER
```

```
C-SECOND ORDER SECTIONS:  K<=(L+1)/2
C-ODD ORDER L: LAST SECTION WILL CONTAIN SINGLE POLE
C-LN DEFINES COEFFICIENT ARRAY SIZE
C-EP REGULATES THE PASSBAND RIPPLE
C-TRANSFER FUNCTION SCALING IS INCLUDED IN FIRST SECTION (L EVEN)
C-ANALOG COEFFICIENTS ARE RETURNED IN D AND C
C-IERROR=0     NO ERRORS DETECTED
C-        1      INVALID FILTER ORDER L
C-        2      INVALID SECTION NUMBER K
C-        3      INVALID RIPPLE PARAMETER EP
      DIMENSION D(0:LN),C(0:LN)
      PI=4.*ATAN(1.)
      IERROR=0
      IF(L.LE.0) IERROR=1
      IF(K.GT.INT((L+1)/2).OR.K.LE.0) IERROR=2
      IF(EP.LE.0.) IERROR=3
      IF(IERROR.NE.0) RETURN
      GAM=((1.+SQRT(1.+EP**2))/EP)**(1./L)
      SIGMA=.5*(1./GAM-GAM)*SIN((2*K-1)*PI/(2*L))
      OMEGA=.5*(GAM+1./GAM)*COS((2*K-1)*PI/(2*L))
      DO 1 LL=0,LN
        D(LL)=0.
        C(LL)=0.
    1 CONTINUE
      IF(INT(L/2).NE.INT((L+1)/2).AND.K.EQ.(L+1)/2) THEN
        D(0)=-1.*SIGMA
        C(0)=D(0)
        C(1)=1.
        RETURN
      ENDIF
      C(0)=SIGMA**2+OMEGA**2
      C(1)=-2.*SIGMA
      C(2)=1.
      D(0)=C(0)
      IF(INT(L/2).EQ.INT((L+1)/2).AND.K.EQ.1) D(0)=D(0)/SQRT(1.+EP**2)
      RETURN
      END
C-
      SUBROUTINE SPCBII(L,K,LN,WS,ATT,D,C,IERROR)
C-LATEST DATE: 11/13/85
C-GENERATES KTH SECTION COEFFICIENTS FOR LTH ORDER NORMALIZED
C-      LOWPASS CHEBYSHEV TYPE II ANALOG FILTER
C-SECOND ORDER SECTIONS:  K<= (L+1)/2
C-ODD ORDER L:  FINAL SECTION WILL CONTAIN SINGLE POLE
C-LN DEFINES COEFFICIENT ARRAY SIZE
C-WS AND ATT REGULATE STOPBAND ATTENUATION
C-      MAGNITUDE WILL BE 1/ATT AT WS RAD/SEC
C-ANALOG COEFFICIENTS ARE RETURNED IN ARRAYS D AND C
C-IERROR=0     NO ERRORS DETECTED
C-        1      INVALID FILTER ORDER L
C-        2      INVALID SECTION NUMBER K
C-        3      INVALID STOPBAND FREQUENCY WS
C-        4      INVALID ATTENUATION PARAMETER
      DIMENSION D(0:LN),C(0:LN)
      PI=4.*ATAN(1.)
      IERROR=0
      IF(L.LE.0) IERROR=1
      IF(K.GT.INT((L+1)/2).OR.K.LT.1) IERROR=2
      IF(WS.LE.1.) IERROR=3
      IF(ATT.LE.0.) IERROR=4
      IF(IERROR.NE.0) RETURN
      GAM=(ATT+SQRT(ATT**2-1.))**(1./L)
      ALPHA=.5*(1./GAM-GAM)*SIN((2*K-1)*PI/(2*L))
      BETA=.5*(GAM+1./GAM)*COS((2*K-1)*PI/(2*L))
      SIGMA=(WS*ALPHA)/(ALPHA**2+BETA**2)
      OMEGA=(-1.*WS*BETA)/(ALPHA**2+BETA**2)
      DO 1 LL=0,LN
        D(LL)=0.
        C(LL)=0.
    1 CONTINUE
      IF(INT(L/2).NE.INT((L+1)/2).AND.K.EQ.(L+1)/2) THEN
        D(0)=-1.*SIGMA
        C(0)=D(0)
        C(1)=1.
```

```
            RETURN
            ENDIF
            SCLN=SIGMA**2+OMEGA**2
            SCLD=(WS/COS((2*K-1)*PI/(2*L)))**2
            D(0)=SCLN*SCLD
            D(2)=SCLN
            C(0)=D(0)
            C(1)=-2.*SIGMA*SCLD
            C(2)=SCLD
            RETURN
            END
C-
            SUBROUTINE SPBSSL(L,WSCL,LN,D,C,IERROR)
C-LATEST DATE: 11/13/85
C-GENERATES ANALOG FILTER COEFFICIENTS FOR LTH ORDER
C-      NORMALIZED LOWPASS BESSEL FILTER
C-COEFFICIENTS ARE RETURNED IN ARRAYS D AND C
C-LN SPECIFIES ARRAY SIZE (LN>=L)
C-WSCL CONTROLS FREQUENCY SCALING SUCH THAT RESPONSE AT 1 RAD/SEC
C-      IS EQUAL TO THAT OF UNSCALED H(S) AT WSCL RAD/SEC
C-IERROR=0        NO ERRORS DETECTED
C-      1         INVALID FILTER ORDER (L<=0 OR L>=LN)
C-      2         INVALID SCALE PARAMETER (WSCL<=0)
            DIMENSION D(0:LN),C(0:LN)
            IERROR=0
            IF(L.LE.0.OR.L.GT.LN) IERROR=1
            IF(WSCL.LE.0.) IERROR=2
            IF(IERROR.NE.0) RETURN
            DO 1 LL=0,LN
               D(LL)=0.
               C(LL)=0.
          1 CONTINUE
            DO 2 K=0,L
               DENOM=(2**(L-K))*SPBFCT(K,K)*SPBFCT(L-K,L-K)
               C(K)=(WSCL**K)*SPBFCT(2*L-K,2*L-K)/DENOM
          2 CONTINUE
            D(0)=C(0)
            RETURN
            END
C-
            SUBROUTINE SPIIRD(IFILT,IBAND,NS,LS,F1,F2,F3,F4,DB,B,A,IERROR)
C-LATEST DATE: 09/25/86
C-IIR LOWPASS, HIGHPASS, BANDPASS, AND BANDSTOP DESIGN OF CHEBYSHEV 1,
C-      CHEBYSHEV 2, AND BUTTERWORTH DIGITAL FILTERS IN CASCADE FORM.
C-IFILT=1(CHEB1-PASSBAND RIPPLE), 2(CHEB2-STOPBAND RIPPLE), OR
C-      3(BUTTERWORTH-NO RIPPLE).
C-IBAND=1(LOWPASS), 2(HIGHPASS), 3(BANDPASS), OR 4(BANDSTOP).
C-NS  =NUMBER OF SECTIONS IN CASCADE.
C-LS  =ORDER OF EACH SECTION: USUALLY 2(IBAND=1,2) OR 4(IBAND=3,4).
C-F1-F4=FREQ. IN HZ-SEC. (SAMPLING FREQ.=1.0) AS IN PLOTS BELOW.
C
C      LOWPASS         HIGHPASS        BANDPASS        BANDSTOP
C
C        F  F            F  F           F F  F F        F F F F
C        1  2            1  2           1 2  3 4        1 2 3 4
C    0 XXX------     0 +------XXX    0 +----XX----   0 XX--------X
C    I X . .         I .   .X       I . .XX. .      IX. . . .X
C    I X.  .         I .  .X        I . .XX. .      IX. . . .X
C    I...X  .        I....X         I...X..X .      I.X.....X
C    I    X .        I .X           I .X  X .       I  X. .X
C    I     X.        I .X           I .X   X        I  X. .X
C DB I......X     DB I.X         DB I.X......X    DB I...XXX
C    I       X       IX             IX        X      I
C
C-      F3 AND F4 ARE NOT USED WITH ANY LOW OR HIGHPASS.
C-      F2 IS NOT USED WITH LOWPASS BUTTERWORTH.
C-      F1 IS NOT USED WITH HIGHPASS BUTTERWORTH.
C-      F1 AND F4 ARE NOT USED WITH BANDPASS BUTTERWORTH.
C-      F2 AND F3 ARE NOT USED WITH BANDSTOP BUTTERWORTH.
C
C-DB  =DB OF STOPBAND REJECTION.  APPLIES TO CHEB. FILTERS ONLY.
C-      NOT USED WITH BUTTERWORTH.  MUST BE GREATER THAT 3 DB.
C-B   =NUMERATOR COEFFICIENTS, ALWAYS DIMENSIONED B(0:LS,NS).
```

```
C-A     =DENOMINATOR COEFFICIENTS, ALWAYS DIMENSIONED A(LS,NS).
C-IERROR=0    NO ERRORS.
C-         1-5  SEE SPFBLT ERROR LIST.
C-         6    IFILT OR IBAND OUT OF RANGE.
C-         7    F1-F4 NOT IN SEQUENCE OR NOT BETWEEN 0.0 AND 0.5,
C-              OR DB NOT GREATER THAN 3.
C-         11+  SEE SPCHBI, SPCBII, OR SPBWCF ERROR LIST.
C-
      DIMENSION B(0:LS,NS),A(LS,NS),D(0:4),C(0:4),WORK(0:4,0:4)
      COSH(X)=0.5*(EXP(X)+EXP(-X))
      COSH1(X)=ALOG(X+SQRT(X*X-1.))
      WARP(F)=TAN(PI*F)
      BPSUB(OM)=(OM*OM-WARP(FH)*WARP(FL))/((WARP(FH)-WARP(FL))*OM)
      OMIN(OM1,OM2)=MIN(ABS(OM1),ABS(OM2))
      PI=4.*ATAN(1.)
      IERROR=6
      IF(0.GE.NS) RETURN
      IF(IFILT.LT.1.OR.IFILT.GT.3) RETURN
      IF(IBAND.LT.1.OR.IBAND.GT.4) RETURN
      IERROR=7
      IF(IBAND.EQ.1.OR.IBAND.EQ.4) FL=F1
      IF(IBAND.EQ.2.OR.IBAND.EQ.3) FL=F2
      IF(IBAND.LE.3) FH=F3
      IF(IBAND.EQ.4) FH=F4
      IF(IBAND.LT.3) THEN
        IF(IFILT.LT.3.AND.(0.0.GE.F1.OR.F1.GE.F2.OR.F2.GE.0.5)) RETURN
        IF(IFILT.EQ.3.AND.(0.0.GE.FL.OR.FL.GE.0.5)) RETURN
      ELSEIF(IFILT.LT.3) THEN
        IF(0.0.GE.F1.OR.F1.GE.F2.OR.F2.GE.F3) RETURN
        IF(F3.GE.F4.OR.F4.GE.0.5) RETURN
      ELSEIF(IFILT.EQ.3) THEN
        IF(0.0.GE.FL.OR.FL.GE.FH.OR.FH.GE.0.5) RETURN
      ENDIF
      IF(IFILT.LT.3.AND.DB.LE.3.) RETURN
      IF(IFILT.LT.3) THEN
        IF(IBAND.LE.2) OMEGA=WARP(F2)/WARP(F1)
        IF(IBAND.EQ.3) OMEGA=OMIN(BPSUB(WARP(F1)),BPSUB(WARP(F4)))
        IF(IBAND.EQ.4) OMEGA=OMIN(1./BPSUB(WARP(F2)),1./BPSUB(WARP(F3)))
        ALAMDA=10.**(DB/20.)
        EPSLON=ALAMDA/COSH(2*NS*COSH1(OMEGA))
      ENDIF
      DO 1 K=1,NS
        IF(IFILT.EQ.1) CALL SPCHBI(2*NS,K,4,EPSLON,D,C,IERROR)
        IF(IFILT.EQ.2) CALL SPCBII(2*NS,K,4,OMEGA,ALAMDA,D,C,IERROR)
        IF(IFILT.EQ.3) CALL SPBWCF(2*NS,K,4,D,C,IERROR)
        IF(IERROR.NE.0) THEN
          IERROR=10*IFILT+IERROR
          RETURN
        ENDIF
        CALL SPFBLT(D,C,LS,IBAND,FL,FH,B(0,K),A(1,K),WORK,IERROR)
        IF(IERROR.NE.0) RETURN
    1 CONTINUE
      RETURN
      END




C-***********************************************************************
C-*                     CHAPTER 8 ROUTINES                             *
C-***********************************************************************
C-
      SUBROUTINE SPFIRL(L,FCN,IWNDO,B,IERROR)
C-LATEST DATE: 05/17/86
C-FIR LOWPASS FILTER DESIGN USING WINDOWED FOURIER SERIES
C-L=FILTER LENGTH = L+1
C-FCN=NORMALIZED CUT-OFF FREQUENCY IN HERTZ-SECONDS
C-IWNDO=WINDOW USED TO TRUNCATE FOURIER SERIES
C-        1-RECTANGULAR; 2-TAPERED RECTANGULAR; 3-TRIANGULAR
C-        4-HANNING; 5-HAMMING; 6-BLACKMAN
C-B(0:L)=DIGITAL FILTER COEFFICIENTS RETURNED
C-IERROR=0       NO ERRORS DETECTED
```

```
C-       1         INVALID FILTER LENGTH (L<=0)
C-       2         INVALID WINDOW TYPE IWNDO
C-       3         INVALID CUT-OFF FCN; <=0 OR >=0.5
      DIMENSION B(0:L)
      IERROR=0
      IF(L.LE.0) IERROR=1
      IF(IWNDO.LT.1.OR.IWNDO.GT.6) IERROR=2
      IF(FCN.LE.0..OR.FCN.GE.0.5) IERROR=3
      IF(IERROR.NE.0) RETURN
      DO 1 I=0,L
        B(I)=0.0
    1 CONTINUE
      PI=4.*ATAN(1.)
      WCN=2.*PI*FCN
      DLY=L/2.
      LIM=INT(L/2)
      IF(DLY.EQ.L/2) THEN
        LIM=LIM-1
        B(L/2)=WCN/PI
      ENDIF
      DO 2 I=0,LIM
        B(I)=((SIN(WCN*(I-DLY)))/(PI*(I-DLY)))*SPWNDO(IWNDO,L+1,I)
        B(L-I)=B(I)
    2 CONTINUE
      RETURN
      END
C-
      SUBROUTINE SPFIRD(L,IBAND,FLN,FHN,IWNDO,B,IERROR)
C-LATEST DATE: 05/18/86
C-FIR DIGITAL FILTER DESIGN USING WINDOWED FOURIER SERIES
C-L=LENGTH OF FILTER = L+1
C-IBAND=1(LOWPASS); 2(HIGHPASS); 3(BANDPASS); 4(BANDSTOP)
C-FLN=NORMALIZED LOW CUT-OFF FREQUENCY IN HZ-SEC
C-FHN=NORMALIZED HIGH CUT-OFF (BP,BS) IN HZ-SEC
C-IWNDO=1(RECTANGULAR); 2(TAPERED RECTANGULAR); 3(TRIANGULAR)
C-      4(HANNING); 5(HAMMING); 6(BLACKMAN)
C-DIGITAL FILTER COEFFICIENTS ARE RETURNED IN B(0:L)
C-IERROR=0       NO ERRORS DETECTED
C-       1         INVALID LENGTH   (L<=0)
C-       2         INVALID WINDOW TYPE
C-       3         INVALID FILTER TYPE
C-       4         INVALID CUT-OFF FREQUENCY
      DIMENSION B(0:L)
      PI=4.*ATAN(1.)
      DO 1 I=0,L
        B(I)=0.
    1 CONTINUE
      IERROR=0
      IF(L.LE.0) IERROR=1
      IF(IWNDO.LT.1.OR.IWNDO.GT.6) IERROR=2
      IF(IBAND.LT.1.OR.IBAND.GT.4) IERROR=3
      IF(FLN.LE.0..OR.FLN.GT.0.5) IERROR=4
      IF(IBAND.GE.3.AND.FLN.GE.FHN) IERROR=4
      IF(IBAND.GE.3.AND.FHN.GE.0.5) IERROR=4
      IF(IERROR.NE.0) RETURN
      DLY=L/2.
      LIM=INT(L/2)
      MID=0
      IF(DLY.EQ.LIM) THEN
        LIM=LIM-1
        MID=1
      ENDIF
      WC1=2.*PI*FLN
      IF(IBAND.GE.3) WC2=2.*PI*FHN
      GO TO (5,10,15,20) IBAND
C-LOWPASS DESIGN
    5 CONTINUE
      DO 6 I=0,LIM
        S=I-DLY
        B(I)=((SIN(WC1*S))/(PI*S))*SPWNDO(IWNDO,L+1,I)
        B(L-I)=B(I)
    6 CONTINUE
      IF(MID.EQ.1) B(L/2)=WC1/PI
      RETURN
C-HIGHPASS DESIGN
```

```
   10 CONTINUE
      DO 11 I=0,LIM
        S=I-DLY
        B(I)=((SIN(PI*S)-SIN(WC1*S))/(PI*S))*SPWNDO(IWNDO,L+1,I)
        B(L-I)=B(I)
   11 CONTINUE
      IF(MID.EQ.1) B(L/2)=1.-WC1/PI
      RETURN
C-BANDPASS DESIGN
   15 CONTINUE
      DO 16 I=0,LIM
        S=I-DLY
        B(I)=((SIN(WC2*S)-SIN(WC1*S))/(PI*S))*SPWNDO(IWNDO,L+1,I)
        B(L-I)=B(I)
   16 CONTINUE
      IF(MID.EQ.1) B(L/2)=(WC2-WC1)/PI
      RETURN
C-BANDSTOP DESIGN
   20 CONTINUE
      DO 21 I=0,LIM
        S=I-DLY
        B(I)=((SIN(WC1*S)+SIN(PI*S)-SIN(WC2*S))/(PI*S))*
     +       SPWNDO(IWNDO,L+1,I)
        B(L-I)=B(I)
   21 CONTINUE
      IF(MID.EQ.1) B(L/2)=(WC1+PI-WC2)/PI
      RETURN
      END



C-*******************************************************************
C-*                    CHAPTER 9 ROUTINES                           *
C-*******************************************************************
C-
      SUBROUTINE SPCONV(X,Y,L,NMIN,IERROR)
C-LATEST DATE: 11/27/85
C-FAST CONVOLUTION OF SEQUENCES X AND Y.
C-X AND Y SHOULD BE DIFFERENT VECTORS, DIMENSIONED X(0:L) AND Y(0:L).
C-L=LAST INDEX IN BOTH X AND Y.  MUST BE (POWER OF 2)+1 AND AT LEAST 5.
C-NMIN=MINIMUM SHIFT OF INTEREST IN THE CONVOLUTION FUNCTION.
C-FFT LENGTH ,N, USED INTERNALLY, IS L-1.
C-LET K=INDEX OF LAST NONZERO SAMPLE IN Y(0)---Y(N-1).  THEN X(0)
C- THRU X(N-1) MUST INCLUDE PADDING OF AT LEAST K-NMIN-1 ZEROS.
C-CONVOLUTION FUNCTION (OUTPUT) REPLACES X(NMIN) THRU X(N-1).
C-IERROR=0  NO ERROR DETECTED
C-        1  L-1 NOT A POWER OF 2
C-        2  NMIN OUT OF RANGE
C-        3  INADEQUATE ZERO PADDING
      DIMENSION X(0:L),Y(0:L)
      COMPLEX CX
      IERROR=1
      N=L-1
      TEST=N
    1 TEST=TEST/2.
      IF(TEST-2.) 8,2,1
    2 IERROR=2
      IF(NMIN.LT.0.OR.NMIN.GE.N) RETURN
      IERROR=3
      DO 3 K=N-1,0,-1
        IF(Y(K).NE.0.) GO TO 4
    3 CONTINUE
    4 DO 5 J=N-1,0,-1
        IF(X(J).NE.0.) GO TO 6
    5 CONTINUE
    6 IF(N-J.LE.K-NMIN) RETURN
      CALL SPFFTR(X,N)
      CALL SPFFTR(Y,N)
      DO 7 M=0,N/2
        CX=CMPLX(X(2*M),X(2*M+1))*CMPLX(Y(2*M),Y(2*M+1))
        X(2*M)=REAL(CX)/N
        X(2*M+1)=AIMAG(CX)/N
```

```
      7 CONTINUE
        CALL SPIFTR(X,N)
        IERROR=0
      8 RETURN
        END
C-
        SUBROUTINE SPCORR(X,Y,L,ITYPE,NMAX,IERROR)
C-LATEST DATE: 09/18/86
C-FAST CORRELATION OF X(0:L) AND Y(0:L).  FINDS RXY(0) THRU RXY(NMAX).
C-L=LAST INDEX IN BOTH X AND Y.  MUST BE (POWER OF 2)+1 AND AT LEAST 5.
C-ITYPE=TYPE OF CORRELATION=0 IF X AND Y ARE THE SAME VECTOR (AUTO-
C-        CORRELATION), OR NOT 0 IF X AND Y ARE DIFFERENT VECTORS.
C-NMAX=MAXIMUM LAG OF INTEREST IN THE CORRELATION FUNCTION.
C-FFT LENGTH ,N, USED INTERNALLY, IS L-1.
C-LET K=INDEX OF FIRST NONZERO SAMPLE IN Y(0)---Y(N-1).  THEN X(0)
C- THRU X(N-1) MUST INCLUDE PADDING OF AT LEAST NMAX-K ZEROS.
C-CORRELATION FUNCTION, RXY, REPLACES X(0) THRU X(NMAX).
C-Y(0) THRU Y(L) IS REPLACED BY ITS FFT, COMPUTED USING SPFFTR.
C-IERROR=0  NO ERROR DETECTED
C-        1  L-1 NOT A POWER OF 2
C-        2  NMAX OUT OF RANGE
C-        3  INADEQUATE ZERO PADDING
        DIMENSION X(0:L),Y(0:L)
        COMPLEX CX
        IERROR=1
        N=L-1
        TEST=N
      1 TEST=TEST/2.
        IF(TEST-2.) 8,2,1
      2 IERROR=2
        IF(NMAX.LT.0.OR.NMAX.GE.N) RETURN
        IERROR=3
        DO 3 K=0,N-1
          IF(Y(K).NE.0.) GO TO 4
      3 CONTINUE
      4 DO 5 J=N-1,0,-1
          IF(X(J).NE.0.) GO TO 6
      5 CONTINUE
      6 IF(N-1-J.LT.NMAX-K) RETURN
        CALL SPFFTR(X,N)
        IF(ITYPE.NE.0) CALL SPFFTR(Y,N)
        DO 7 M=0,N/2
          CX=CMPLX(X(2*M),-X(2*M+1))*CMPLX(Y(2*M),Y(2*M+1))
          X(2*M)=REAL(CX)/N
          X(2*M+1)=AIMAG(CX)/N
      7 CONTINUE
        CALL SPIFTR(X,N)
        IERROR=0
      8 RETURN
        END




C-******************************************************************
C-*                CHAPTER 10 ROUTINES                             *
C-******************************************************************
C-
        SUBROUTINE SPDECI(X,LX,RATIO,LX2,IERROR)
C-LATEST DATE: 11/20/85
C-LINEAR DECIMATION OF A SEQUENCE OF EQUALLY-SPACED SAMPLES.
C-X(0:LX)=ORIGINAL DATA VECTOR WITH STEP SIZE T1, TO BE INCREASED
C-        TO T2.
C-LX=LAST INDEX OF ORIGINAL SEQUENCE, X(0) --- X(LX). (INPUT.)
C-RATIO=T2/T1=STEP SIZE RATIO.  MUST BE GE 1.0 AND LE LX.  (INPUT.)
C-LX2=LAST INDEX OF DECIMATED SEQUENCE=LX/RATIO.  (OUTPUT.)
C-IERROR=0 IF NO ERROR DETECTED, 1 IF RATIO IS OUT OF RANGE.
C-COMPUTATION IS IN PLACE.  NEW SEQUENCE REPLACES X(0) THROUGH X(LX2).
C-THE REMAINING ELEMENTS IN X, X(LX2+1) THROUGH X(LX), ARE SET TO ZEROS.
        DIMENSION X(0:LX)
        IERROR=1
        IF(RATIO.LT.1.0.OR.RATIO.GT.FLOAT(LX)) RETURN
        LX2=LX/RATIO
        DO 1 K=1,LX
          K1=K*RATIO
```

```
        IF(K.LE.LX2) X(K)=X(K1)+(K*RATIO-K1)*(X(K1+1)-X(K1))
        IF(K.GT.LX2) X(K)=0.
    1 CONTINUE
      IERROR=0
      RETURN
      END
C-
      SUBROUTINE SPLINT(X,LX,LX1,RATIO,LX2,IERROR)
C-LATEST DATE: 11/20/85
C-LINEAR INTERPOLATION BETWEEN EQUALLY SPACED SAMPLES.
C-X(0:LX)=VECTOR CONTAINING THE ORIGINAL DATA VECTOR, X(0:LX1) WITH
C-         STEP SIZE T1, TO BE INTERPOLATED.
C-LX1=LAST INDEX OF INPUT SEQUENCE, X(0) --- X(LX1).  (INPUT.)
C-RATIO=T2/T1=STEP SIZE RATIO.  MUST BE GT 0.0 AND LT 1.0.  (INPUT.)
C-LX2=LAST INDEX IN INTERPOLATED SEQUENCE=LX1/RATIO.  (OUTPUT.)
C-IERROR=0  NO ERROR DETECTED
C-        1   RATIO OUT OF RANGE
C-        2   LX TOO SMALL FOR INTERPOLATED RESULT
C-COMPUTATION IS IN PLACE.  LX MUST BE AT LEAST LX2=LX1/RATIO.
C-THE REMAINING ELEMENTS, X(LX2+1) --- X(LX), ARE SET TO ZERO.
      DIMENSION X(0:LX)
      IERROR=1
      IF(RATIO.LE.0.0.OR.RATIO.GE.1.0) RETURN
      IERROR=2
      LX2=LX1/RATIO
      IF(LX2.GT.LX) RETURN
      DO 1 K=LX,1,-1
        IX=K*RATIO
        IF(K.LE.LX2) X(K)=X(IX)+(K*RATIO-IX)*(X(IX+1)-X(IX))
        IF(K.GT.LX2) X(K)=0.
    1 CONTINUE
      IERROR=0
      RETURN
      END
C-
      SUBROUTINE SPZINT(X,LX,LX1,RATIO,LX2,IERROR)
C-LATEST DATE: 11/27/85
C-INTERPOLATION BETWEEN EQUALLY-SPACED SAMPLES USING ZERO PADDING.
C-X(0:LX)=DATA VECTOR CONTAINING ORIGINAL SEQUENCE, X(0) THRU X(LX1).
C-LX1=LAST INDEX OF ORIGINAL DATA SEQUENCE.  (NOTE: X(LX1+1) THRU
C-   X(LX) CAN BE ANYTHING INITIALLY.)  COMPUTAION IS IN PLACE.
C-RATIO (R) MUST BE 1./(POWER OF 2.).
C-ROUTINE WORKS BY TAKING FFT OF X(0:LX1) PADDED WITH N-(LX1+1) ZEROS,
C- WHERE N=SMALLEST POWER OF 2 GT LX1, THEN PADDING THE FFT WITH
C- ZEROS TO INCREASE LENGTH TO N/R, THEN TAKING THE SCALED INVERSE FFT.
C-LX MUST BE BIG ENOUGH TO ACCOMODATE FINAL FFT, I.E.,
C-        LX GT ((SMALLEST POWER OF 2) GT LX1)/R.
C-LX2=LAST INDEX OF OUTPUT SEQUENCE, COMPUTED AS LX1/R.
C-IERROR=0  NO ERROR DETECTED
C-        1   RATIO IS NOT THE RECIPROCAL OF A POWER OF 2
C-        2   LX IS TOO SMALL FOR X(0:LX TO HOLD THE INTERPOLATED RESULT.
      DIMENSION X(0:LX)
      IERROR=1
      BASE=1.0
    1 BASE=BASE/2.
      IF(RATIO-BASE) 1,2,5
    2 IERROR=2
      N=1
    3 N=2*N
      IF(N.LE.LX1) GO TO 3
      N2=N/RATIO+0.5
      IF(N2.GE.LX) RETURN
      LX2=LX1/RATIO+0.5
      DO 4 K=0,LX
        IF(K.GT.LX1) X(K)=0.0
        X(K)=X(K)/N
    4 CONTINUE
      CALL SPFFTR(X,N)
      X(N)=X(N)/2.
      CALL SPIFTR(X,N2)
      IERROR=0
    5 RETURN
      END
```

```
C-****************************************************************
C-*                     CHAPTER 11 ROUTINES                     *
C-****************************************************************
C-
      SUBROUTINE SPNORM(X,Y,N,L,A,C,IERROR)
C-LATEST DATE: 11/13/85
C-SETS UP NORMAL EQN FOR LEAST-SQUARES POLYNOMIAL FIT: AB=C
C-X(0:N-1) & Y(0:N-1) ARE N-POINT INPUT DATA ARRAYS
C-L=ORDER OF POLYNOMIAL
C-A(0:L,0:L) = ARRAY RETURNED      C(0:L) = VECTOR RETURNED
C-IERROR=0      NO ERRORS DETECTED
C-      1        INVALID N: N<=0
C-      2        INVALID L: L<=0 OR L>N
      DIMENSION X(0:N-1),Y(0:N-1),A(0:L,0:L),C(0:L)
      IERROR=0
      IF(N.LE.0) IERROR=1
      IF(L.LE.0.OR.L.GT.N) IERROR=2
      IF(IERROR.NE.0) RETURN
      DO 2 J=1,L
        S1=0.
        S2=0.
        S3=0.
        DO 1 K=0,N-1
          S1=S1+X(K)**J
          S2=S2+X(K)**(L+J)
          S3=S3+Y(K)*(X(K)**J)
    1   CONTINUE
        A(0,J)=S1
        A(J,L)=S2
        C(J)=S3
    2 CONTINUE
      A(0,0)=N
      C(0)=0.
      DO 3 K=0,N-1
        C(0)=C(0)+Y(K)
    3 CONTINUE
      DO 5 I=1,L
        DO 4 J=0,L-1
          A(I,J)=A(I-1,J+1)
    4   CONTINUE
    5 CONTINUE
      RETURN
      END
C-
      SUBROUTINE SPGAUS(A,C,L,B,IERROR)
C-LATEST DATE: 11/13/85
C-USES GAUSSIAN ELIMINATION TO SOLVE AB=C MATRIX EQUATION
C-CAUTION:  DO NOT USE WITH ILL-CONDITIONED SYSTEMS
C-NO PIVOTING IS PERFORMED
C-A(0:L,0:L) = SQUARE MATRIX INPUT     C(0:L) = VECTOR INPUT
C-L=ORDER OF SYSTEM OF EQUATIONS
C-B(0:L) = SOLUTION VECTOR RETURNED
C-IERROR=0      NO ERRORS DETECTED
C-      1        INVALID ORDER   L<=0
C-      2        ILL-CONDITIONED     PIVOT <1.E-10
      DIMENSION A(0:L,0:L),B(0:L),C(0:L)
      IERROR=1
      IF(L.LE.0) RETURN
      DO 1 I=0,L
        B(I)=0.0
    1 CONTINUE
      IERROR=2
      DO 5 I=0,L
        PIV=A(I,I)
        IF(ABS(PIV).LT.1.E-10) RETURN
        C(I)=C(I)/PIV
        DO 2 J=I,L
          A(I,J)=A(I,J)/PIV
    2   CONTINUE
        DO 4 II=I+1,L
          SCL=A(II,I)
          C(II)=C(II)-C(I)*SCL
          DO 3 J=I,L
            A(II,J)=A(II,J)-A(I,J)*SCL
    3     CONTINUE
```

```
 4    CONTINUE
 5 CONTINUE
      DO 8 I=L,0,-1
        B(I)=C(I)
        DO 6 J=I+1,L
          B(I)=B(I)-A(I,J)*B(J)
 6      CONTINUE
 8 CONTINUE
      IERROR=0
      RETURN
      END
C-
      SUBROUTINE SPORTH(Y,N,L,ORTHB,P,IERROR)
C-LATEST DATE: 11/13/85
C-GENERATES COEFFICIENTS FOR LEAST SQUARES FIT VIA ORTHOGONAL
C       POLYNOMIALS - LTH ORDER
C-DATA SAMPLES MUST OCCUR AT REGULAR INTERVALS
C-Y(0:N-1)=DATA ARRAY OF N SAMPLE POINTS
C-ORTHB(0:L)=COEFFICIENTS RETURNED
C-P(0:L,0:L)=WORK ARRAY USED INTERNALLY
C-IERROR=0       NO ERRORS DETECTED
C-      1        N<=0 OR L<=0
C-      2        N<L  CANNOT COMPUTE LEAST SQUARES FIT
      DIMENSION Y(0:N-1),ORTHB(0:L),P(0:L,0:L)
      IERROR=0
      IF(N.LE.0.OR.L.LE.0) IERROR=1
      IF(N.LT.L) IERROR=2
      IF(IERROR.NE.0) RETURN
      CALL SPLSMT(N,L,P)
      DO 5 I=0,L
        SN=0.
        SD=0.
        DO 2 NN=0,N-1
          SUM=0.
          DO 1 J=0,I
            SUM=SUM+P(I,J)*SPBFCT(NN,J)
 1        CONTINUE
          SN=SN+Y(NN)*SUM
          SD=SD+SUM**2
 2      CONTINUE
        ORTHB(I)=SN/SD
 5    CONTINUE
      RETURN
      END
C-
      SUBROUTINE SPPOLY(DX,Y,N,L,B,ORTHB,P,IERROR)
C-LATEST DATE: 11/13/85
C-GENERATES POLYNOMIAL COEFFICIENTS FOR LEAST SQUARES FIT
C-    B(0:L)=COEFFICIENTS RTND: B0+B1*X+....+BL*X**L
C-DX=POINT SPACING IN X DIRECTION (SAMPLE INTERVAL)
C-Y(0:N-1)=ARRAY OF N EQUALLY SPACED DATA SAMPLES
C-WORK ARRAYS:
C-      ORTHB(0:L) - ORTHOGONAL POLYNOMIAL COEFFICIENTS
C-      P(0:L,0:L) - MATRIX USED IN SPORTH AND SPSTRL
C-IERROR=0       NO ERRORS DETECTED
C-      1        N<=0 OR L<=0
C-      2        N<L  CANNOT COMPUTE LEAST SQUARES FIT
C-      3        DX <=0.  INVALID SAMPLE INTERVAL
      DIMENSION Y(0:N-1),B(0:L),ORTHB(0:L),P(0:L,0:L)
      CALL SPORTH(Y,N,L,ORTHB,P,IERROR)
      IF(DX.LE.0.) IERROR=3
      IF(IERROR.NE.0) RETURN
      DO 2 J=0,L
        B(J)=0.
        DO 1 I=J,L
          B(J)=B(J)+ORTHB(I)*P(I,J)
 1      CONTINUE
 2    CONTINUE
      CALL SPSTRL(L,P)
      DO 4 J=0,L
        SUM=0.
        DO 3 I=J,L
          SUM=SUM+B(I)*P(I,J)
 3      CONTINUE
        B(J)=SUM/(DX**J)
```

```
        4 CONTINUE
          RETURN
          END
C-
          SUBROUTINE SPLEVS(AVECT,C,L,B,WK,IERROR)
C-LATEST DATE: 05/27/86
C-LEVINSON'S ALGORITHM - SOLUTION OF AB=C MATRIX EQUATION
C-AVECT(0:L)= 1ST ROW OF SYMMETRIC TOEPLITZ MATRIX
C-C(0:L)= DATA VECTOR
C-B(0:L)=SOLUTION VECTOR RETURNED
C-WK(0:L)=WORK VECTOR USED INTERNALLY
C-IERROR=0      NO ERRORS DETECTED
C-        1       INVALID L<0
C-        2       APPROX DIVIDE BY ZERO - DENOMINATOR .LT. 1.E-10
          DIMENSION AVECT(0:L),C(0:L),B(0:L),WK(0:L)
          DOUBLE PRECISION U,V,W
          IERROR=0
          IF(L.LT.0) IERROR=1
          IF(ABS(AVECT(0)).LT.1.E-10) IERROR=2
          IF(IERROR.NE.0) RETURN
          DO 1 I=0,L
            WK(I)=0.0
            B(I)=0.0
        1 CONTINUE
          B(0)=C(0)/AVECT(0)
          IF(L.EQ.0) RETURN
          V=1.
          IERROR=2
          DO 5 M=0,L-1
            U=AVECT(M+1)/AVECT(0)
            W=(C(M+1)-B(0)*AVECT(M+1))/AVECT(0)
            IF(M.GT.0) THEN
              DO 2 K=1,M
                U=U-WK(M-K)*AVECT(K)/AVECT(0)
                W=W-B(K)*AVECT(M+1-K)/AVECT(0)
        2     CONTINUE
            ENDIF
            WK(M)=U/V
            IF(M.GT.0) THEN
              DO 3 N=0,(M-1)/2
                TEMP=WK(N)
                WK(N)=TEMP-WK(M)*WK(M-N-1)
                IF(M-N-1.NE.N) WK(M-N-1)=WK(M-N-1)-WK(M)*TEMP
        3     CONTINUE
            ENDIF
            V=V-WK(M)*U
            IF(ABS(V).LT.1.E-10) RETURN
            B(M+1)=W/V
            DO 4 K=0,M
              B(K)=B(K)-WK(M-K)*B(M+1)
        4   CONTINUE
        5 CONTINUE
          IERROR=0
          RETURN
          END
C-
          SUBROUTINE SPDURB(AAVECT,L,B,IERROR)
C-LATEST DATE: 05/21/86
C-DURBIN'S ALGORITHM - SPECIAL CASE OF LEVINSON'S ALGORITHM
C-AAVECT(0:L+1) = AUGMENTED A DATA VECTOR
C-B(0:L) = SOLUTION VECTOR RETURNED
C-IERROR=0      NO ERRORS DETECTED
C-        1       INVALID L    <0
C-        2       APPROX DIVIDE BY ZERO   DENOMINATOR .LT.1.E-10
          DIMENSION AAVECT(0:L+1),B(0:L)
          DO 1 I=0,L
            B(I)=0.0
        1 CONTINUE
          IERROR=0
          IF(L.LT.0) IERROR=1
          IF(ABS(AAVECT(0)).LT.1.E-10) IERROR=2
          IF(IERROR.NE.0) RETURN
          B(0)=-AAVECT(1)/AAVECT(0)
          IF(L.EQ.0) GO TO 7
          IERROR=2
```

```fortran
      DO 5 NN=1,L
        GAMMA=0.
        GAMMAP=0.
        DO 2 I=1,NN
          GAMMA=GAMMA-AAVECT(I)*B(NN-I)
          GAMMAP=GAMMAP-AAVECT(I)*B(I-1)
    2     CONTINUE
        SCL=AAVECT(0)-GAMMAP
        IF(ABS(SCL).LT.1.E-10) GO TO 6
        BETA=-(AAVECT(NN+1)-GAMMA)/SCL
        DO 3 I=0,(NN-2)/2
          TMP=B(I)
          B(I)=B(I)+BETA*B(NN-1-I)
          IF(NN.GT.1) B(NN-1-I)=B(NN-1-I)+BETA*TMP
    3     CONTINUE
        IF(INT((NN-2)/2).NE.INT((NN-1)/2))
     +      B((NN-1)/2)=B((NN-1)/2)+BETA*B((NN-1)/2)
        B(NN)=BETA
    5 CONTINUE
      IERROR=0
    6 CONTINUE
    7 DO 8 I=0,L
        B(I)=-B(I)
    8 CONTINUE
      RETURN
      END
C-
      SUBROUTINE SPSTRL(L,S)
C-LATEST DATE: 11/13/85
C-GENERATES MATRIX OF STIRLING #S OF THE FIRST KIND
      DIMENSION S(0:L,0:L)
      DO 2 I=0,L
        S(I,0)=0.
        S(I,I)=1.
        DO 1 J=I+1,L
          S(I,J)=0.
    1   CONTINUE
    2 CONTINUE
      DO 3 I=2,L
        S(I,1)=-(I-1)*S(I-1,1)
    3 CONTINUE
      DO 5 I=3,L
        DO 4 J=2,I-1
          S(I,J)=S(I-1,J-1)-(I-1)*S(I-1,J)
    4   CONTINUE
    5 CONTINUE
      RETURN
      END
C-
      SUBROUTINE SPLSMT(N,L,P)
C-LATEST DATE: 11/13/85
C-GENERATES MATRIX USED FOR ORTHOGONAL POLYOMIALS
      DIMENSION P(0:L,0:L)
      DO 2 I=0,L
        P(I,0)=1.
        DO 1 J=I+1,L
          P(I,J)=0.
    1   CONTINUE
    2 CONTINUE
      DO 5 I=1,L
        DO 4 J=1,I
          P(I,J)=((-1.)**J)*SPBFCT(I,J)*SPBFCT(I+J,J)/
     +           (SPBFCT(N-1,J)*(SPBFCT(J,J)**2))
    4   CONTINUE
    5 CONTINUE
      RETURN
      END
C-
      FUNCTION SPBFCT(I1,I2)
C-LATEST DATE: 02/14/87
C-GENERATES (I1)!/(I1-I2)!=I1*(I1-1)*...*(I1-I2+1).
C-NOTE: 0!=1 AND SPBFCT(I,I)=SPBFCT(I,I-1)=I!.
      SPBFCT=0.
      IF(I1.LT.0.OR.I2.LT.0.OR.I2.GT.I1) RETURN
      SPBFCT=1.
```

```
                IF(I2.EQ.0) RETURN
                DO 1 I=I1,I1-I2+1,-1
                  SPBFCT=SPBFCT*I
              1 CONTINUE
                RETURN
                END




C-*****************************************************************
C-*                      CHAPTER 12 ROUTINES                      *
C-*****************************************************************
C-
        SUBROUTINE SPNLMS(X,N,D,B,L,MU,SIG,AL,PX,IERROR)
C-LATEST DATE: 11/13/85
C-IMPLEMENTS NLMS ALGORITHM B(K+1)=B(K)+2*MU*E*X(K)/((L+1)*SIG)
C-X(0:N-1)=DATA VECTOR        INPUT SENT       OUTPUT RETURNED
C-D(0:N-1)=DESIRED SIGNAL VECTOR
C-N SPECIFIES NUMBER OF DATA POINTS IN X AND D
C-B(0:L)=ADAPTIVE COEFFICIENTS OF LTH ORDER FIR FILTER
C-MU=CONVERGENCE PARAMETER - DECLARE REAL
C-SIG=INPUT SIGNAL POWER ESTIMATE - UPDATED INTERNALLY
C-AL=FORGETTING FACTOR    SIG(K)=AL*(X(K)**2)+(1-AL)*SIG(K-1)
C-IERROR=0         NO ERRORS DETECTED
C-       1         INVALID ORDER    L<0
C-       2         INVALID CONVERGENCE PARAMETER   MU<=0 OR >=1
C-       3         INPUT POWER ESTIMATE   SIG<=0
C-       4         FORGETTING FACTOR    AL<0 OR =>1
C-       5         RESPONSE EXCEEDS 1.E10
        DIMENSION X(0:N-1),D(0:N-1),B(0:L),PX(0:L)
        REAL MU
        IERROR=0
        IF(L.LT.0) IERROR=1
        IF(MU.LE.0..OR.MU.GE.1.) IERROR=2
        IF(SIG.LE.0.) IERROR=3
        IF(AL.LT.0..OR.AL.GE.1.) IERROR=4
        IF(IERROR.NE.0) RETURN
        IERROR=5
        DO 5 K=0,N-1
          PX(0)=X(K)
          X(K)=0.
          DO 1 LL=0,L
            X(K)=X(K)+B(LL)*PX(LL)
      1   CONTINUE
          IF(ABS(X(K)).GT.1.E10) RETURN
          E=D(K)-X(K)
          SIG=AL*(PX(0)**2)+(1-AL)*SIG
          TMP=2*MU/((L+1)*SIG)
          DO 2 LL=0,L
            B(LL)=B(LL)+TMP*E*PX(LL)
      2   CONTINUE
          DO 3 LL=L,1,-1
            PX(LL)=PX(LL-1)
      3   CONTINUE
      5 CONTINUE
        IERROR=0
        RETURN
        END




C-*****************************************************************
C-*                      CHAPTER 13 ROUTINES                      *
C-*****************************************************************
C-
        FUNCTION SPMEAN(X,N)
C-LATEST DATE: 11/13/85
C-COMPUTES MEAN VALUE OF N-POINT DATA VECTOR X(0:N-1)
        DIMENSION X(0:N-1)
        SPMEAN=0.0
        IF(N.LE.0) RETURN
```

```
        DO 1 I=0,N-1
          SPMEAN=SPMEAN+X(I)
      1 CONTINUE
        SPMEAN=SPMEAN/N
        RETURN
        END
C-
        FUNCTION SPVARI(X,N)
C-LATEST DATE: 11/13/85
C-COMPUTES VARIANCE OF N-POINT DATA VECTOR X(0:N-1)
        DIMENSION X(0:N-1)
        SPVARI=0.0
        IF(N.LE.1) RETURN
        DO 1 I=0,N-1
          SPVARI=SPVARI+X(I)**2
      1 CONTINUE
        SPVARI=(SPVARI-N*(SPMEAN(X,N)**2))/(N-1.)
        RETURN
        END
C-
        SUBROUTINE SPLMTS(X,N,XMIN,IMIN,XMAX,IMAX)
C-LATEST DATE: 01/21/87
C-FINDS THE GLOBAL MINIMUM AND MAXIMUM OF THE N-POINT DATA
C-SEQUENCE IN ARRAY X(0:N-1).
C-XMIN AND XMAX ARE THE MINIMUM AND MAXIMUM DATA VALUES.
C-IMIN AND IMAX ARE THE ARRAY LOCATIONS CORRESPONDING TO
C-THE FIRST OCCURRENCE OF THE RESPECTIVE EXTREME.
        DIMENSION X(0:N-1)
        IMIN=0
        IMAX=0
        IF(N.LE.1) GO TO 2
        DO 1 I=1,N-1
          IF(X(I).GT.X(IMAX))IMAX=I
          IF(X(I).LT.X(IMIN))IMIN=I
      1 CONTINUE
      2 XMIN=X(IMIN)
        XMAX=X(IMAX)
        RETURN
        END
C-
        SUBROUTINE SPLFIT(X,Y,N,A,B,IERROR)
C-LATEST DATE: 11/13/85
C-FITS STRAIGHT LINE TO DATA IN ARRAYS X(0:N-1) AND Y(0:N-1)
C-N SPECIFIES NUMBER OF DATA POINTS
C-EQUATION OF LINE:  Y=AX+B    A,B ARE RETURNED
C-IERROR=0      NO ERRORS DETECTED
C-       1      N<=0
C-       2      ARITHMMETIC PROBLEM - DENOMINATOR APPROX 0.
        DIMENSION X(0:N-1),Y(0:N-1)
        A=0.
        B=0.
        IERROR=1
        IF(N.LE.0) RETURN
        XYSUM=0.
        XSUM=0.
        YSUM=0.
        X2SUM=0.
        DO 1 I=0,N-1
          XYSUM=XYSUM+X(I)*Y(I)
          XSUM=XSUM+X(I)
          YSUM=YSUM+Y(I)
          X2SUM=X2SUM+X(I)**2
      1 CONTINUE
        IERROR=2
        DEN=N*X2SUM-XSUM**2
        IF(DEN.LT.1.E-10) RETURN
        A=(N*XYSUM-XSUM*YSUM)/DEN
        B=SPMEAN(Y,N)-A*SPMEAN(X,N)
        IERROR=0
        RETURN
        END
C-
        SUBROUTINE SPEXPN(X,Y,N,A,B,IERROR)
C-LATEST DATE: 11/13/85
C-FITS EXPONENTIAL CURVE TO DATA IN ARRAYS X(0:N-1) AND Y(0:N-1)
C-N SPECIFIES NUMBER OF DATA POINTS
```

```
C-EQUATION: Y=A(B**X)     A,B ARE RETURNED
C-DATA IN ARRAY Y IS CHANGED TO LOG DATA INTERNALLY
C-IERROR=0      NO ERRORS DETECTED
C-        1       N<=0
C-        2       SPLFIT ERROR = 2
C-        3       Y DATA VALUE <=0.  CANNOT COMPUTE LOG
      DIMENSION X(0:N-1),Y(0:N-1)
      A=0.
      B=0.
      IERROR=1
      IF(N.LE.0) RETURN
      IERROR=3
      DO 1 I=0,N-1
        IF(Y(I).LE.0.) RETURN
        Y(I)=ALOG10(Y(I))
    1 CONTINUE
      CALL SPLFIT(X,Y,N,BLG,ALG,IERROR)
      IF(IERROR.NE.0) RETURN
      A=10.**ALG
      B=10.**BLG
      RETURN
      END
C-
      SUBROUTINE SPPWRC(X,Y,N,A,B,IERROR)
C-LATEST DATE: 11/13/85
C-FITS POWER FUNCTION CURVE TO DATA IN ARRAYS X(0:N-1) AND Y(0:N-1)
C-N SPECIFIES NUMBER OF DATA POINTS
C-EQUATION:  Y=A(X**B)    A,B ARE RETURNED
C-DATA IN X AND Y IS CONVERTED TO LOG INTERNALLY
C-IERROR=0      NO ERRORS DETECTED
C-        1       N<=0
C-        2       SPLFIT ERROR = 2
C-        3       X OR Y DATA <=0.  CANNOT COMPUTE LOG
      DIMENSION X(0:N-1),Y(0:N-1)
      A=0.
      B=0.
      IERROR=1
      IF(N.LE.0) RETURN
      IERROR=3
      DO 1 I=0,N-1
        IF(X(I).LE.0..OR.Y(I).LE.0.) RETURN
        X(I)=ALOG10(X(I))
        Y(I)=ALOG10(Y(I))
    1 CONTINUE
      CALL SPLFIT(X,Y,N,B,ALG,IERROR)
      IF(IERROR.NE.0) RETURN
      A=10.**ALG
      RETURN
      END
C-
      SUBROUTINE SPXEXP(X,Y,N,A,B,IERROR)
C-LATEST DATE: 11/13/85
C-FITS CURVE OF FORM Y=BX*(E**AX) TO X(0:N-1) AND Y(0:N-1)
C-N SPECIFIES NUMBER OF POINTS     A,B ARE RETURNED
C-DATA IN X AND Y IS MODIFIED INTERNALLY
C-IERROR=0      NO ERRORS DETECTED
C-        1       N<=0
C-        2       SPLFIT ERROR = 2
C-        3       NEGATIVE DATA - CANNOT COMPUTE LN
      DIMENSION X(0:N-1),Y(0:N-1)
      A=0.
      B=0.
      IERROR=1
      IF(N.LE.0) RETURN
      IERROR=3
      DO 1 I=0,N-1
        IF(X(I).LE.0..OR.Y(I).LE.0.) RETURN
        Y(I)=ALOG(Y(I))-ALOG(X(I))
    1 CONTINUE
      CALL SPLFIT(X,Y,N,A,BLN,IERROR)
      IF(IERROR.NE.0) RETURN
      B=EXP(BLN)
      RETURN
      END
```

```
C-
       SUBROUTINE SPRFTM(IDIR,X,N,T0,T10,T90,T100,IERROR)
C-LATEST DATE: 10/06/86
C-FINDS RISE/FALL TIME PARAMETERS FOR N-POINT DATA VECTOR X(0:N-1)
C-IDIR=1: RISE TIME       IDIR=-1: FALL TIME
C-LINEAR INTERPOLATION IS USED TO FIND T10 AND T90
C-TIMES RETURNED ASSUME NORMALIZED SAMPLE INTERVAL  T=1
C-T0: MINIMUM    T10: 10%       T90: 90%      T100: MAXIMUM
C-IERROR=0    NO ERRORS DETECTED
C-       1       N<=0
C-       2       INVALID IDIR PARAMETER  .NE.1,-1
C-       3       T0=T100   MINIMUM AND MAXIMUM ARE SAME
       DIMENSION X(0:N-1)
       IERROR=0
       IF(N.LE.0) IERROR=1
       IF(IDIR.NE.1.AND.IDIR.NE.-1) IERROR=2
       IF(IERROR.NE.0) RETURN
       CALL SPLMTS(X,N,XMN,IMN,XMX,IMX)
       K100=IMX
       IF(IDIR.EQ.1)K0=0
       IF(IDIR.EQ.-1)K0=N-1
       DO 1 I=K100-IDIR,K0,-IDIR
         IF(X(I).GE.X(I+IDIR).AND.X(I).LT.X(K100)) THEN
           K0=I+IDIR
           GO TO 2
         ENDIF
     1 CONTINUE
     2 IERROR=3
       IF(X(K0).EQ.X(K100)) RETURN
       T0=K0
       T100=K100
       A90=.9*(X(K100)-X(K0))+X(K0)
       DO 5 I=K100,K0,-IDIR
         IF(X(I).LT.A90) GO TO 6
     5 CONTINUE
       I=K0
     6 T90=I+IDIR*(A90-X(I))/(X(I+IDIR)-X(I))
       A10=.1*(X(K100)-X(K0))+X(K0)
       DO 7 I=K100,K0,-IDIR
         IF(X(I).LT.A10) GO TO 8
     7 CONTINUE
       I=K0
     8 T10=I+IDIR*(A10-X(I))/(X(I+IDIR)-X(I))
       IERROR=0
       RETURN
       END




C-******************************************************************
C-*                    CHAPTER 14 ROUTINES                        *
C-******************************************************************
C-
       FUNCTION SPWNDO(ITYPE,N,K)
C-LATEST DATE: 11/13/85
C-THIS FUNCTION GENERATES A SINGLE SAMPLE OF A DATA WINDOW.
C-ITYPE=1(RECTANGULAR), 2(TAPERED RECTANGULAR), 3(TRIANGULAR),
C-     4(HANNING), 5(HAMMING), OR 6(BLACKMAN).
C-     (NOTE:  TAPERED RECTANGULAR HAS COSINE-TAPERED 10% ENDS.)
C-N=SIZE (TOTAL NO. SAMPLES) OF WINDOW.
C-K=SAMPLE NUMBER WITHIN WINDOW, FROM 0 THROUGH N-1.
C-  (IF K IS OUTSIDE THIS RANGE, SPWNDO IS SET TO 0.)
       PI=4.*ATAN(1.)
       SPWNDO=0.
       IF(ITYPE.LT.1.OR.ITYPE.GT.6) RETURN
       IF(K.LT.0.OR.K.GE.N) RETURN
       SPWNDO=1.
       GO TO (1,2,3,4,5,6), ITYPE
     1 RETURN
     2 L=(N-2)/10
       IF(K.LE.L) SPWNDO=0.5*(1.0-COS(K*PI/(L+1)))
       IF(K.GT.N-L-2) SPWNDO=0.5*(1.0-COS((N-K-1)*PI/(L+1)))
       RETURN
```

```
      3 SPWNDO=1.0-ABS(1.0-2*K/(N-1.0))
        RETURN
      4 SPWNDO=0.5*(1.0-COS(2*K*PI/(N-1)))
        RETURN
      5 SPWNDO=0.54-0.46*COS(2*K*PI/(N-1))
        RETURN
      6 SPWNDO=0.42-0.5*COS(2*K*PI/(N-1))+0.08*COS(4*K*PI/(N-1))
        RETURN
        END
C-
        SUBROUTINE SPMASK(X,LX,ITYPE,TSV,IERROR)
C-LATEST DATE: 02/20/87
C-THIS ROUTINE APPLIES A DATA WINDOW TO THE DATA VECTOR X(0:LX).
C-ITYPE=1(RECTANGULAR), 2(TAPERED RECTANGULAR), 3(TRIANGULAR),
C-     4(HANNING), 5(HAMMING), OR 6(BLACKMAN).
C-     (NOTE:  TAPERED RECTANGULAR HAS COSINE-TAPERED 10% ENDS.)
C-TSV=SUM OF SQUARED WINDOW VALUES.
C-IERROR=0 IF NO ERROR, 1 IF ITYPE OUT OF RANGE.
        DIMENSION X(0:LX)
        IERROR=1
        IF(ITYPE.LT.1.OR.ITYPE.GT.6) RETURN
        TSV=0.
        DO 1 K=0,LX
          W=SPWNDO(ITYPE,LX+1,K)
          X(K)=X(K)*W
          TSV=TSV+W*W
      1 CONTINUE
        IERROR=0
        RETURN
        END
C-
        SUBROUTINE SPUNWR(X,LX,IRD)
C-LATEST DATE: 11/13/85
C-SIMPLE PHASE UNWRAPPING ROUTINE, USEFUL WHERE PHASE WAS COMPUTED
C-  USING ATAN2 FUNCTION AND IN SIMILAR SITUATIONS WHERE THE PHASE
C-  IS WRAPPED INTO THE RANGE FROM -PI TO PI RADIANS.
C-X(0:LX)=SEQUENCE OF PHASE ANGLES IN RADIANS OR DEGREES, ALL
C-        ASSUMED TO BE OUTPUTS OF ATAN2 IN THE RANGE (-PI,PI) RAD.
C-IRD=1 TO INDICATE X IS IN RADIANS, OR 2 TO INDICATE DEGREES.
C-THE ROUTINE INSERTS A CORRECTION OF 2*PI OR 360 DEGREES WHEREVER
C-  THE PHASE JUMPS MORE THAN PI RADIANS.
        DIMENSION X(0:LX)
        ANGL=180.
        IF(IRD.EQ.1) ANGL=4.*ATAN(1.)
        COR=0.
        DO 2 K=1,LX
          DX=X(K)-(X(K-1)-COR)
          IF(ABS(DX).LE.ANGL) GO TO 1
          COR=COR-SIGN(2.*ANGL,DX)
      1   X(K)=X(K)+COR
      2 CONTINUE
        RETURN
        END
C-
        SUBROUTINE SPHILB(X,LX)
C-LATEST DATE: 12/17/86
C-GENERATES THE WEIGHTS OF AN FIR HILBERT TRANSFORMER.
C-AFTER EXECUTION THE WEIGHTS ARE IN X(0) THROUGH X(L-1), WHERE
C-  L=LX IF LX IS ODD OR L=LX+1 IF LX IS EVEN.
C-WHEN USED AS A CAUSAL FILTER, THE TRANSFORMER HAS APPROXIMATELY
C-  UNIT GAIN, A GROUP DELAY OF (L-1)/2 SAMPLES, PLUS
C-  APPROXIMATELY 90 DEGREES PHASE SHIFT AT ALL FREQUENCIES.
        DIMENSION X(0:LX)
        PI=4.*ATAN(1.)
        L2=LX/2
        X(L2)=0.
        X(LX)=0.
        DO 1 K=1,L2
          X(L2+K)=2.*(K-2*INT(K/2))/(PI*K)
          X(L2-K)=-X(L2+K)
      1 CONTINUE
        CALL SPMASK(X,2*L2,5,TSV,K)
        RETURN
        END
```

```
C-
      SUBROUTINE SPCHRP(X,LX,LX1,F1,F2,WORK,LX2,IERROR)
C-LATEST DATE: 11/27/85
C-COMPUTES THE CHIRP-Z TRANSFORM OF A COMPLEX SEQUENCE.
C-X=COMPLEX INPUT ARRAY CONTAINING THE COMPLEX SEQUENCE.
C-LX=LAST INDEX IN THE ARRAY X(0:LX).  LX+1 MUST BE A POWER OF 2.
C-LX1=LAST INDEX OF THE COMPLEX INPUT SEQUENCE, X(0)...X(LX1).
C-      LX1 MUST BE LESS THAN LX.
C-F1=FREQUENCY OF FIRST DFT COMPONENT.  (SAMPLING FREQ.=1.0.)
C-F2=FREQUENCY OF LAST DFT COMPONENT, GREATER THEN F1.
C-WORK=COMPLEX WORK ARRAY, DIMENSIONED COMPLEX WORK(0:LX).
C-AFTER EXECUTION DFT COMPONENTS, SPACED EVENLY FROM F1 THRU F2,
C-  ARE STORED IN X(0)...X(LX2).  THE ORIGINAL TIME SERIES IS LOST.
C-LX2=LAST FREQUENCY INDEX AS ABOVE, COMPUTED AS LX2=LX-LX1 DURING
C-      EXECUTION OF THE ROUTINE.
C-IERROR=0  NO ERRORS
C-      1  LX IS NOT GREATER THAN LX1
C-      2  F2 IS NOT GREATER THAN F1
C-      3  LX+1 IS NOT A POWER OF 2
      COMPLEX X(0:LX),WORK(0:LX),AM1,W2
      TP=8.*ATAN(1.)
      IERROR=1
      LX2=LX-LX1
      IF(LX2.LT.1) RETURN
      IERROR=2
      IF(F1.GE.F2) RETURN
      IERROR=3
      LXC=1
    1 LXC=2*LXC+1
      IF(LXC-LX) 1,2,6
    2 AM1=EXP(CMPLX(0.,-TP*F1))
      W2=EXP(CMPLX(0.,-TP*(F2-F1)/(2.*LX2)))
      DO 3 K=0,LX
        IF(K.LE.LX1) WORK(K)=(AM1*W2**K)**K*X(K)
        IF(K.GT.LX1) WORK(K)=0.
        IF(K.LE.LX2) X(K)=W2**(-K*K)
        IF(K.GT.LX2) X(K)=W2**(-(LX+1-K)**2)
    3 CONTINUE
      CALL SPFFTC(WORK,LX+1,-1)
      CALL SPFFTC(X,LX+1,-1)
      DO 4 K=0,LX
        X(K)=X(K)*WORK(K)
    4 CONTINUE
      CALL SPFFTC(X,LX+1,+1)
      DO 5 K=0,LX2
        X(K)=X(K)*W2**(K*K)/(LX+1)
    5 CONTINUE
      IERROR=0
    6 RETURN
      END
C-
      FUNCTION SPWLSH(LOGN,NSEQ,K)
C-LATEST DATE: 02/24/86
C-GENERATES A WALSH-ORDERED WALSH COEFFICIENT.  THE INPUTS ARE
C-      LOGN=LOG BASE 2 OF TRANSFORM ARRAY SIZE.
C-      NSEQ=SEQUENCY INDEX (0 THRU 2**LOGN-1).
C-      K=TIME INDEX (0 THRU 2**LOGN-1).
C-SPWLSH IS THE WALSH COEFFICIENT, EITHER -1.0 OR 1.0.
C-REF. -- AHMED AND RAO(SPRINGER-VERLAG,1975), PAGES 90-91.
C-
      X(M,N)=MOD(M/N,2)
      IS1=2**(LOGN-1)
      IS2=2
      SPWLSH=X(NSEQ,IS1)*X(K,1)
      DO 1 I=2,LOGN
        SPWLSH=SPWLSH+(X(NSEQ,IS1)+X(NSEQ,IS1/2))*X(K,IS2)
        IS1=IS1/2
        IS2=IS2*2
    1 CONTINUE
      SPWLSH=1.-2.*MOD(SPWLSH,2.)
      RETURN
      END
```

Plotting Routines

Several plotting subroutines are used in the programming examples in this text to produce different types of plots. The comments in these subroutines are listed in this appendix so the reader can understand the calling sequences. Since plotting routines are designed for specific equipment, we assume that the reader will want to supply his or her own versions of one or more of these routines.

```
        SUBROUTINE PXY(X,Y,N,ITYPE,MODE,ISEQ,ITIC,NMBR,X1,Y1,X2,Y2)
C-XY PLOT OF X(N) VERSUS Y(N).
C-ITYPE=1(LINEAR), 2(LOGX), 3(LOGY), OR 4(LOGLOG).
C-MODE=0(LINE), +N(MARKER EVERY N PTS. W/LINE), -N(MARKERS ONLY).
C-ISEQ=0(ONLY PLOT), 1(FIRST PLOT), 2(IN BETWEEN), 3(LAST PLOT).
C-ITIC=0,2(TIC MARKS) OR 1,3(NO TIC MARKS).
C-     (NOTE: ITIC=2 OR 3 CHANGES ASPECT RATIO FROM 1.0 TO 1.5.)
C-NMBR=0(SOFT AXIS NUMBERS), 1(NO NUMBERS), OR 2(HARD NUMBERS).
C-X1,Y1=LOWER LEFT CORNER=(.1,.1) OR MORE.
C-X2,Y2=UPPER RIGHT CORNER=(.9,.9) OR LESS, ASSUMING ASPECT RATIO=1.0.
C-


        SUBROUTINE PY(XSTART,XINCR,Y,N,ITYPE,ISEQ,WORK,X1,Y1,X2,Y2)
C-PLOT ROUTINE SIMILAR TO PXY.  PRODUCES PLOT OF ARRAY Y(N) VS X.
C-XSTART=STARTING X VALUE, X(1).
C-XINCR=X INCREMENT.
C-ITYPE,ISEQ=SAME AS IN PXY.
C-WORK(N)=WORK ARRAY, SAME SIZE AS Y(N).
C-X1,Y1,X2,Y2=SAME AS IN PXY.
C-


        SUBROUTINE MCPLOT(X0,DX,Y,N,IPAT,ICHR,NCV,NC,NXL,NYL,XSZ,YSZ)
C-PLOT ROUTINE SIMILAR TO PY.  PRODUCES PLOT OF ARRAY Y(N) VS X.
C-X0=X-AXIS STARTING POINT.
C-DX=X-INCREMENT.
C-Y=ARRAY OF Y DATA.
C-N=NUMBER OF POINTS IN Y TO PLOT.
C-IPAT=LINE PATTERN: 0(SOLID),1(DOTS),3(DASHES),9(SPARSE DOTS).
C-ICHAR=DATA POINT SYMBOL: 0(NONE),4(+),7(O),9(*).
C-NCV=NUMBER OF CURVES ON PLOT.
C-NXL=X-LABEL: 1(TIME),2(SAMPLE NO.),3,4,5(FREQ. IN HZ,KHZ,RAD/S).
C-NYL=Y-LABEL: 1(AMPL.),2(MAG.),3(PWR. IN DB),4,5(PHASE IN DEG.,RAD.)
C-XSZ,YSZ=HORIZONTAL,VERTICAL SIZE IN INCHES.
C-
```

```
      SUBROUTINE HP(Y,N,ISEQ,X1,Y1,X2,Y2)
C-HISTOGRAM PLOT OF Y(N) OVER (0.0,0.5).
C-ISEQ=0(ONLY PLOT), 1(FIRST PLOT), 2(IN BETWEEN), 3(LAST PLOT).
C-IF ISEQ=0 OR 1, X1,Y1=LOWER LEFT CORNER=(.1,.1) OR MORE.
C-IF ISEQ=0 OR 1, X2,Y2=UPPER RIGHT CORNER=(.6,.8) OR LESS.
C-Y SCALES ARE RETAINED FOR ISEQ=2 OR 3.
C-
```

Index